Microbial Growth Kinetics

Microbial Growth Kinetics

Nicolai S. Panikov
Institute of Microbiology, Russian Academy of Sciences, Russia

CHAPMAN & HALL
London · Glasgow · Weinheim · New York · Tokyo · Melbourne · Madras

Published by Chapman & Hall, 2–6 Boundary Row, London SE1 8HN, UK

Chapman & Hall, 2–6 Boundary Row, London, SE1 8HN, UK

Blackie Academic & Professional, Wester Cleddens Road, Bishopbriggs, Glasgow G64 2NZ, UK

Chapman & Hall GmbH, Pappelallee 3, 69469 Weinheim, Germany

Chapman & Hall USA., One Penn Plaza, 41st Floor, New York NY 10119, USA

Chapman & Hall Japan, ITP-Japan, Kyowa Building, 3F, 2–2–1 Hirakawacho, Chiyoda-ku, Tokyo 102, Japan

Chapman & Hall Australia, Thomas Nelson Australia, 102 Dodds Street, South Melbourne, Victoria 3205, Australia

Chapman & Hall India, R. Seshadri, 32 Second Main Road, CIT East, Madras 600 035, India

First edition 1995

Typeset in 10/12pt Times by Florencetype Ltd, Stoodleigh, Devon

Printed in Great Britain by Hartnolls Ltd, Bodmin, Cornwall

ISBN 0 412 56630 3

A catalogue record for this book is available from the British Library

Library of Congress Catalog Card Number: 94-68794

∞ Printed on permanent acid-free text paper, manufactured in accordance with ANSI/NISO Z39.48-1992 and ANSI/NISO Z39.48-1984 (Permanence of Paper).

Contents

Preface

Kinetics (from the greek χίνε τίχος, forcing to move) is a branch of natural science that deals with the rates and mechanisms of any processes, physical, chemical or biological. Microbiological kinetics studies all dynamic manifestations of microbial life: growth itself, survival and death, adaptations, mutations, product formation, cell cycles, interactions with environment and other organisms. Contrary to simple dynamic recording, kinetic studies require the perception of the underlying mechanisms by the combination of experimental measurements and mathematical modelling. The model formalizes the postulated mechanism of the studied reaction, so that the comparison of observation and the model's prediction allows us to discard an incorrect hypothesis.

This book pursues general kinetic principles and ecological applications. Nowadays, ecological problems have acquired one of the highest priorities and the prospects of resolving them (monitoring and prediction of global climate changes, development of sustainable agriculture, remedying of a polluted environment) are usually associated with system approaches which rely on quantitative analysis and mathematical simulations of natural processes. However, every microbiologist who takes the time to digest contemporary mathematical simulations of this kind will be greatly disappointed by the very limited attention generally paid to the 'microbial compartment'. This dreadful contradiction to the real contribution of the microbial community to biospheric functions is in no way the fault of computing people. It is simply that the knowledge of microbial life in the natural environment is very poorly understood in quantitative terms. To date, we have only fragmentary data on the real growth rates and survival of microbial populations *in situ*, and the total biomass and abundance of some particular species. It is highly problematic to follow microbial dynamics in a natural habitat, and almost impossible to understand and explain the observed data in an unambiguous way.

Most of the mentioned 'blank spots' belong to the field of microbial kinetics. Slow implementation of biokinetics to microbial ecology may be explained by the complexity of natural objects, but more important is the inadequacy of the theoretical background and principles of kinetic analysis which has evolved independently of ecological problems. The contemporary kinetic theory of microbial growth has been developed from pressure to solve urgent problems in the fermentation industry. Most of the kinetic studies originated from chemical kinetics and are focused on the solution of particular well-defined problems, such as the optimization of cell biosynthesis under steady-state and the fully controlled continuous fermenter. This has resulted in the domination of models which have a narrow range of applicability and fail in the prediction of the diverse adaptive reactions under changeable environmental conditions (of particular importance for ecologists!). In addition, most efforts have been directed to studies of the limited range of industrial microorganisms (enterobacteria, baker and fodder yeast, bacilli) whose growth properties differ considerably from populations that dominate in natural habitats.

The objective of the experimental and simulation studies summarized in this book was to develop new principles of kinetic analysis which would allow us to understand the complex dynamic behaviour of microorganisms, both in a laboratory culture and in nature. I have made an attempt to show that biokinetics may be efficiently used to resolve a number of urgent problems in microbial ecology.

These objectives require a careful examination of the knowledge already accumulated in this particular field, and hence the first two chapters of the book concentrated on an historical survey and the organization of the known facts. I have tried to follow, step-by-step, how the facts were discovered, and how new ideas have been developing from its very origin. The results of original studies ware presented in the second half of the book. They have been arranged in an order of increasing complexity of the examined systems, from homogeneous growth (Chapter 3) to colonies and biofilms (Chapter 4), mixed cultures and competitive analysis (Chapter 5), and, finally, to microbial growth in soil (Chapter 6).

The experimental data presented were obtained mainly during the course of my work at Moscow University, Department of Soil Biology and at the Institute of Microbiology, Russian Academy of Sciences. It may be attributed to efforts of a research team of enthusiastic young scientists and students of which I have the pleasure to be the formal leader.

I am greatly indebted to Professor John S. Pirt, my first and the only mentor in microbial kinetics, and to Professor Michael J. Bazin for valuable discussions and support in the publishing of this book.

I appreciate very much the assistance of Dr Elizabeth Scott who had the difficult task of improving my English, as well as Maria Syzova, Alexander Dorofeyev and Svetlana Dedysh for friendly help in the preparation of the manuscript.

I would like to acknowledge also the financial support of this work by

NATO International Scientific Exchange Programmes and the Royal Society/Russian Academy of Sciences sponsored collaborative project. Part of the performed research was sponsored by the International Science Foundation, Grant No. MIQ 000, and the Russian Fund of Fundamental Research, Grant No. 93–04–7202.

Historical development of microbial growth theory

1

Wagner

Pardon! a great delight is granted
When, in the spirit of the ages planted,
We mark how, ere our times, a sage has thought,
And then, how far his work, and grandly, we have brought.

Faust

O yes, up to the stars at last!
Listen, my friend: the ages that are past
Are now a book with seven seals protected . . .
So, of tentimes, you miserable mar it!
At the first glance who sees it runs away.
An offal-barrel and a lumber-garret,
Or, at the best, a Punch-and-Judy play,
With maxims most pragmatical and hitting,
As in the mouths of puppets are befitting!

Johann Wolfgang von Goethe

The history of microbiology is fairly short and yet it is full of dramatic events and instructive illustrations. It has been discussed in numerous writings, including biographies of prominent microbiologists, popular scientific issues and traditional 'historical introductions' to textbooks and monographs. Our excursion into history is by no means an introduction. Rather, it belongs intrinsically to the main body of the book and we shall concentrate on aspects which have not been adequately reviewed elsewhere. The general aim of this chapter is to trace the origin of contemporary concepts of microbial growth. To define further the boundary of this survey we will list the key questions:

1. How did our current views on the nature of microbial culture evolve?
2. How have cultivation techniques been improved over the years to attain the present-day level of fully controlled fermentation?

3. What was the contribution of the exact sciences (i.e. physics, chemistry), as well as mathematical modelling to the understanding of microbial processes?
4. How have theoretical microbiological concepts been applied to microbial communities in natural habitats, such as sediments, soils and waters?

There is a tradition to start an historical survey on microbial kinetics quoting Jacob Monod (1942). There can be no doubt that this outstanding French scientist did make significant contributions to the development of the theory of microbial growth. However, anyone who chooses to undertake an even cursory historical enquiry would not be sceptical about the earliest achievements. So let us start *ab ovo*.

1.1 THE BIRTH OF MICROBIOLOGY

People were cultivating microorganisms for the production of wine, vinegar and sour dairy products long before their discovery, developing skills and methodology using a trial-and-error approach. However, real progress in microbiology, including the fundmental discovery of the existence of microorganisms was critically dependent on achievements in the fields of the exact sciences. An important contribution to the invention of the microscope was made by Galileo Galilei, who was in fact a creator of contemporary natural sciences. In 1610, he built not only a telescope but also an optical device, called an **occialino**, to examine minute objects.

The optical instrument constructed by Antony van Leeuwenhoek (1680–1720) was less sophisticated than the occialino, but due to perfect lens grinding, intuitive use of dark-field microscopy, and surely, due to the phenomenal research abilities of the Dutchman, it allowed the observation of yeasts and bacteria. Leeuwenhoek made accurate observations of bacilli, cocci and spirochetes, recorded bacterial motility, noticed proliferation and made approximate measurements using sand grains as a standard of comparison.

At the same time the truly scientific study of microorganisms was made possible by advances in chemistry. This science had been ridded of mediaeval alchemy and had become objective and precise. In the 19th century it was already based on reliable experimental technique and could be used to explain and predict certain phenomena. A decisive factor in the switching to quantitative methodology was the introduction of new research tools – dry weight determinations and gasometric analysis. This led to the establishment of the mass conservation law (formulated by Lomonosov in 1748 and by Lavoisier in 1789), of Dalton's law of multiple proportions (in 1803) and of Proust's law of constant composition (in 1808). These three laws had formed the quantitative basis of chemistry, the stoichiometry of chemical reactions. Simultaneously, an oxygen theory of combustion was developed by Lavoisier (1774–1777), followed by the experimentally tested atomic theory of Dalton (in 1803), and the molecular theory of Avogadro (in 1811). Berzelius, Liebig, and Dumas founded the analytical chemistry of natural compounds and in 1861 Butlerov advanced the structural theory of organic compounds. After this time chemistry was converted from being a predominantly analytical science to a synthetic one.

The research subjects of the new generation of chemists and of biologists partly overlapped in the studies of fermentations. These processes were viewed as being purely chemical. Lavoisier used alcoholic fermentation as an example of a chemical reaction to verify his mass conservation law, and Gay Lussac in 1810 established its stoichiometry as

$$C_6H_{12}O_6 = 2CO_2 + 2C_2H_5OH$$

According to Berzelius, fermentation proceeded through abiotic contact catalysis, while Liebig interpreted it as chemical interaction of sugars with products of animal and plant degradation.

The discovery made in 1837 by Cagniard-Latour in France and Schwann and Kützing in Germany of yeast development during fermentation was ignored by chemists because this biological study was based on observation, while the accepted standards of methodology at that time already called for quantitative experimentation. It was very indicative that justice was restored due to Pasteur, who was a chemist adherent to exact science rather than to a descriptive one.

Louis Pasteur

The birth of microbiology as a science is usually dated back to 1857, when Pasteur's first paper, *Mémoire de la Fermentation appelée lactique*, was published. Omitting the exhaustive commentaries available on this work (Stephenson, 1949), we would like to stress two points important for our review.

1. Pasteur was the first to use growth dynamic data combined with chemical assays as experimental evidence for his conclusions. Thus, to prove that chemical transformations were caused by microbial activity, he demonstrated the increase of microbial biomass during the course of this particular process.
2. Pasteur put microbial cultivation onto a solid scientific footing by introducing defined nutrient media, with each component having its own well-specified function. For example, *Penicillium* moulds were grown on a medium containing (per l of water) 20 g of sucrose, 2 g of ammonium tartrate, and 0.5 g of bakery yeast ash as a source of mineral elements.

This research direction was continued by Raulin (1869), a former student of Pasteur. Using liquid culture of *Aspergillus niger* as an example, he accurately measured the amount of mycelium mass formed (x) and residual concentrations (s) of individual chemical compounds in the defined medium to calculate the trophic coefficient, α (subsequently, it was replaced by its reciprocal, Y, and called the growth yield):

$$\alpha = \frac{1}{Y} = \frac{s_0 - s}{x - x_0}, \qquad [1.1]$$

where x_0 is the inoculum biomass and s_0 is the initial amount of substrate in the medium. Raulin arrived at the conclusion that α values were almost constant, i.e. to get a 1 g increase in biomass, a microbial culture should consume 4.6 g sugar and 0.2 g NH_4NO_3. Raulin determined α values not only for carbon and nitrogen, but also for P, K, Mg, Fe and Zn. Corresponding compounds were added not in the form of 'ash' but as individual substances.

S.N. Winogradsky and M. Beijerinck

Along with Pasteur, these famous scientists should be honoured as founders of general microbiology. Like the ingenious Frenchman, they were graced with flushes of scientific insight and the ability to scrupulously test their ideas through laborious experimental work. They should be credited not only for their discoveries but also for shaping the methodology of scientific research in general.

The contribution made by Winogradsky to microbial growth theory can be summarized as follows:

1. The development of cultivation technique. Winogradsky introduced to microbiological practice the method of continuous flow cultivation in its microscopic version. This technique was applied to sulphur bacteria. The bacteria were grown in a droplet of hydrogen sulphide solution held under a cover glass on a microscope slide, with nutrient solution being replaced manually many times a day. The flow of nutrient turned out to be absolutely essential to maintain the growth of bacteria. In addition, this methodology contributed to the discovery of chemolithotrophy. Winogradsky made microchemical determinations in the output solution and found a smooth increase in sulphate ion concentration which was closely related to bacterial growth and the oxidation of sulphide. The elegance and simplicity of this technique was striking: 'this is such kind of noble simplicity which allowed to solve the most difficult problems' (Timirjaziev). Subsequently, continuous-flow methodology was neglected for nearly sixty years, and only from the 1950s onwards

did it become extremely popular as one of the most effective methods of controlled cultivation.

2. Physiology and growth stoichiometry of bacteria oxidizing inorganic compounds. Chemolithotrophs were the most rewarding objects for the quantitative studies of the end of the 19th century, because at that time inorganic chemistry was in the ascendant. The growth dynamics of nitrification revealed that the ratio between oxidized NH_4^+ and assimilated CO_2 was constant and equal to 35.4 : 1 (Winogradsky, 1949, pp.170–174). Later, determinations made by Meyerhof fully confirmed not only the general conclusions but also the numeric stoichiometric values obtained by Winogradsky.
3. Microbial growth under laboratory conditions and in natural habitats. Winogradsky and Beijerinck raised, for the first time, the problem of distinction between indigenous microbial populations *in situ*, and domesticated microbial cultures. Winogradsky considered laboratory cultivation conditions as 'abnormal' in the sense that they distorted microbial properties as compared with natural habitats. Adherence to this idea eventually led Winogradsky to the formulation of an essentially new research strategy in soil microbiology.

1.2 EVOLUTION OF VIEWS ON MICROBIAL GROWTH, 1900–1930

Robert Koch

During this period microbiology was developing mainly as an applied branch of science and was concerned primarily with the fighting of infectious diseases. The mentality of microbiologists of that time was undoubtedly and to a large degree shaped by Koch. His famous postulates were equally well understood by researchers and by physicians. The microbial cultivation technique developed by Koch and his followers was straightforward and reliable. Also simplified were theoretical concepts of growth, activity and compositional structure of bacteria. Microbiology as an applied science has benefited from Koch's deliberate primitivism. 'He was the right man at the right time and medicine probably owes as much to his limitations as to his great gifts' (Stephenson, 1949, p.8). On the other side of the coin, the 'average' microbiologist at the beginning of the 20th century lost the broad-mindedness and attachment to exact sciences which had been so typical for the initial historical period. The development of quantitative microbiology was not altogether stopped but was certainly relegated away from the main stream of scientific progress. It was continued by a few enthusiasts, now almost forgotten.

Growth dynamics of microbial culture

One of the first biometric observations of microbial growth was made by the English bacteriologist Ward (1895). As a research object, he chose the exceptionally large bacteria *Bacillus ramosus* whose cells form filaments up to 1 mm in length. Ward was the first to present growth observation data graphically. He also introduced the measure of growth rate in terms of generation time and formulated the basic concepts of microbial kinetics by identifying two groups of factors affecting growth rate, internal factors (filament age, viability, germination strength of spores) and external factors (temperature, illumination, amount of nutrients).

Concurrently, a detailed study of the growth dynamics of typhoid bacteria was undertaken in Germany by Müller (1895). He established the existence of the lag phase in the development of bacterial cultures, and distinguished logarithmic and decelerating growth. For many following years it was growth phase differentiation that was the favourite subject of theoretical analysis. The growth phases were identified on the grounds of the dynamic pattern of microbial cell number. Such dynamics were recorded by plating technique or sometimes by direct microscopic count. Most clearly, phases were recognized by the use of the following approximation function (Buchanan, 1918):

$$N = N_0 \exp(\mu F(t)\, t) \qquad [1.2]$$

where N is an instant number of microorganisms, N_0 is the value of N at the start of the respective growth phase, and t is time. The empirical function $F(t)$ was allowed to have different forms for the seven consecutive growth phases:

- Initial stationary $F(t) = 0$;
- Lag $F(t) = t^{n-1}$;
- Logarithmic growth $F(t) = 1$;
- Negative growth acceleration $F(\mathrm{t}) = t^{-t-1}$;
- Maximum stationary $F(t) = 0$;
- Accelerating death $F(t) = t^{n-1}$;
- Logarithmic decrease $F(\mathrm{t}) = -1$, $1.56 \leqslant n \leqslant 2.7$.

Many features of microbial cultures were found to be related to growth phases; metabolic activity of cells, their resistance to unfavourable factors, electrophoretic mobility, agglutination ability, etc. (Müller, 1895; Sherman and Albus, 1923). However, main attention was paid to growth phase-dependent morphological variations.

'Life cycles' and 'cytomorphosis' in bacteria

The interest in this issue was stirred by a long and uncompromising dispute between monomorphists and pleomorphists. The founder of monomorphism was Cohn, who was also the author of a well-known morphological classification of bacteria. The pleomorphists (Zopf, Löhnis, Enderlein, Mellon, Almquist) were obviously right in their criticism of the oversimplification and primitivism of monomorphists, who denied variability of shape and size of bacterial cells.

However, pleomorphists went too far in assuming that all the bacterial diversity originated from just one or a few species, to produce an enormous variety of forms, depending on cultivation conditions and the stage of their 'life cycle'. A well-substantiated criticism of the views of pleomorphists was given by Winogradsky in his papers of 1887 and 1937 (Winogradsky, 1949, pp. 25–47, 123–141). He showed that their fallacy stemmed basically from the failure to isolate pure cultures of some difficult 'capricious' bacteria. Thereby, the truth was eventually restored. However, even more important to our story was the consequence of this scandalous dispute since it provoked a research interest in morphology and the development of bacteria. A clear-cut trend emerged identifying microbial growth phases with ontogenetic stages in the development of higher organisms. Most openly this was found in the works of the American bacteriologist Henrici (1928). Instead of a 'life cycle', he introduced the concept of 'cytomorphosis' as 'the cell of bacteria undergo a regular metamorphosis during the growth of a culture similar to the metamorphosis exhibited by the cells of a multicellular organism during its development, each species presenting three types of cells, a young form, an adult form and a senescent form . . .'.

Lag phase

It is natural that, having adopted such concept, the microbiologist would attach the greatest attention to the lag phase since this 'embryonic' phase seemed to be decisive for the fate of a microbial culture. It was during this phase that internal conditions were created which would definitely determine the future growth and death of the cell population. Prolonged latency was explained: (a) as being the time needed to synthesize some intermediate metabolite; (b) as a 'preadjustment' of the environment required for cell growth; (c) as cell recovery after stress caused by metabolic products; and (d) as being the results of some kind of cell inertia (Penfold, 1914; Sherman and Albus, 1924). Already at that time, mathematical modelling was used to test special hypotheses. Thus, it was suggested by Buchanan (1918) that, like higher organisms, bacteria did have a resting phase even though they may not form any specific dormant structures (spores or cysts). Active metabolism may be restored in response to an external signal in the form of transfer to a fresh nutrient medium. During the lag phase, latent cells, y, were assumed to germinate into active cells, x, with germination times having a Pearson distribution. Then

$$\dot{x} = \mu(t)x\,, \quad \mu(t) = \frac{\mu_m x}{x+y}\,, \quad y = x_0 - x\,,$$

$$x(t) = \int_{-a_1}^{t} c\left(1+\frac{t}{a_1}\right)^{m_1}\left(1-\frac{t}{a_2}\right)^{m_2} \exp(\mu t)\, dt\,, \qquad [1.3]$$

where μ is the specific growth rate of microorganisms, μ_m is the maximum value of μ, and a_1, a_2, m_1, m_2 are Pearson distribution parameters.

Other growth phases

In many textbooks and monographs of that era one can find a derivation of the exponential growth equation for binary dividing bacteria, based on the geometric progression 2, 2^2, 2^3 . . . (Rahn, 1932). Then

$$N = N_0 2^n\,, \qquad [1.4]$$

where n is the number of divisions. Interestingly, true exponential growth was thought to occur only in the case of symmetrically dividing bacteria with equal probability of subsequent division for the mother and daughter cells. Growth of other microorganisms, such as yeasts and especially of mycelium fungi, appeared to be very complicated and unpredictable. In the case of budding yeasts, however, the exponential law was thought to give a reasonable approximation (Rahn, 1932, p.398).

Initially, the only measure of the growth rate was the generation time, g, calculated by the Pedersen formula,

$$g = (t - t_0)\frac{\ln 2}{(\ln N - \ln N_0)}\,. \qquad [1.5]$$

Later Slator (1916) introduced a 'growth constant', μ. It was defined as the proportionality coefficient in the differential equation relating the growth rate of cells to their instant number:

$$\frac{dN}{dt} = \mu N \qquad [1.6]$$

The solution of this equation is given by an exponential function,

$$N = N_0 \exp(\mu t)\,. \qquad [1.7]$$

Comparison of Eqns [1.4] and [1.7] gives the following relationship between μ, n, and g,

$$\mu = \frac{\ln 2}{g} = \frac{n \ln 2}{t} \qquad [1.8]$$

A remarkable feature of the theoretical views at that time on the exponential growth of microorganisms was that the value of μ was presumed to be in general a constant for the given species. A drop in μ during post-exponential growth was treated as a deviation from the normal development or as a result of the expression of some 'internal development programme'. For this reason, little attention was paid to studying how μ was affected by different environmental factors. The only exception was temperature. Its effect on the rate of a microbial process, k, was described by the Arrhenius equation

$$\ln k = \ln C - \frac{E_a}{RT} \,.$$

where T is absolute temperature, C is an integration constant, R is the gas constant, and E_a is the activation energy.

Deceleration and eventual arrest of microbial growth in the population was attributed to the action of the following factors (Rahn, 1932): (a) depletion of substrate; (b) accumulation of toxic metabolic products; and (c) formation of specific auto-regulatory metabolites. The self-poisoning of cultures by metabolites was already at that time described by simple empirical equations. For example, in the case of inhibition of growth of saccharomycetes by ethanol

$$-\frac{ds}{dt} = kx\left(1 - \frac{p}{K_i}\right), \qquad [1.9]$$

where s and p are concentrations of substrate (sugar) and ethanol, respectively, x is the yeast biomass, k is the metabolic activity and K_i is the inhibition constant.

The phase of decline was usually approximated by a first-order equation (Eqn. [1.1]), without making too much effort to trace the exact survival dynamics. Of more interest was the dynamics of microbial death caused by the action of an antiseptic or by unfavourable environmental factors. The approximating equations in this case were chosen as the simplest possible form, such as the following (Buchanan, 1918):

$$N - N_0 \exp(-KC^n t)\,, \qquad [1.10]$$

where K and n are empirically fitted constants and C is the concentration of the biocide.

Simulation of colonial growth

There were several attempts to apply kinetics analysis to situations more complicated than homogeneous microbial growth. The Russian microbiologist Egunov (1914) proposed the following equation for the growth of fungal or bacterial colonies on agar media:

$$\frac{dR}{dt} = K_R - kR\,, \qquad [1.11]$$

where R is the radius of the colony, K_R is its linear growth rate (assumed to be constant), and k is an empirically fitted constant characterizing the growth inhibition by metabolic products. Some microorganisms (fungi on all media and *Bacillus subtilis* on gelatin) were found to increase the radius of their colonies at a constant rate, in this case $k = 0$. This phenomenon was explained as follows:

> '. . . the colony can grow only by its very thin peripheral zone. For a uniform growth, this zone must always stay beyond the influence of increasing amount of metabolic products' (Egunov, 1914, p.5).

The work of Egunov was, however, neglected. Only some 50 years later were these kinds of studies resumed, and the early findings on the constancy of K_R for fungi, due to constancy of peripheral zone were brilliantly confirmed by much more advanced and sophisticated experimental methods (Pirt, 1975). Apart from Eqn. [1.11], Egunov came up with several other successful simulations of dynamic microbial behaviour. His conclusive words sound almost poetic:

> '. . . for a biologist, microbial cultures on the plate are just like the stars in the sky for an astronomer; they conceal biological laws and mathematical analysis is the only means to discern them' (Egunov, 1914, p.20).

The problem of 'infinite' vegetative growth

A large series of quantitative investigations of microbial growth dynamics was performed on monoxenic protozoan cultures. In prolonged experiments involving daily transfers of these unicellular organisms to a fresh nutrient medium, the specific growth rate, μ, was measured as a function of different medium factors (temperature, acidity, composition of the nutrient broth, etc.) and the internal properties of the culture (Woodruff and Baitsell, 1911a,b; Galadzhiev, 1932). The major problem addressed in these essentially continuous-flow experiments was whether there existed any limit to the reproductive capacity of protozoa. Would they age in the course of vegetative propagation or would they always remain young? The answer was unambiguous as no degeneration of paramecia was detected even for 20–22 years (!) of continuous vegetative growth, the μ-values remained more or less constant and showed oscillations between 0.5 and 1.5 day^{-1}. Initially these oscillations were attributed entirely to the effects of temperature. Subsequent experiments under constant temperature, however, revealed that it was only the seasonal (winter, summer) bias that vanished, whereas autonomous sustained oscillations of μ with a period of several days were still evident. Their nature is obscure until now. There were some attempts to explain periodic acceleration of vegetative propaga-

Table 1.1 The first models of mathematical ecology

Simulated process	Equation	Eqn. no.	References
Unlimited growth of population	$\dot{N} = (\mu - a)N = rN$	[1.12]	Malthus (1798)
Limited growth of population	$\dot{N} = rN \dfrac{\ln(N/K)}{\ln K}$	[1.13]	Gompertz (1825)
	$\dot{N} = rN(1 - N/K)$	[1.14]	Verhulst (1838)
Interspecific competition	$\dot{N}_1 = N_1[r_1 - \gamma_1 F(N_1, N_2)]$, $\dot{N}_2 = N_2[r_2 - \gamma_2 F(N_1, N_2)]$	[1.15]	Lotka (1925); Volterra (1931)
Prey–predator interactions	$\dot{N}_1 = r_1 N_1 - \gamma_1 N_1 N_2$, $\dot{N}_2 = r_2 N_1 N_2 - \gamma_2 N_2$	[1.16]	Lotka (1925); Volterra (1931)
	$\dot{N}_1 = r_1 N_1 - \gamma_1 N_2 \sqrt{N_1}$, $\dot{N}_2 = \begin{cases} r_2 N_2 \sqrt{N_1}, & \text{if } N_1 \neq 0, \\ -aN_2, & \text{if } N_1 = 0 \end{cases}$	[1.17]	Gause and Vitt (1934)

N, population density; N_1, population of prey; N_2, population of predator; K, environment capacity; μ, true growth rate; a, death rate; r, apparent growth rate; γ, coefficient of interspecific interaction.

tion in protozoa by endomixis or by allelopathic effects (Robertson, 1924), but these hypotheses were rejected shortly after their advancement.

The rise of mathematical ecology and demography

The mathematical relationships we have discussed so far were purely descriptive, i.e. they characterized microbial growth only in empirical terms. We may guess that this approach originated from mathematical demography of that time and from mathematical models of population dynamics of higher organisms. An interest in these areas arose as early as the last century in connection with vigorous industrial development, the active migration of people and a dramatic increase in the exploitation of natural resources. Listed in Table 1.1 are models of population dynamics which have been and are still used in mathematical ecology. Most often population dynamics were approximated by the so-called logistic equation [1.14].

Its derivation is based on the following assumptions:

1. The population size is confined by a parameter, K, (the capacity of the environment).
2. The apparent specific growth rate of population, r, ($r = \mu - a$, where a is the death rate) is directly related to the difference between K and the instant population density, N.

The logistic equation is of little value for prediction because it is never possible to know the values of K and r in advance, but when used *post factum,* it does allow a reasonably good fit to a wide range of dynamic curves. For this reason, the logistic equation has been frequently used to describe the growth of microorganisms under various circumstances, in homogeneous cultures and on solid media, in continuous-flow columns and in soil (Pearl, 1925; Henrici, 1928; Saunders and Bazin, 1973; Koch, 1975).

A major step forward in population dynamics studies was made between 1920 and 1930 when a mere registration and formalistic description of time series gave way to a deeper theoretical analysis of the driving forces of particular dynamic processes. The pioneers in this field were the American physicist Alfred J. Lotka and the Italian mathematician Vito Volterra. In Lotka's classical work, *Elements of Physical Biology* (Lotka, 1925), an attempt was made to solve various biological problems by using approaches borrowed from mechanics, molecular physics and physical chemistry. Thus, he 'derived' Eqn. [1.14] from the principles of chemical kinetics, analysed interactions in 2–3-species communities, and developed the first models of epidemic disease.

An even more important contribution, as is generally accepted (Svirezhev, 1976), was made by Volterra, who developed the first mathematical theory of biological communities. This theory describes the interaction of competing populations, as well as the interaction of populations at different trophical levels. It takes into account the time-lag phenomena (time delay) and can be extended to so-called conservative and dissipative biological communities (Volterra, 1931).

In works of the Italian mathematician and his followers (Gause and Vitt, 1934; Kostitzin, 1934), the qualitative theory of dynamical systems was expanded to apply to biological objects. This theory was originally developed by Lyapunov and Poincaré for solving problems of celestial mechanics. The central idea was to analyse the behaviour of the system under small perturbations away from steady-state solutions. This approach is still recognized as one of the most effective in mathematical biology (Romanovsky *et al.*, 1984). Let us discuss, for illustration, its historically first biological application, which by happy chance was related to a microbiological object, the protozoa. In Gause's experiments, paramecia *P. caudatum* were grazed by *Didinium nasutum* (Gause, 1934). The population dynamics of each species exhibited aperiodic damped oscillations before elimination of both. The predator consumed all the prey and then perished from starvation. However, the classical Volterra model (Eqn. [1.15], Table 1.1) contradicts these experimental data, predicting undamped periodic oscillations of N_1 and N_2. In a theoretical study, Gause and Vitt (1934) modified Eqn. [1.15] to improve the agreement with observed dynamics (Eqn. [1.17]). It was enough to assume firstly, that the predator mortality was significant only at small N_1, and secondly, that its growth rate was proportional to the square root of N_1. Thus, a purely mathematical operation allowed the gaining of new insight into the biology of the studied object. Information of this kind is all the more valuable because it can suggest new experiments that otherwise would not have occurred to a traditionally minded protozoologist.

The emergence of microbiological kinetics

Van Niel wrote in 1949: 'Growth is the expression *par excellence* of the dynamic nature of living organisms. Among the general methods available for the scientific investigation of dynamic phenomena, the most useful ones are those which deal with the kinetic aspects' (Van Niel, 1949, p.102). These words proved to be prophetic, because in the next year, 1950, the chemostat theory was published, and microbiological kinetics began its vigorous development. But what about earlier times? Were kinetic investigations with microbial cultures performed by anyone? As a matter of fact not all of the dynamic studies, discussed above, could be considered as kinetic ones *sensu stricto,* because very few were aimed at the analysis of the mechanism of the growth phenomenon.

To make clear the distinction between the terms 'dynamic' and 'kinetic', one needs to recall the origin of microbiological kinetics. We have no doubts that it was derived from kinetics of chemical and enzymatic reactions.

Chemical kinetics, which studies mechanisms and rates of chemical reactions, developed as a separate branch of physical chemistry by the end of the last century. Guldberg and Vaage, Menshutkin, Arrhenius and van't Hoff built up the foundations of chemical kinetics as they invented and refined the technique for reaction rate measurements, introduced the concept of the kinetic order and formulated the general principles governing the rates of simple chemical reactions. The study of temperature effects led then to an understanding of the role of active molecules in chemical interactions and, eventually, to the formulation by Eyring of transition-state theory based on thermodynamics and quantum mechanics (1930–1935). In the first third of the century the greatest attention was paid to complex chemical reactions, including peroxide oxidation (Bach,

Ehngler), as well as to linear (Bodenstein) and branched chain reactions (Hinshelwood, Semenov).

The study of kinetics of enzymatic reactions was born at the turn of the century. Experimental methods available for the founders (O'Sullivan, Thompson, Wurtz, Buchner, 1880–1897) were far from perfect. There were no buffers, no purified enzymes, and rather cumbersome analytical techniques to record the full reaction dynamics (later to be substituted for more convenient and exact measurement of initial reaction rates). Nevertheless, many important facts were established. Thus, it was found that enzymes were true catalysts (being regenerated after each reaction event) and that catalysis occurred via the formation of enzyme–substrate complexes (ESCs). The presence of ESCs was proven by kinetic data alone. The thermoinactivation of pure enzyme solution (exemplified by invertase) was much higher than that of this enzyme in the presence of substrate (sucrose). Enzymatic reactions were shown by Brown (1890–1902) to be of mixed kinetic order, being first-order at small substrate concentrations and zero-order at high ones. In 1902–1903 Henri discovered that, besides ESCs, there were other enzyme complexes, in particular with the reaction products (actually the explanation of a competitive product inhibition) and developed the first mathematical model of the enzymatic reaction.

A fundamental step forward was made by the German biochemists Michaelis and Menten, who advanced a technique for measuring initial reaction rates under fully controlled conditions, i.e. at constant temperature and pH values by using acetate buffer (Michaelis and Menten, 1913). Another of their accomplishments, of a theoretical nature, was the derivation of the equation describing the rate of an enzymatic reaction, v, as a function of substrate concentration, s:

$$v = V\frac{s}{K_s + s}, \qquad [1.18]$$

where V is maximal reaction rate and K_s is a parameter characterizing enzyme affinity to substrate (formally, the dissociation constant for the enzyme–substrate complex; see Table 1.2).

This equation predicts a hyperbolic relationship between v and s, which has been confirmed by numerous experiments of the authors themselves and by others over a period of more than 70 years. Admittedly, there are several deviations from the simple Michaelis–Menten kinetics, which occur when an enzymatic reaction is complicated by cooperative phenomena, non-trivial inhibition, inactivation of enzymes, etc. However, what we would like to stress are not the exceptions but the way in which this equation was obtained and verified. Unlike empirical equations [1.2], [1.3] and [1.9–1.17], Eqn. [1.18] was deduced from well-defined and clearly stated assumptions about the catalytic mechanism. This mechanism is best

Table 1.2 The first kinetic models in enzymology (after Cornish-Bowden, 1976)

Mechanism of enzymatic reaction	Respective kinetic equation	Originally suggested by
$E + S \xrightarrow{k_1} ES \xrightarrow{k_2} E + P$	$v = \dfrac{Vs}{K_m + s}$ $V = k_2 e_0,\ K_m = k_2/k_1$	Van Slyke and Cullen (1914)*
$E + S \overset{K_s}{\rightleftarrows} ES \xrightarrow{k_2} E + P$	$v = \dfrac{Vs}{K_s + s}$ $V = k_2 e_0$	Michaelis and Menten (1913)
$E + S \underset{k_{-1}}{\overset{k_1}{\rightleftarrows}} ES \xrightarrow{k_2} E + P$	$v = \dfrac{Vs}{K_m + s}$ $K_m = (k_{-1} + k_2)/k_1$	Briggs and Haldane (1925)*
$E + S \overset{K_s}{\rightleftarrows} ES \xrightarrow{k_2} E + P$ $K_{ss} \updownarrow$ ES_2	$v = \dfrac{Vs}{K_s + s + s^2/K_{ss}}$	Haldane (1930)*
$E + S \overset{K_s}{\rightleftarrows} ES \xrightarrow{k_2} E + P$ $E + I \overset{K_i}{\rightleftarrows} EI$	$v = \dfrac{Vs}{K_s(1 + i/K_i) + s}$	Michaelis and Menten (1913)

E, enzyme; e_0, total enzyme concentration; S, substrate; s, substrate concentration; ES, enzyme–substrate complex (ESC); ES_2, double non-productive ESC; I, inhibitor; i, inhibitor concentration; k_1, k_2, k_3, kinetic constants of elementary reactions; K_s, K_{ss}, K_i, dissociation constants of respective dynamic complexes.

* Reference cited in Cornish-Bowden (1976).

characterized by a flow diagram showing the interaction between substrate, S, and free enzyme, E. Examples of such diagrams along with corresponding steady-state equations are shown in Table 1.2. To verify the postulated mechanism, a prediction given by equation should be compared with experimental data. If there is no agreement, the hypothesis is, obviously, rejected. When there is agreement, however, no unambiguous conclusion can be made. As often happens, several equations produce the same or close residual errors of fitting to one particular set of experimental points. Moreover, several basically different mechanisms can lead to one and the same equation. For instance mechanisms 1–3 and 5 (Table 1.2) produce hyperbolic relationships between v and s; hence measurements of initial reaction rates at several substrate concentrations do not allow the differentiation of these mechanisms at all. Instead we may follow non-steady-state transient dynamics of the enzymatic reaction or record the dependence of v on s at various inhibitor concentrations. In this way the described methodology based on a combination of experimentation, together with mathematical modelling permits the discarding of wrong alternatives, but not approval of fair ones. This is well-known scientific methodology, and the kinetic approach proved to be quite fruitful. As early as 1920–1930, the kinetics of enzymatic reactions became an advanced discipline in its own right (Haldane, 1930).

Returning to microbiology, when and how was the outlined kinetic approach applied to microbial cultures and microbial populations? Did somebody use dynamic data to clarify the mechanisms of the growth process? It would be hard to answer this question unequivocally, because the notion of a mechanism is in a way like a Russian matrioshka doll in that you keep opening it and there is always something inside. For the sake of clarity, we shall define mechanistic studies as those which explain some complicated process via several simpler reactions, e.g. cell growth should be explained by the activity of subcellular elements and complexes of enzymes, microbial population dynamics by the behaviour of individual cells, etc. If so, the emergence of microbial kinetics can be dated back to the publication of a paper, *Optima and limiting factors*, written by Blackman (1905), a botanist from Cambridge University. It is interesting, that although this work dealt with plants, it is much more often cited by microbiologists than by botanists.

To begin with, Blackman advanced the notion of a limiting factor as one that controls the rate of the studied process. The limiting factors include the availability of nutrient substances and energy (for phototrophs this is CO_2, moisture, illumination, the content of chlorophyll and mineral substances) as well as environmental characteristics (temperature, acidity and tonicity of the medium). It can be easily seen that in doing so Blackman gave a dynamic interpretation of Liebig's 'law of minimum'.

For the sake of simplicity, Blackman supposed a linear relationship between the growth rate of phototrophs and the availability of a growth-limiting nutrient (like the CO_2 concentration in the air). When the CO_2 concentration goes over a certain threshold value, there is no further increase in the rate of the process. Hence, according to Blackman, the concentration dependence of the growth rate is described by a broken line, which can be treated as a rough approximation to the Michaelis–Menten hyperbola. Blackman went even further. Admitting the presence of a huge variety of enzymes ('congeries of enzymes') in any cell, he supposed that the growth rate of a cell was on the whole determined by a single enzymatic reaction (the so-called **master reaction**), which can be identified kinetically as the slowest one or 'bottle-neck'. The simplicity of Blackman's approach is so attractive that it is still invoked as a justification for the use of enzymatic kinetics equations to describe microbial growth (Varfolomeyev, 1987). It is worth mentioning, however, that Blackman's contemporaries tended to oversimplify the bottle-neck concept and to interpret it too literally. Thus, kinetic parameters of a microbial culture were thought to be completely identical to those of the responsible cell enzymes. For example, temperature dependence of the growth rate was naively used for calculation of the activation energy parameter attributed to particular cell enzymes.

At the same time, the bottle-neck postulate was questioned on several grounds. Specifically, it was argued that: (a) the longer the metabolic pathway, the larger may be the difference between the rates of the bottle-neck reaction and the end product formation; and (b) the truly master reaction would not necessarily be the slowest, much more important should be its position in the sequence between reversible and irreversible steps, the real bottle-neck being localized just before the first irreversible reaction (Burton, 1936).

In order to substantiate the outlined arguments, their advocates resorted to growth 'mechanisms' like the following ones:

$$A \xrightarrow{k_1} B \xrightarrow{k_2} C \ldots \xrightarrow{k_n} M, \qquad A \underset{k_{-1}}{\overset{k_1}{\rightleftarrows}} B \underset{k_{-2}}{\overset{k_2}{\rightleftarrows}} C \ldots \overset{k_n}{\rightleftarrows} M,$$

where A, B, C, . . . are the substrate and intermediates of intracellular metabolic reactions, M denotes metabolic products, and k_1, k_2, . . . are first-order rate constants. Such blunt conventional and artificial 'schemes' of microbial metabolism can hardly be expected in reality, and although they can still be found in one form or another, even in recent publications (Ierusalimsky, 1967; Varfolomeyev, 1987), they are unable either to confirm or disprove the bottle-neck postulate.

Earlier works in quantitative microbiology were not based on the combination of the experimentation with mathematical modelling, an approach firmly established by that time in chemistry and enzymology. Moreover, it seems that a remarkable progress in enzymology was generally beyond the scope of even the most educated microbiologists of that time. Thus, neither the Michaelis–Menten equation nor any other equations from enzymatic kinetics can be found in concurrent books on microbial physiology (Henrici, 1928; Rahn, 1932). Instead, the dependence of rates of enzymatic reactions upon substrate concentrations was described by equations borrowed from chemical kinetics (mostly, by the first-order reactions). Experimental studies of specific growth rate, μ, dependence on limiting substrate concentration, s, were entirely devoid of mathematical symbolism (Penfold and Norris, 1912; Meyerhof, 1917; Henrici, 1928). Surprisingly, even Hinshelwood, highly appreciated for his numerous discoveries in chemical kinetics and perfect exercising of mathematical tools, refrained from the kinetic analysis of his experimental studies of the μ dependence on s (Dagley and Hinshelwood, 1938). The only point common to many earlier works was the assertion of a linear relationship between the limiting substrate concentration in a fresh medium and the overall cell yield of biomass. However, the latter finding was obviously related to growth stoichiometry but by no means to growth kinetics.

The most prominent kinetic investigations that are similar to modern ones were carried out between 1930 and 1935 by the Russian microbiologist Tauson (see Tauson, 1933, 1937). By studying the growth of *Aspergillus* on different substrates, he came to the important conclusion that, even in the exponential phase, the specific growth rate, μ, was not constant and varied as a function of limiting substrate concentration. The dynamics of fungal growth in a batch culture was described by the following set of differential equations:

$$\frac{dx}{dt} = \mu(s)\,x\,, \qquad \frac{ds}{dt} = -\frac{\mu(s)\,x}{Y} - mx\,, \qquad [1.19]$$

where m is the maintenance coefficient, to be discussed below. An analysis of the kinetics described by equation set [1.19] allowed Tauson to develop the new method of 'retarded' culture, which was the first example of a continuously controlled cultivation technique in the history of microbiology. Nowadays, we would call it a **fed-batch cultivation** since at regular intervals small amounts of limiting substrate were added into the submerged fungal culture and as a result, the substrate supply fully determined the microbial growth rate. Today, we also know that residual substrate concentrations, s, should be extremely low in this particular type of continuous culture. This is why Tauson obtained a reasonable approximation of the function $\mu(s)$ in the form of a first-order equation.

Bioenergetics of microbial growth

It was Lavoisier who put studies of bioenergetics onto a scientific footing. Together with Laplace, in 1783, he had designed the first calorimeter, and then found a direct relationship between muscular work of human and oxygen uptake during respiration. In 1842, the German physician and physicist Mayer made a rough estimate of the mechanical equivalent of heat and formulated the generalized energy conservation law. Mayer extended his conclusions to biological objects as he explained the different content of oxygen in venous blood of people living in the tropics and at high latitude by additional energy requirements for thermogenesis under cold climate conditions.

The chemical interpretation of animal respiration was brought to completion by the German physiologist Max Rubner. He formulated the law of nutrients isodynamy, which was based on the estimation of energy content in food products. In 1891, he succeeded R. Koch as the Director of the Hygienics Institute (later to be known as the Microbiological Institute) in Berlin and worked there until 1909. Rubner was, in fact, the first to undertake quantitative studies of the energy requirements of microorganisms. In addition to the recording of growth substrate uptake (a common practice at the time), he also used direct as well as indirect calorimetric measurements (Rubner, 1903). Direct calorimetry involves the measurement of growth-related heat production in the culture, whereas indirect determinations are made from the difference in combustion enthalpy of substrates and end products of microbial growth. Both approaches yielded similar results, as was demonstrated with alcoholic fermentation. By comparing metabolic activity of microorganisms and animals, Rubner concluded that microorganisms possessed higher specific rates and higher efficiency of energy conservation.

Bioenergetic research was continued by Terroine in France (Terroine and Wurmser, 1922) and by Tauson in the USSR (Tauson, 1933). As a measure of the efficiency of microbial growth, they introduced the coefficient of energy utilization, CEU, or energetic yield ('le rendement energetique'):

$$\mathrm{CEU} = \frac{\text{energy content of biomass}}{\text{energy content of substrate consumed}}.$$

CEU was determined by indirect calorimetry; that is, by measuring in a calorimeter the heat released from combustion of the fresh nutrient medium, Q_0, the total cultural liquid, Q_1, and its filtrate, Q_f. Then

$$\mathrm{CEU} = \frac{Q_1 - Q_f}{Q_0 - Q_f}. \qquad [1.20]$$

By the first law of thermodynamics, any change in the internal energy of an isolated system, ΔE, is caused by the transfer of heat, Q, and by work done, W: $\Delta E - Q + W$. For microbial growth, we have $\Delta E = Q_{\text{growth}} - p\Delta V$, where Q_{growth} is the heat production associated with growth and $p\Delta V$ is the work done by the microbial system on its environment (p is the pressure and ΔV is the change in volume of the gas phase). The growth of microorganisms usually proceeds at a constant atmospheric pressure, p = constant. Also, under aerobic conditions, the change in volume can be neglected, $\Delta V = 0$, because the uptake of O_2 is counterbalanced by evolving CO_2. Then, the heat release is the only pathway for the dissipation of energy in utilizing substrate:

$$\mathrm{CEU} = \frac{\Delta H_s - Q_{growth}}{\Delta H_s}, \qquad [1.21]$$

where ΔH_s is the combustion enthalpy per unit of substrate.

Maintenance energy

It had already been noticed by the pioneer researchers that the biomass yield per unit mass of utilized energy source was determined by the partition of energy expenditures between two routes, for growth and for so-called maintenance purposes, including the turnover of macromolecules, osmoregulation and cell motility. The concept of maintenance energy was borrowed from general physiology, where there is a notion of basic metabolism, estimated as respiration rate of an organism at rest and on an empty stomach. Basic metabolism occurs not only in multicellular organisms but also in a single cell. An example is the erythrocyte, which does not multiply but does respire to maintain its viability. Pfeffer (1904) compared a microbial cell with a steam engine. Even when it is idle, you have to keep on burning coal in the fire-box in order to have it ready to start off on signal. This analogy was refined by Rahn (1932), who compared a cell with an electric machine. Both depend on a continuous supply of energy to maintain the required redox potential and compensate for degradation processes. According to Warburg, the degradation was caused by diffusion, whereas Meyerhof assumed it to be caused by spontaneous chemical processes tending to increase the entropy of the system. If so, the goal of maintenance metabolism would be to bring the living system back to the original non-equilibrium state. First estimates of energy requirements for microbial motility were obtained by von Angerer (1919). By using Stokes' equation, he evaluated the energy needed to keep a cell in motion at $(10/216) \times 10^{-16}$ cal h^{-1}, which amounts to as little as 0.01% of the total energy flux.

A major contribution to the quantitative description of maintenance energy requirements was made by Terroine and Tauson. Terroine introduced the coefficient of basic metabolism, m, which was the specific rate of energy consumption at $\mu = 0$ (subsequently this parameter came to be called the maintenance coefficient). A direct measurement of m in a population of viable cells, which at the same time do not multiply, is very difficult, if at all possible. The reason is that bacterial cells normally either grow and divide ($\mu > 0$), or die and undergo lysis ($\mu < 0$), depending on substrate concentration, the presence of inhibitory products and other factors. The only way to overcome this problem is by an indirect estimation of the maintenance coefficient from the relationship between the energy consumption and growth rate. French researchers (Terroine and Wurmser, 1922) made microorganisms grow at different rates by changing the pH of the medium. From Tauson's point of view, this technique was unsatisfactory, since metabolic activity of microorganisms *per se* could well be affected by the acidity of the medium. Instead, he suggested the simultaneous recording of the energy source uptake, Δs, and the increment in microbial biomass, Δx (Tauson, 1934). These quantities are related by the following equation, which includes the desired parameters m and $Y^{\max}$ (the 'true' or maximal yield coefficient),

$$\Delta s = s_0 - s = \frac{x - x_0}{Y^{\max}} + m \int_0^t x \, dt \; , \qquad [1.22]$$

where s_0 and x_0 are the values of s and x at $t = 0$.

In another work (Tauson, 1937) he described a more accurate technique involving the use of a 'retarded' culture. With a constant rate of a limiting substrate supply, the microbial biomass should increase linearly with time, $x = kt$. By taking into account maintenance requirements, the energy source consumption is described by the following differential equation

$$\dot{s} = \frac{-k}{Y^{\max}} - mx = \frac{-k}{Y^{\max}} - mkt \; . \qquad [1.23]$$

By integration one obtains

$$\Delta s = s_0 - s = x\left(\frac{1}{Y^{\max}} + \frac{mt}{2}\right) . \qquad [1.23a]$$

This equation can be used to find the values of parameters m and $Y^{\max}$ by one of the two experiments: (a) by growing microorganisms at two different feeding rates, or (b) by cultivation at one feeding rate but long enough to attain significant changes in the biomass yield. Tauson used both methods to measure the values of m and $Y^{\max}$ for submerged cultures of *Aspergillus niger* and *A. flavus*. The numerical values he obtained (m = 0.016–0.055 cal/cal day^{-1}, $Y^{\max}$ = 50–55%) are in agreement with recent measurements in chemostat (Pirt, 1975; Roels, 1980). Essentially, Tauson also observed the non-constancy of m, which declined in old cultures. He explained this effect not only by the accumulation of dead cells (for which $m = 0$), but also by a reduction in the activity of old starving cells. In other words, he explained the variation in m by changes in the physiological state of a microbial culture.

Unfortunately, the works of Terroine and Tauson, carried out between 1920 and 1940, went unnoticed and were not repeated or cited by their contemporaries. Their findings were rediscovered only some 40 to 50 years later by Pirt, Herbert, Marr and others.

1.3 JACQUES LUCIEN MONOD

By devoting an entire section of our survey solely to Monod, we in no way wish to undermine the adopted chronological principle. The simple reason is that the efforts of other researchers during the indicated period were overshadowed by Monod's enormous contribution to the development of quantitative microbiology. For several decades, from the 1940s to 1960s, his works were the focus of interest of the scientific community. They were cited and discussed, interpreted and popularized, experimentally tested and applied to particular cases. It is difficult at this point to avoid several biographic extractions, which are based mainly on Stanier's reminiscences (1977).

Main events of Monod's life (1910–1976) and career

Monod was born in France, was Swiss by origin and Protestant by religious conviction. His father was a professional artist, art historian, and a man of great erudition. Many other ancestors belonged to the intellectual part of society (professors, pastors, public servants). He studied at the Sorbonne, but his real scientific education was informal. It was obtained from active communication with four scientists. As Monod put it later himself, he owed 'to Georges Teissier the taste for quantitative descriptions, to André Lwoff the initiation to the powers of microbiology, to Boris Ephrussi the discovery of physiological genetics, to Louis Rapkine the idea that only chemical and molecular descriptions can give a complete interpretation of the functioning of living organisms'.

Upon graduation from the Sorbonne, Monod spent several years in the 'search for a problem'. During this time he worked on the growth and differentiation of ciliates, took part in a long expedition to Greenland, and trained in genetics at the California Institute of Technology. He then returned to France, where his search for a problem eventually led him to the analysis of growth. The first experimental object was axenic cultures of protozoa *Tetrahymena pyriformis* and the effect of nutrient concentration on its growth rate. André Lwoff gave Monod two valuable pieces of advice that shaped the rest of his scientific life. The first one was to use the bacterium *Escherichia coli* instead of protozoa, since it could grow in a synthetic medium of simple composition. The second, given by Lwoff somewhat later, was to study the mechanism of enzymatic adaptation in bacteria.

The first results were not long in coming. With striking ease and elegance Monod conducted his innovative experimental studies on the kinetics and stoichiometry of bacterial growth, developed a comprehensive and consistent theory of microbial growth and discovered the phenomenon of **diauxie**. All these findings were included in his doctoral thesis, *Recherches sur la croissance des cultures bactériennes*, which appeared as a book in Nazi-occupied Paris (Monod, 1942).

After the start of the Second World War, Monod, having two sons and a Jewish wife, fled to the country to find relative safety before joining the underground resistance in Paris. Eventually he reached a position of considerable responsibility as a resistance leader and was very nearly lost at the hands of the Gestapo. Nevertheless, he did manage to continue his experimental work under the cover of the Pasteur Institute during occasional and secret visits to Lwoff's laboratory.

During the occupation he joined the French communist party. This can be considered as an act of solidarity with other Nazi-fighters rather than an expression of Marxist faith. Shortly after the end of the war he broke publicly with the French communists over the issue of their unquestioning support for Lysenko's doctrines and repressions against Russian geneticists.

The war over, Monod returned to the Sorbonne and was then invited by Lwoff to the Pasteur Institute. By the end of the 1940s, he had made his second important contribution to microbial growth theory by developing the principle of continuous culture and by designing a continuous fermenter, the 'bactogéne', intended for the mass production of bacterial vaccines. However, this technical innovation was given a cold reception by his colleagues at the Institute, who preferred to stay with Roux flasks.

In the 1950s, Monod continued his previous work on enzymatic adaptation but this time employed new research tools, including sensitive assays of β-galactosidase activity with a chromogenic substrate, enzyme isolation, purification and immunological detection. These experiments were conducted with *E. coli* mutants. Eventually these sparkling studies led to the discovery of the *lac*-operon, the next great contribution of Monod to biological theory, which was formally appreciated by the Nobel Prize in 1965 (together with F. Jacob and A. Lwoff). Concurrently Monod, together with Jacob and Changeux, finished the works on allosteric effects and gave an explanation of the enzyme activity regulation (Monod *et al.*, 1965).

In 1970 Monod published his second book, *Le Hasard et la Nécessité (Chance and Necessity)*, devoted to the philosophy of science. He offered great biological generalization, and showed with skill and elegance the connection between classical Darwinism and modern molecular biology.

In 1971, Monod became the Director of the Pasteur Institute and plunged into unrewarding administrative work, aggravated by the Institute's peculiar financial status. This work consumed the last years of Monod's life. His career as an active scientist was over. He suffered a severe attack of viral hepatitis in 1972 and developed an aplastic anaemia. In 1976, Monod retired from the Institute and this year was the last of his life.

Recherches sur la croissance des cultures bactériennes (Monod, 1942)

With respect to citation frequency, this book is probably a 'champion' among those on microbiology. However, it is probably often cited without careful reading (the reason being that it was not translated into English, and most of its readers are unable to cope with large portion of French). This is unfortunate, because the original ideas come to be distorted as the same excerpt is rewritten from one survey to another. The book consists of two parts, the first dealing with the general laws of microbial growth, and the second describing the diauxie phenomenon.

We are mostly interested in its first part. It begins with a survey of previous investigations, primarily those which deal with quantitative descriptions of phases in the development of batch culture. Monod was not aware of many of the contributions discussed above, specifically, he did not know anything of the works of Tauson, Hinshelwood and Michaelis, but was able to obtain similar results by pursuing his own line of investigation. His contribution and main results can be summarized as follows.

1. First of all, we should mention **the development of experimental technique.** Monod did for microbial kinetics as much as Michaelis and Menten did for enzymology. That is, he invented a **system** of precise methods for the dynamic measurement of microbial growth, introduced, defined and controlled cultivation conditions and made experimental procedure simple and reproducible.

To determine bacterial biomass he used turbidimetry in combination with preceding dry-weight calibration (before Monod, turbidimetry was viewed as a convenient and quick but relative measure of biomass density; growth was expressed in arbitrary units of turbidity or even 'galvanometer deflections').

The cultivation of two bacterial species, *Escherichia coli* and *Bacillus subtilis* was carried out in liquid synthetic media at a constant temperature and with agitation by shaking or air sparging. The concentration of the limiting substrate (sugar) did not exceed $0.5\ \mathrm{g\,l^{-1}}$, this being the essential condition to overcome self-inhibition and to keep a constant acidity by means of phosphate buffer alone.

2. Measurements of **biomass yield,** Y, per unit of substrate consumed, revealed that Y was almost a constant and was not influenced either by the culture age, the chemical form of the substrate or by its initial concentration in the medium. Thus, for the 15 different substrates tested (including mono-, di- and polysaccharides, pentoses, hexoses and alcohols), the biomass yield varied only between 0.17 and $0.25\ \mathrm{g\,g^{-1}}$. The maximum content of biomass, x_m, in the culture occurred at the moment of complete utilization of the substrate and was a linear function of s_0,

$$x_\mathrm{m} = x_0 + Ys_0 \ . \tag{1.24}$$

Monod, however, failed to establish a minimum threshold concentration of the substrate in the medium, at which $\mu = 0$. He also did not detect any difference in the values of x_m for bacteria grown with and without agitation (although growth rates in these two cases were significantly different). On the grounds of these observations he concluded that the consumption of substrate for maintenance (*ration d'entretien*) by the studied microorganisms was negligible. Today we should consider this statement as an erroneous one. In reality, enterobacteria and bacilli are characterized by m values as high as 40–100 mg glucose h^{-1} g^{-1} biomass. It is still puzzling why Monod's experiments failed to reveal such intensive maintenance substrate consumption. Another mysterious result of Monod's studies was that he found abnormally low values of Y, lying in the range of 0.17–0.25 g dry weight g^{-1} glucose. This is to be compared with the range 0.4–0.6, reported later by others. One possible explanation for both of these discrepancies could be double limitation, of which Monod was unaware. In fact, he did not use any chelating components of the nutrient media (such as EDTA, citrate), so precipitation of $MgNH_4PO_4$ was likely to occur. It would be sufficient to cause Mg^{2+} deficiency and a decline in biomass yield per organic substrate as a result of its non-productive oxidation and accumulation of exometabolites. With a reduced growth rate, these products could be reutilized, concealing, thereby, a drop in Y associated with maintenance.

3. The heart of Monod's theory is **the concept relating microbial growth rate,** μ, with limiting substrate concentration, s. To begin with, he proved (by cultivation of a stationary culture on a filtrate) that it was the depletion of nutrients but not accumulation of toxic metabolites which caused deceleration and subsequent cessation of bacterial growth in a batch culture. Then, by using the material balance equation, he calculated the instant residual substrate concentration in the course of microbial growth, $s = s_0 - (x - x_0)/Y$. The obtained values of s were plotted versus instant values of $\mu = d(\ln x)/dt$. It transpired that the obtained dependence of μ on s could be approximated by an equation with two parameters, μ_m and K_s

$$\mu = \mu_m \frac{s}{K_s + s} \qquad [1.25]$$

It is interesting that originally Monod did not draw any parallel between Eqn. [1.25] and the Michaelis–Menten equation. It was only in a later paper (Monod, 1949) that he mentioned this similarity. At the same time he stressed that stringent compliance with Blackman's 'bottle-neck' principle was unlikely. According to Monod, 'microbial growth rate should be controlled by a large number of different rate-determining steps', rather than by a single 'master reaction'. The saturation constant, K_s, characterizes the affinity of microbial cells to substrate. The value of K_s should be expected to bear some relation to the apparent dissociation constants of the enzyme involved in the first step of substrate conversion. However, this interpretation of Eqn. [1.25] developed in 1950, whereas in 1942 the choice of a hyperbolic function was entirely intuitive and empirical, except that a remark was made on the similarity between Eqn. [1.25] and the adsorption isotherm.

At first glance, Eqn. [1.25] has no real impact on the development of microbial growth theory, it is just another empirical formula! However, the last chapter of the first part of Monod's book shows that this is not the case. Here Monod integrates the differential equation for exponential microbial growth taking into account relationship [1.25]:

$$\frac{dx}{dt} = \mu(s)\,x = \mu_m \frac{s}{K_s + s}\,x\,. \qquad [1.26]$$

By substitution of s by x and separating the variables, he obtains the following relationship

$$\mu_m t = (1 + P)\ln\frac{x}{x_0} - P\ln\left(Q - \frac{x}{x_0}\right) + P\ln(Q - 1)\,, \qquad [1.27]$$

where

$$P = \frac{YK_s}{Ys_0 + x_0} = \frac{YK_s}{x_m} \quad \text{and} \quad Q = \frac{Ys_0 + x_0}{x_0} = \frac{x_m}{x_0}\,.$$

Equation [1.27] describes the S-shaped growth dynamics of a batch culture and includes the three parameters, Y, μ_m, and K_s, which can be thought of as 'passport' data for a particular organism (e.g. for *E. coli* grown on glucose at 30°C, $Y = 0.23$, $\mu_m = 1.35\ h^{-1}$ and $K_s = 4\ mg\ l^{-1}$). The initial conditions, s_0 and x_0 are set by the experimenter, who selects the inoculation dose and medium composition. Thus, knowing the values of all these entities, the growth dynamics can be calculated **before the experiment.** This was a great advancement in the theory development. Similar to ballistics, where the trajectory of a projected body with a given mass can be calculated from its initial momentum, in microbiology it became possible to predict the dynamic pattern of microbial growth, provided there was information on growth characteristics and cultivation conditions.

Monod's book demonstrated an excellent agreement between calculation by Eqn. [1.27] and experimental data for different s_0. It was convincing, although not a rigorous test of the relevance of the model. As we have already mentioned, a good fit is not a sufficient criterion for a model's validity, because low residuals may also be obtained with many empirical expressions, e.g. with the logistic equation. However, the principal distinction between Monod's model and the logistic equation is that the latter could not be used for prediction. We will never find the list of r and K parameters, characteristic for a particular species as in the case of Monod's model.

At the same time we should not overestimate the importance of Eqn. [1.27] as a practical tool for predicting the **real** dynamics of microbial processes. Let us recall that Monod deliberately restricted the range of examined cultivation conditions by low s_0 values, and excluded from consideration the lag and death phases. It is only recently that the explanation and prediction of the dynamics of microbial growth in these phases has become possible.

Development of the chemostat theory

The advancement of this theory should not be confused with the invention of the continuous-flow cultivation technique. As previously mentioned, this technique was first used by Winogradsky to maintain the microculture of sulphur bacteria. At that time he supplied fresh nutrient manually several times per day. Also manually and discontinuously, by small pulses, fresh nutrients were delivered to protozoa culture during the 10–30-year long experiments of Woodruff and Baitsell (1911a,b), and Galadzhiev (1932), as well as in the case of 'retarded' cultures realized by Tauson. The first fermentations with a truly continuous supply of the nutrient were set in motion between 1920 and 1950 (Haddon, 1928; Utenkov, 1941; Jordon and Jacobs, 1944; McClung, 1949). The medium solution was delivered into the fermentation vessel either by means of a pump, or by 'self-flow' through a resistance capillary under constant hydrostatic pressure. The culture volume was either allowed to increase with the inflow of fresh medium or was maintained constant by the continuous removal of cell suspension. In the latter case, cultivation was carried out with agitation (in a continuous fermenter with complete mixing) or without it (as in a plug-flow culture). It can be concluded from one of the first reviews on continuous cultivation (McClung, 1949) that this method was regarded merely as a technical means to replenish depleted nutrient elements or to replace the 'spoiled' medium. The distinct properties of a continuous flow system as compared with simple batch cultures were not fully realized. The credit for their discovery undoubtedly belongs to Monod (1950), who provided a theoretical analysis of a continuous cultivation.

The theory of the continuous-flow culture was a natural harvest of his preceding work, where two basic dynamic variables were introduced, the concentration (density) of microbial biomass, x, and the concentration of a growth-limiting substrate, s. Monod derived conservation equations tracing main sources and sinks for these variables. Below we present them in exactly the same form as they appeared in his original paper of 1950:

$$\frac{dx_B}{dt} = \frac{dx_T}{dt} - \frac{dx_P}{dt} = (\mu - D)\,x_B\,, \qquad [1.28]$$

$$\frac{ds}{dt} = D(s_0 - s) - \frac{x_B}{R}\mu, \quad \mu = \mu_0\frac{s}{s_K + s}\,,$$

where x_B and s are, respectively, biomass and substrate concentrations in the fermenter, x_P is the biomass concentration in the product bottle, x_T is the total biomass $x_T = x_B + x_P$, s_0 is the substrate concentration in fed nutrient, R is yield ($\equiv Y$), μ_0 is maximal growth rate ($\equiv \mu_m$) and D is the dilution rate defined as the ratio between the nutrient flow rate, F, ($cm^3\,h^{-1}$) and the culture volume, V (cm^3): $D = F/V$ (h^{-1}).

It is especially important that the described open system can attain so-called **steady-state**, when variables x and s do not change with time, being equal to respectively $\bar{x}$ and $\bar{s}$. Under steady-state conditions both derivatives of Eqn. [1.28] are set to zero, $dx/dt = 0$, and $ds/dt = 0$. From the first equation it follows that $\mu = D$, i.e. the specific rate of microbial growth is equal to dilution rate, which is under the full control of the experimenter. From the second equation, we find

$$\bar{x} = Y(s_0 - \bar{s})\,, \qquad [1.29]$$

i.e. the steady-state biomass concentration depends only on the yield factor, Y, and the difference, $s_0 - \bar{s}$. In contrast to batch culture, it is independent of the initial cell concentration, x_0 (compare with Eqn. [1.1]). In other words, a steady-state culture 'does not remember' its prehistory, and its properties are determined solely by current cultivation parameters.

Other conclusions which may be drawn from an analysis of equation set [1.28] are as follows:

1. The dilution rates permitting stable microbial growth (i.e. those D at which $\bar{x} > 0$), are confined between 0 and wash-out point $D_{wash} = \mu_m s_0/(K_s + s_0)$.
2. The specific growth rate μ does not depend on either s_0 or x and is governed solely by the substrate concentration in the cultural liquid ($\bar{s}$ or s).
3. However, the effect of $\bar{s}$ upon μ may be evaluated from the dependence of $\bar{x}$ on D, as soon as $\bar{x}$ and $\bar{s}$ are linked by the conservation condition [1.29].

The most important biological implication of the chemostat theory was the discovery of the fact that microorganisms can grow endlessly with any rate between 0 and μ_m. Such substrate-limited growth is nevertheless exponential, because two subsequent acts of cell division will be separated by a constant time interval. Earlier, substrate limitation had been observed by Monod only transiently at the end of the exponential phase of batch growth. In this work, however, it was shown both theoretically and experimentally that substrate-limited growth can be stable and sustained. Today, this idea may seem absolutely trivial but it was not easily taken by Monod's contemporaries, who believed that in general there could be only one particular rate of microbial exponential growth. Curiously, some interpreters of Monod's paper failed to recognize a distinction between μ, a variable quantity, and a constant parameter μ_m. From this misunderstanding a ridiculous conclusion was made that 'there is only one rate of medium flow . . . at which steady-state conditions will be maintained' (Golle, 1953).

Another inference hard to absorb for some conventionally educated microbiologists was the lack of immediate dependence of μ on s_0. As a result, there were a myriad of attempts to determine the values of parameters μ_m and K_s from measurements of μ as a function of s_0 rather than of s (for references, see Guady *et al.*, 1971).

In the second part of his paper, Monod outlined a wide range of problems which could be studied both experimentally and theoretically with the help of continuous culture. Specifically, he discussed the possibility of testing alternative hypotheses on the regulation of enzyme biosynthesis. An hypothesis could be expressed as a differential equation describing the formation and elimination (decay and removal with the flow) of particular enzyme. Then, the activity of the enzyme observed in a continuous-culture experiment at various D could be compared with the model calculation. Interestingly, it was in this paper that Monod noted that apart from steady-state microbial cultivation, the non-equilibrium, transient growth may give very valuable information on mechanisms studied. Apart from enzyme biosynthesis, Monod envisaged the use of continuous cultures for quantitative description of population dynamics, in particular the selection of mutants with fast growth.

The characteristic trait of Monod's kinetic studies is well illustrated by the above examples. Such investigations were never aimed at the discovery of formalized growth laws as an end in itself. Instead, Monod used his dynamic data along with other facts to clarify actual problems of microbiology and biochemistry. Since his main interests were focused on problems of enzyme synthesis and adaptation, it was actually these problems that he tried to solve by biokinetic methods.

In all current commentaries and reviews on continuous culture, the honour for the development of this theory is shared between Monod and the American researchers Novick and Szilard (1950a,b). Without belittling the real scientific contribution of the prominent geneticist and the world famous nuclear physicist, we would like to point out that their studies on continuous culture were not as deep and consistent as Monod's investigation. That is, they did not present explicitly a system of differential equations for x and s, did not suggest an analytical expression for the relationship between μ and s (just indicating that it was a monotonous function with saturation at high s), and neither did they outline the prospects of chemostat application in microbiology. The largest paper, which appeared in *Proceedings of the National Academy of Sciences* (Novick and Szilard, 1950b) concentrated on one particular field related to microbial population genetics. The study of tryptophan-limited growth of an auxotrophic strain of *E. coli* allowed the clarification of the driving forces of spontaneous mutation in a continuous culture. Another paper, published in *Science* (Novick and Szilard, 1950a) was devoted to methodology. In this paper general chemostat theory was outlined in an extremely condensed form and verbally, without any use of mathematical symbolism; nevertheless it was a masterpiece. Using several phrases the authors delineated the most essential points of substrate-limited growth and made a happy choice in suggesting the term 'chemostat' for a substrate-limited, continuous-flow cultivator with complete mixing. Chemostat is now a well-established word in the scientific dictionary.

1.4 VERIFICATION, REFINEMENT AND DEVELOPMENT OF THE CHEMOSTAT THEORY (1950–PRESENT)

Developed by J. Monod, the chemostat theory made a first breach in an impregnable citadel. At last, the first relevant mathematical theory of microbial growth was constructed, and although still in its infancy it was able to describe, predict and control microbial processes. The ingenious French scientist did not glorify the victory, but went on to other challenges, never returning to the problems of microbial growth. Meanwhile new centres of kinetic investigations emerged in Great Britain (The Microbiological Research Establishment, Porton Down), Czechoslovakia (The Institute of Microbiology, Prague) and the Soviet Union (The Institute of Microbiology, Moscow, Institute of Physics, Krasnoyarsk).

1.4.1 'CHEMOSTAT RUSH'

For about three decades following 1950, scientific progress in the field was associated with advances in continuous cultivation. The interest in this method stemmed partly from the possibility of its implementation on an industrial scale for controlled microbial fermentation. As early as the 1950s, continuous cultivation began to be used for the production of ethanol and baker's yeast. Somewhat later, in the 1960s, the mass-scale production of single cell protein (SCP) gained top priority, and was viewed as an alternative to the 'green revolution' in providing proteins for cattle breeding. Concurrently, the microbial production of antibiotics, vitamins and amino acids continued to attract considerable attention. To bridge the gap between laboratory experiments and industrial requirements, new techniques of continuous-flow cultivation were developed, including chemostats in series, fermenters with feedback of biomass, and the turbidostat (Table 1.3). Each type of fermenter came with its own operational theory which was, in fact, an extension and adjustment of the general chemostat theory for the given specific case (Herbert, 1961a).

Besides practical needs, the interest in continuous cultivation was fuelled by influential and prominent scientists. The introduction and wide use of the new methodology was actively encouraged and organized by Ierusalimsky, Rabotnova, Malek, Herbert, Pirt and others. They vigorously promoted the idea that any microbiological study, be it ecological or genetic, biochemical or cytological, should involve the cultivation of microorganisms under controlled conditions, preferably with the use of a chemostat. The magnitude of this 'promotional campaign' becomes obvious from the number of meetings devoted specifically to continuous cultivation. There were as many as eight international conferences, and in the USSR and Czechoslovakia, such meetings were held almost annually from 1960 to 1980 (for reviews, see [Řičica, 1973; Rabotnova, 1974; Cooney, 1979). The number of publications advancing the use of continuous-flow cultivation or dealing with its specific problems grew annually and reached its peak in the mid-1970s. Today, the 'chemostat rush' is subsiding, and more reserved assessments of its possibilities have become common. In any case, statements made 30 years ago about the absolute supremacy of continuous over batch (static) cultures appear now to be too categorical: '. . . static culture and the development changes observed with it are *artifacts*. . . . The natural conditions which most fully correspond to the dynamics of microbial multiplication exist *only in continuous culture*. Therefore, *only* by using this method can we produce cultures which are really physiological, and correspond to the real physiological and biochemical characteristics of microorganisms' (Malek, 1958, p.16). At the same time, such conviction in the power of continuous cultures had a positive effect in encouraging the solution of severe technical problems and by attracting attention to the quantitative aspects of growth, all of which made a great contribution to the advancement of microbiology and biotechnology.

1.4.2 GROWTH STOICHIOMETRY

Experimental verification and elaboration of the original chemostat model

Monod, strictly speaking, did not verify his continuous-culture model. Equation set [1.28] was derived by logical reasoning using experimental

Table 1.3 Elaborations on the chemostat

<table>
<tr><th>Diagram</th><th>Equation of mass balance for biomass</th><th>The field of application</th></tr>
<tr><td colspan="3">Turbidostat (Bryson, 1952)</td></tr>
<tr><td>
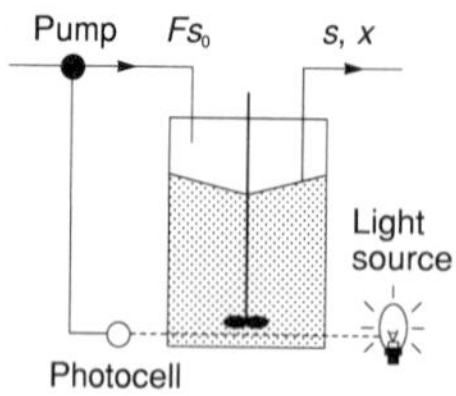

</td><td>$\dot{x} = \mu x - Dx$
$x = \text{constant}$</td><td>Cultivation at maximal growth rate, under substrate inhibition, selection of strains resistant to toxins</td></tr>
<tr><td colspan="3">pH-stat (Pechurkin and Terskov, 1973)</td></tr>
<tr><td>
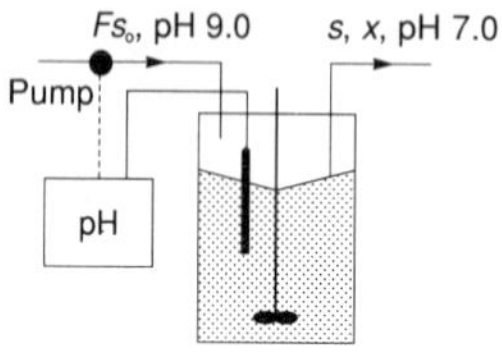

</td><td>$\dot{x} = \mu x - Dx$
$x = \text{constant}$</td><td>The same application field; technically this is more reliable</td></tr>
<tr><td colspan="3">Chemostat with biomass feedback (Herbert, 1961a)</td></tr>
<tr><td>
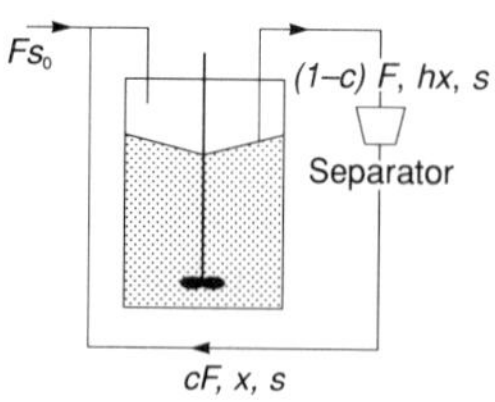

</td><td>$\dot{x} = \mu x - cDx - (1 - C)Dhx,$
$\mu = AD,\ A = h + c(1 - h)$
$0 \leqslant c < 1,\ 0 \leqslant h < 1$</td><td>Microbial processes aimed at complete substrate removal, e.g. in waste treatment</td></tr>
<tr><td colspan="3">Chemostats in series (Herbert, 1961a)</td></tr>
<tr><td>
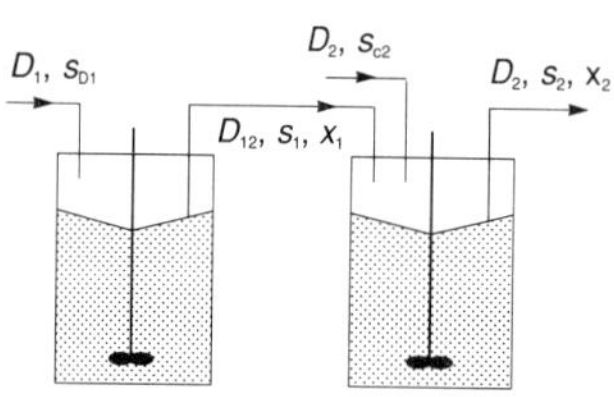

</td><td>$\dot{x}_1 = \mu_1 x_1 - D_1 x_1,$
$\dot{x}_2 = \mu_2 x_2 + D_{12} x_1 - D_2 x_2,$
$\mu_2 = D_2 - D_{12}\dfrac{x_1}{x_2}$</td><td>The second stage allows growth under inhibitory conditions, at $D_2 > \mu_m$ and at very low dilution rates $D_2 \to 0$.
Used for water purification and production of secondary metabolites</td></tr>
</table>

data previously obtained from batch culture (Monod, 1942). The testing of the theory was limited to the assembling of a simple apparatus, the 'bactogéne', and by the demonstration that it could indeed operate in a steady-state mode. A major contribution to the solution of the many and diverse technical problems associated with continuous culture and to the experimental verification of Monod's chemostat model was made by British microbiologists at the Porton Down Research Centre. They succeeded in designing a highly advanced and reliable fermenter, which is still commercially manufactured. In contrast to all previous continuous-flow fermenters, the nutrient medium was dispensed by mechanical peristaltic pumps (rather than by 'self-flow' through a resistance capillary). The apparatus was equipped with a disc impeller for vortex aeration and mixing (instead of air sparging), it had an automatic pH-control system and the capability of continuous pO_2 and temperature monitoring. Special attention was given to those 'petty details' which could ensure reliable and easy sterile operation, e.g. tube connectors and joints, inoculation ports, sampling devices, rotameters and stabilizers of air flows, air filters and sterilization facilities.

From the 1950s until the present day, the main object of British researchers has been *Aerobacter* (*Klebsiella*) *aerogenes*. The testing of the chemostat theory was performed mainly by measuring the steady-state concentrations of microbial biomass and substrate for different values of D. The first studies of glycerol-limited growth of *A. aerogenes* showed an almost complete agreement with the Monod model (Herbert *et al.*, 1956), the only deviation being that the critical dilution rate in the chemostat D_{wash} exceeded the value of μ_m determined in a batch culture. This was explained by an 'apparatus effect', being the result of imperfect mixing, biomass accretion on the walls of the culture vessel and short-term fluctuations of flow rate or liquid volume. However, more accurate measurements have subsequently revealed a first significant deviation from the Monod model. That is, **a non-constancy of biomass yield per substrate consumed**.

Variation in biomass yield from energy source: maintenance requirements

Variation of the yield, Y, in chemostat culture limited by energy source, was shown to fit the following pattern. The maximum yield occurred at subcritical D with a decline at lower dilution rates. Denis Herbert, an acknowledged leader of the Porton Down group of microbiologists, explained observed Y variation as being due to the endogenous degradation of cellular components occurring at a constant specific rate, a:

$$\dot{x} = \mu x - Dx = (\mu_{true} - a)x - Dx \,, \qquad [1.30]$$

where μ_{true} is the true growth rate of microorganisms. Then, at steady-state, we have

$$D = \mu = \mu_{true} - a\,. \qquad [1.31]$$

It is worth noting that the concept of endogenous degradation of microbial cell components was not at all new. Rubner was already aware of this process and described it by a first-order equation. The chemostat just provided a convenient method allowing the determination of a from the dependence of $\bar{x}$ on D. This possibility was fully exploited by John S. Pirt, another outstanding member of the Porton Down team (Pirt, 1965). In order to account for the variation in Y, Pirt modified the second, rather than the first, equation of the Eqn. [1.28] to obtain

$$\dot{s} = D(s_0 - s) - \frac{\mu x}{Y^{max}} - mx\,, \qquad [1.32]$$

where m is the maintenance coefficient, Y^{max} is the 'true', or the maximal, yield of biomass (both parameters were originally introduced by Terroine and Tauson). Under steady-state, the apparent biomass yield, Y, becomes a function of D:

$$Y = \frac{\bar{x}}{s_0 - s} = \frac{Y^{max} D}{D + mY^{max}}\,. \qquad [1.33]$$

Using reciprocal transformation, Eqn. [1.33] can be rearranged as

$$\frac{1}{Y} = \frac{1}{Y^{max}} + \frac{m}{D} \qquad [1.34]$$

or, upon multiplying by $\mu = D$,

$$q_e = \frac{\mu}{Y^{max}} + m\,, \qquad [1.35]$$

where q_e is the specific rate of energy source consumption ($q_e = \mu/Y$). Transformations [1.34] and [1.35] are used for the graphical determination of m and Y^{max} from plots of $\bar{x}$ and $\bar{s}$ against D.

It should be remembered that the first experimental measurements of m in continuous culture (to be exact, in fed-batch culture) were made by Tauson as early as 1937. Later this method was actually reinvented by Marr *et al.* (1963). However, the results obtained with fed-batch culture are less reliable than those obtained using a chemostat. In the fed-batch method, maintenance requirements are evaluated from the curvature of the line describing x dynamics. With infrequent sampling, one may fail to observe the deviation from linearity and accept the wrong result, $m = 0$.

Thus, there are two approaches to describe the D-dependent variation of Y, one by Herbert–Rubner, given by Eqn. [1.30], and another by Pirt–Tauson, Eqn. [1.32]. As shown by Pirt (1965), both approaches are mathematically equivalent, provided $m = a/Y^{max}$. In both cases, a decrease of Y at low D was explained by the partial energy diversion from cell biosynthesis, but the biological interpretation of this diversion was different. In the first approach, non-productive energy consumption was associated with turnover and endogenous degradation of cell constituents, whereas in the second approach it included the direct consumption of energy source for so-called maintenance functions such as osmoregulation, compensation of decayed cell components and support of cell motility. A rough estimate presented by Pirt (1975) showed that a major portion of the overall maintenance energy can be accounted for by osmotic work. This conclusion was supported by the discovery of a linear relation between the respiration rate and the magnitude of the transmembrane gradient of K^+, which is the main ionic component of the intercellular pool (Hueting *et al.*, 1979).

The described concept of maintenance requirements was a subject of a severe criticism (Tempest and Neijssel, 1977). One of the strongest arguments against it was an apparent increase in Y^{max} observed in chemostat cultures, limited by P, N and other 'conserved' substrates under conditions of energy excess. To preserve the constancy of the 'true' yield, Pirt (1982) had to modify Eqn. [1.35] in the following way:

$$q_e = \frac{\mu}{Y^{max}} + m + m'\left(1 - \frac{\mu}{\mu_m}\right) \qquad [1.36]$$

where $m'(1 - \mu/\mu_m)$ is the second μ-dependent component of maintenance energy which operates under excess of energy substrate. This equation is, however, of an *ad hoc* nature and as such is of restricted applicability.

In the opposing case, under severe energy limitation, as in chemostat at extremely low D, we observe a different type of deviation from the predictions of Eqn. [1.35]. As $\mu = D$ approaches zero the decrease in q_e was so strong that it became less than the value of m measured at higher D (Pirt, 1972). This effect has not yet been adequately understood. At any rate, it became clear that the concept of maintenance energy requirements, assuming the m-constancy, was valid only for a narrow range of cultivation conditions.

The minimum growth rate

Studies of extremely slow growth in the chemostat led to the discovery of an intriguing effect which could not be explained with conventional chemostat theory. When $D = \mu$ was reduced to extremely low values

($< \mu_m$ by two to three orders of magnitude), a substantial fraction of cells in the population of *A. aerogenes* lost viability (Tempest *et al.*, 1967). The death terms should be taken into account in the material balance equation for x,

$$\dot{x}_{via} = \mu x_{via} - a' x_{via} - D x_{via} ,$$

$$\dot{x}_{d} = a' x_{via} - D x_{d} , \qquad [1.37]$$

$$\dot{x} = \dot{x}_{via} + \dot{x}_{d} = \mu x_{via} - D x = (\mu\theta - D) x ,$$

where x_{via} and x_d are concentrations of viable and dead cells, respectively, $\theta = x_{via}/x$ is the fraction of viable cells, and a' is the specific death rate. (Note that parameters a' in [1.37] and a in [1.30] are differently interpreted. The term a' describes the death rate of cells, whereas a is the rate of cell component degradation to CO_2, H_2O and other waste products.) Under steady-state, we have $\mu = D/\theta$, so that the specific growth rate of surviving microorganisms augments the dilution rate, and the lower the fraction of viable cells θ, the higher the μ. It follows that, with D gradually decreasing, θ will also decline, but in this case μ would approach some lower limit, which was called the **minimum growth rate**. In the case of an *A. aerogenes* culture limited by glycerol, this limit was found to be 0.009 h^{-1} and under ammonium limitation was 0.007 h^{-1} (Tempest *et al.*, 1967).

Variation in biomass yield from conserved substrates: the concept of 'cell quota'

The yield from nitrogen in NH_4^+-limited culture of *A. aerogenes* was found to increase at low D due to accumulation of storage material. This cell compartment was termed the 'inactive biomass' (Herbert, 1959). By subtracting it from the total biomass, the constancy of yield of 'active biomass' could be restored, in formal agreement with Monod's model. Bacteriologists were quite satisfied by this explanation. A further detailed study of the problem was undertaken by algologists (Caperon, 1968; Droop, 1974). They established a similar pattern of D-dependent Y variation in chemostat cultures of algae limited by various biogenic elements. However, they did not try to fit their data to the Monod model. Instead, they advanced a concept of 'cell quota'. The quintessence of this concept was a premise that microbial growth rate is controlled by the intracellular concentration of the 'conserved substrate' rather than by the availability of substrate in the environment.

Let us first inspect the meaning of the term 'conserved substrate'. In fact, this is a antipode of substrates which are sources of energy. Catabolic substrates provide the cell with energy, their consumption being accompanied by a dissipation of chemical substances into waste products outside the cells (CO_2, H_2O), while anabolic or conserved substrates are incorporated into *de novo* synthesized cell components, being conserved in biomass. Conserved substrates include nearly all the non-carbon sources of biogenic elements (N, P, K, Mg, Fe and trace elements), CO_2 for autotrophic microbes, as well as the indispensable amino acids and growth factors. (In some special cases one and the same substrate may be a source of both energy and conserved elements: organic substrates for chemoorganotrophic microbes, NH_4^+ for nitrificators, NO_3^- for denitrificators, etc. These cases will be considered below.) The cell quota, σ, was originally defined as the intracellular concentration of the conserved substrate, but in practice σ was found as a total content of the corresponding element in the biomass. Now, if the amount of extracellular products is small, then for chemostat culture we have

$$\sigma \approx \frac{s_0 - \bar{s}}{\bar{x}} = \frac{1}{Y}. \qquad [1.38]$$

Mathematically, the cell quota concept was expressed as one of the straightforward empirical equations, relating $\mu = D$ and σ, e.g.

$$\mu = \mu_m \left(1 - \frac{\sigma}{\sigma_0}\right), \qquad [1.39]$$

where σ_0 is the minimum quota (low limit of σ attained when $D \to 0$).

Unfortunately, the relationship between μ and σ was either neglected totally (a trend observed in kinetic studies of bacteria) or overstated (a prevailing tendency in algology). The fallacy of exaggeration is more curious and menacing, so let us make this point clear. The μ–σ dependence approximated by Eqn. [1.39] was erroneously recognized by algologists as a primary fundamental relationship rather than a secondary one derived from another basic function. Even today, in some publications, the cell content of the utilized element σ is treated as the primary cause of variations in the microbial growth rate *in situ*. Later in this book (see Chapter 3) we will show that both μ and σ are actually functions of one common independent variable, the limiting substrate concentration in the medium, s. Thus, the main fallacy of the cell quota concept is the negligence of the most essential point of chemostat theory relating μ with s.

Yield of microbial biomass on organic substrates of various chemical nature: the concept of mass–energy balance

The first chemostat studies of chemoorganotrophic microorganisms were usually performed with carbohydrates as a main organic substrate. Since the 1960s, the range of carbon and energy sources in use was markedly broadened to meet the demands for industrial biomass production on natural gas, oil, methanol, ethanol, organic acids and other substrates.

This, in turn, raised the problem of the estimation of the microbial growth efficiency with substrates of diverse chemical nature.

This problem was best solved within the framework of the so-called mass–energy balance (MEB) theory developed by the Russian microbiologists Minkevich and Eroshin (1976). In this theory, aerobic chemoorganotrophic microbial growth is described by the following stoichiometric equation:

$$\underbrace{CH_mO_l + aNH_3 + bO_2}_{\text{substrates}} = \underbrace{Y_cCH_pO_nN_q}_{\text{biomass}} + \underbrace{Y_pCH_rO_sN_t}_{\text{products}} + (1 - Y_c - Y_p)\,CO_2 + cH_2O\,. \qquad [1.40]$$

Here the elemental composition of organic substrate, extracellular product and microbial biomass are expressed by gross formulae, carbon always being unity. This kind of stoichiometric expression has an obvious advantage since the fraction of carbon in dry biomass of various microorganisms is relatively constant, $\sigma_x = 0.46 \pm 0.05$. By contrast, the content of carbon in utilized substrates, σ_s, may vary over a broad range. To characterize substrate and biomass by a single common measure, it was suggested to include in the MEB an index **degree of carbon reduction**, γ, related to the internal energy of organic compounds. This suggestion is based on the following empirical consideration.

The heat liberated by biological or purely chemical oxidation is proportional to oxygen uptake or, equally, to the number of electrons gained by oxygen from oxidized substrates, according to Payne's terminology 'available electrons', AE, (Mayberry *et al.* 1967). The heat production from an oxidation reaction averages at 27 kcal per AE equivalent. A carbon reduction degree, γ, is defined as the number of AE per carbon atom. Its numeric value can be determined from the stoichiometry of the oxidation reaction, e.g.

$$CH_pO_nN_q + bO_2 = CO_2 + 0.5(p - 3q)\,H_2O + qNH_3, \qquad [1.41]$$

$$\gamma = 4b = 4 + p - 2n - 3q\,. \qquad [1.42]$$

The AE balance for Eqn. [1.40] can be written as

$$\gamma_s + b(-4) = Y_c\gamma_x + Y_p\gamma_p\,, \qquad [1.43]$$

where γ_s, γ_x, and γ_p are the carbon reduction degree of respectively substrate, biomass and extracellular product. Dividing both sides of Eqn. [1.43] by γ_s we obtain the relationship delineating the AE distribution between oxygen (i.e. AE used for respiration), biomass and the intracellular product:

$$\underset{\text{oxygen}}{\frac{4b}{\gamma_S}} + \underset{\text{biomass}}{\frac{Y_c\gamma_X}{\gamma_S}} + \underset{\text{product}}{\frac{Y_P\gamma_P}{\gamma_S}} = 1\,. \qquad [1.44]$$

The second term in this equation is the fraction of AE transferred to biomass from utilized substrate, termed the **energetic growth yield**

$$\eta = \frac{Y_c\gamma_X}{\gamma_S}, \qquad [1.45]$$

where Y_c is the biomass yield per unit carbon of utilized substrate. The third term designates that fraction of total substrate internal energy which is transferred to the product; it was called the **energetic product yield**,

$$\xi = \frac{Y_P\gamma_P}{\gamma_S}, \qquad [1.46]$$

where Y_p is the product yield per unit carbon of utilized substrate. Energetic yield η is related to other stoichiometric parameters by the following relationships:

$$\eta = \frac{Y\sigma_X\gamma_X}{\sigma_S\gamma_S}, \qquad [1.47]$$

$$Y_{O2} = \frac{Y_c}{b} = \frac{4\eta}{\gamma_X(1-\eta-\xi)}, \qquad [1.48]$$

$$Y_O = \frac{Y_c}{b}\,\frac{12/32}{\sigma_X} = \frac{3\eta}{2(1-\eta-\xi)\gamma_X\sigma_X}, \qquad [1.49]$$

where biomass yields are expressed as g biomass g^{-1} of substrate (Y), g biomass C g^{-1} substrate C (Y_c), gram-atom biomass C mol^{-1} O_2 (Y_{O2}), and g biomass g^{-1} O_2 (Y_O).

It can be readily seen that η and Tauson's CEU have very similar meanings (strictly speaking, they would be completely identical if the growth heat effect per AE were an absolute constant). The advantage of using η is that it can be found not only thermochemically, but also from any other components of the growth mass balance. This is, by the way, the most attractive feature of the developed MEB approach.

This approach is applicable not only for aerobic chemoorganotrophic microbes but also for anaerobic, autotrophic and methylotrophic organisms, and even for microbial communities of soil or activated sludge. The MEB relies on several assumptions and approximations. Examples of cases where these assumptions do fail were outlined by Manakov (1981) and Roels (1980).

Microscopic approach in studies of growth stoichiometry

Equations [1.40–1.49] exemplify the so-called **macroscopic approach** in studying microbial growth stoichiometry. Its typical features are the use of gross formulae for biomass and metabolic products, evaluation of total mass balance for some biogenic elements (C, N and P), and the use of simple pseudostoichiometric equations involving hypothetical pseudo-compounds. By contrast, the **microscopic approach** focuses on real, rather than hypothetical, metabolic reactions, and attempts to take into account a reasonably limited but still quite large number of individual metabolic intermediates. The final aim of this approach is to organize the biochemical information into a consistent picture of microbial metabolism at the cellular or population level.

The microscopic approach has become possible by virtue of advances in biochemistry, which have succeeded in establishing a sufficiently full picture of metabolic processes in certain microorganisms. The pioneering work in this area was done by Bauchop and Elsden (1960), who were able to sum up the balance of ATP for anaerobically grown microorganisms. As a result, a relation was established between the biomass yield (a macroscopic quantity) and the number of generated ATP moles (a stoichiometric characteristic of real catabolic reactions). Subsequently, the concept of biomass molar yield Y_{ATP} was refined and extended to apply also to aerobic growth, a more detailed picture of ATP expenses for particular biosynthetic reactions and 'membrane energization' being obtained (Stouthamer, 1973).

Another example of advanced microstoichiometric developments is the technique of microbial yield calculation from 'characteristic tables' (Skurida *et al.*, 1983, 1984). The cellular metabolism was divided by the authors into four steps:

- formation of key metabolites from exosubstrates;
- formation of the array of monomeric precursors from key metabolites;
- formation of macrocomponents from monomers; and
- provision of biosynthesis by energy.

Characteristic tables were designed to reduce the enormous complexity of the stoichiometric equation set by means of matrix notation. The essence of catabolism is viewed by this approach to be a compensation of a disbalance between the main energy donors since the consumption of ATP and NADPH in anabolic reactions exceeds their production, whereas FADH and NADH are generated in greater quantities than utilized. A constant proportion between these four donors is maintained by means of catabolic reactions. To identify the single metabolic pathway from a variety of possible catabolic reactions, the authors used the optimization approach and linear programming technique, the only optimization criteria being minimization of energy source expenditure.

We have so far discussed stoichiometry of microbial growth. In this area, chemostat theory is used mainly to address problems of a static nature. It can provide answers to questions like 'how much?' and 'in what proportion?'. Now we are considering the dynamics and will have to deal with questions of the type, 'at what rate?' and 'by which pathway?' We start with the central problem of microbiological kinetics, that is the dependence of μ upon s.

1.4.3 GROWTH DEPENDENCE ON SUBSTRATE CONCENTRATION

Experimental technique

The chemostat provides the opportunity for more rigorous and detailed investigations of this problem. By running an experiment at different dilution rates, D, the corresponding $\bar{s}$ values may be measured, and, therefore, the dependence of $\mu = D$ on $\bar{s}$ can be obtained, in principle, as accurately and in as much detail as desired. Such an approach, however, may and frequently does encounter serious technical problems due to high affinity to limiting substrate of some microorganisms. There is the need first, to select highly sensitive analytical techniques to measure extremely low residual concentrations of particular substances, second, to develop instant sampling procedures to minimize substrate loss, and third, to eliminate apparatus-related artifacts such as non-perfect mixing and fluctuations in nutrient medium supply. This can always be accomplished, although considerable time and effort may be required (Button, 1969, 1985; Droop, 1974; Robertson and Button, 1979).

Verification of the Monod equation

For many years the Monod equation has been tacitly recognized as 'fundamentally true'. This fact may be regarded as a result of the inertia of mind trained on the examples from classical physics. Here the major laws (Newton's mechanics, Coulomb's laws of electricity, etc.), look almost as simple (i.e. are presented by short mathematical formulae); nevertheless, they remain fundamental in nature and of uniform applicability. It was not accidental that many microbiologists viewed the Monod equation as something with a deep inherent meaning rather than simply as an empirical formula. Below, we shall outline a number of naive attempts to deduce this equation logically from the conjectured growth mechanisms.

Even in the 1950s, there were a few works devoted to the experimental testing of this equation (Moser, 1958; Contois, 1959). Such works are still occasionally published nowadays (Powell, 1967; Shehata and Marr, 1971; Varfolomeyev, 1987). It was found that not all experimental data could be reasonably well fitted by Eqn. [1.25]. The calculated hyperbola often passed above experimental points at small s and below them at large s. A better fit could be obtained by using the following, entirely empirical, equations:

$$\mu = \mu_m(1 - \exp(-K_s)) \qquad \text{(Teissier, 1956)} \qquad [1.50]$$

$$\mu = \frac{\mu_m s^n}{K_s + s^n} \qquad \text{(Moser, 1958)} \qquad [1.51]$$

$$\mu = \frac{\mu_m s}{K_s x + s} \qquad \text{(Contois, 1959)} \qquad [1.52]$$

Inclusion of a third parameter into Eqn. [1.25], like the parameter n in Eqn. [1.51], or introduction of the $K_s x$ term, as in Eqn. 1.52, led to a predictable improvement of approximation capability. Nevertheless, these equations failed to gain any recognition because of their entirely *ad hoc* nature.

Biologically justified modifications of Monod's equation

One recognized advantage of the Monod equation was its similarity with the Michaelis–Menten equation. Bearing in mind that, contrary to enzyme molecules, bacterial catalysts do multiply, we should draw a parallel between enzymatic reaction rate v and some specific metabolic rate in microbial culture, e.g. substrate uptake rate, $q_s = (-ds/dt)/x$. In such a way Powell (1958) derived from Eqn. [1.18]

$$q_s = \frac{Q_s s}{K_s + s}, \qquad [1.53]$$

where Q_s is the maximum rate of substrate consumption. If Y = constant, then a substitution $q_s = \mu/Y$ and $Q_s = \mu_m/Y$ into Eqn. [1.53] results in the Monod equation. If Y is D-dependent owing to maintenance requirements, then $q_s = \mu/Y^{max} + m$ (Powell, 1967; Pirt, 1975) and hence,

$$\mu = Y^{max}(q_s - m) = \frac{\mu'_m s}{K_s + s} - mY^{max}, \qquad [1.54]$$

where $\mu'_m = Y^{max}Q_s$. In contrast to the Monod equation, Eqn. [1.54] predicts the occurrence of a threshold substrate concentration, $s^* = mY^{max}K_s/(\mu'_m - mY^{max})$, below which growth is impossible. Equation

[1.54] yields a substantially better fit to experimental data and avoids the above-mentioned systematic simulation errors at modestly small s. It can be also be rewritten in the following form

$$\mu = \frac{(\mu'_{m} - mY^{max})(s - s^{*})}{K_{s} + s} = \mu_{m}\frac{s - s^{*}}{K_{s} + s}. \qquad [1.55]$$

Here, the parameter $\mu_{m} = Y^{max}(Q_{s} - m)$ is the specific growth rate of a microbial population attained asymptotically as s. (Unlike μ'_{m}, the value of μ_{m} is really attainable growth rate.)

A modification of the Monod equation, similar to Eqn. [1.55], was also proposed for the case of conserved substrates. However, the biological meaning of offered modification is entirely different since it was introduced to account for leakage of the limiting substrate from cells. If the specific leakage rate is assumed to be constant, then a decrease in s down to a threshold value s^{*} will lead to the counterbalance of the two reverse processes (uptake and leakage), so that the net consumption of the limiting substrate will be zero. The available publications, unfortunately, skipped this simple algebra (otherwise Eqn. [1.55] would have been obtained), and took instead a straightforward substitution of s for the difference $s - s^{*}$ in the Monod equation (Caperon and Meyer, 1972);

$$\mu = \frac{\mu_{m}(s - s^{*})}{K_{s} + s - s^{*}}. \qquad [1.56]$$

A few valuable refinements of the Monod equation were borrowed from enzymology, in particular from the kinetic analysis of inhibitory effects. Most often it was applied to the non-competitive and substrate inhibition of growth. The former type of inhibition was first observed by Ierusalimsky and his colleagues (Ierusalimsky and Neronova, 1965; Chernavsky and Ierusalimsky, 1966). The combined effects on the growth of propionic bacteria of the substrate (lactate) concentration and metabolic products (propionate or acetate) were described by the so-called Monod–Ierusalimsky equation:

$$\mu = \mu_{m}\frac{s}{K_{s} + s}\,\frac{K_{p}}{K_{p} + p}, \qquad [1.57]$$

where p is the concentration of the inhibitor in the medium, and K_{p} is the inhibition constant.

The retardation of static growth was frequently observed with an excess of such substrates as phenols, methanol, ethanol, etc. These kinetic effects could be approximated by an analogue of Haldane's equation in enzymatic kinetics (Andrews, 1968):

$$\mu = \frac{\mu_m s}{K_s + s + s^2/K_{ss}}, \qquad [1.58]$$

where K_{ss} is the substrate inhibition constant. Equation [1.58] describes a single-peak curve, and so one and the same μ-value may be obtained at two different s, one in the substrate-limiting range, $d\mu/ds > 0$, and the other in substrate inhibition range, $d\mu/ds < 0$. Applied to chemostat cultivation, there is a temptation to speculate that two different steady-states could occur for the same value of D. Such a phenomenon was observed in reality and was termed 'Bistabilität' by the German microbiologists Bergter (1969) and Guthke and Knorre (1980). However, as shown by qualitative analysis of the chemostat model, the steady-state in the substrate inhibition range is unstable. A sustainable maintenance of a population under conditions of substrate inhibition is possible either in the second stage of a two-stage chemostat or in the case of plentiful wall growth in a conventional chemostat (Jones *et al.*, 1973; Pawlowsky *et al.*, 1973).

A new interpretation of the bottle-neck concept

All the discussed modifications of the Monod equation were based on direct borrowing of relationships originally derived in enzymology. A theoretical substantiation for such 'plagiarism' is the Blackman's bottle-neck postulate. Microbial metabolism is symbolized as a unidirectional chain of reactions of substrate S conversion to cell biomass X′ via some hypothetical intermediates $P_1, P_2, \ldots, P_n$ (Varfolomejev, 1987):

$$S + X' \xrightarrow{k_1} P_1 \xrightarrow{k_2} P_2 \rightarrow \ldots \xrightarrow{k_n} P_n \xrightarrow{k_{n+1}} 2X' \qquad [1.59]$$

It can be readily shown that under quasi-steady-state approximation (i.e. assuming steady-state with respect to intermediate concentration p_i, $dp_i/dt = 0$, but not to the total biomass x, which is the sum $x = x' + p_1 + p_2 + \ldots p_n$) microbial growth proceeding in this fashion can indeed be described by the Monod equation, composite parameters μ_m and K_s being expressed via 'elementary' kinetic constants of individual enzymatic reactions:

$$\frac{dx}{dt} = \mu_m \frac{s}{K_s + s} x; \quad \mu_m = \frac{1}{\sum_{i=2}^{n+1} 1/k_i}; \quad K_s = \frac{1}{k_1 \sum_{i=2}^{n+1} 1/k_i} \qquad [1.60]$$

The 'bottle-neck' idea can now be formulated as follows. If one of the constants $k_j \ll k_i$, $i = 2, \ldots, n+1$, $i \neq j$, then $\mu_m = k_j$ and $K_s = k_j/k_1$, and so the jth enzymatic reaction is the master reaction. This constraint is actually stronger than that implied by the original Blackman formulation. That is, the master reaction should be slower as compared with the others by several order of magnitude, rather than merely the slowest one.

The obvious shortcomings of reaction scheme [1.59] is the unjustified oversimplification of the cell metabolism (which is a network, rather than a simple unidirectional sequence of reactions) as well as the lack of a definite interpretation of variables x' and p in real biochemical terms. One of the first attempts to overcome this drawback was made by Ierusalimsky and his physicist colleagues from the Moscow University (Chernavsky and Ierusalimsky, 1966; Stepanova *et al.*, 1967). As a result of their efforts, the chemostat theory was furnished with a powerful mathematical tool, the qualitative (stability) analysis of a dynamic system. An evaluation of the characteristic times of major intracellular metabolic reactions showed that the slowest reaction was the synthesis of ribosomes. This was in accord with fundamental early studies of the 'Copenhagen school' (Maaloe and Kjeldgaard, 1966) on control of macromolecular synthesis in bacterial cells. Thus, altogether, two possible 'bottle-necks' were identified: (a) uptake of limiting substrate (processes of active transport, transphosphorylation of sugars, etc.) (Monod, 1949); and (b) the formation of intracellular protein-synthesizing structures such as rRNA and the ribosomal proteins (Ierusalimsky, 1967).

'Derivations' of $\mu(s)$ from mechanistic considerations

A characteristic feature of the recent past of microbiological kinetics were the dreams to deduce a 'magic' powerful equation which would be able to account for all different types of dynamic behaviour of microorganisms. Such derivations were based on particular assumptions on the growth mechanism. Below we briefly discuss a few typical examples.

1. **Mechanistic considerations at the population level**. Microbial populations were assumed to consist of active cells (which utilize the substrate) and inactive cells (Shvytov, 1974). The probability of a transition from the inactive to the active state was thought to be proportional to s, whereas the probability of the reverse transition was independent of s. From these assumptions, a relationship between μ and s can be inferred, similar to Eqn. [1.25].

2. **Mechanistic considerations at the cellular level**. The growth of a bacterial cell was described by the following pseudostoichiometric equation (Verhoff *et al.*, 1972):

$$\mathrm{A} + a_1\mathrm{S} \underset{r_2}{\overset{r_1}{\rightleftarrows}} \mathrm{B} \xrightarrow{r_3} (1 + a_2)\mathrm{A} + a_3(a_1 - a_2)\mathrm{S}\,, \qquad [1.61]$$

where the symbols A and B denote the 'absorbing' and 'digesting' cell components, respectively. The former are capable of a reversible interaction with the substrate, yielding B-type cell components. The growth, as such, is described by a reaction having a kinetic constant r_3, which reconverts B components back to A, producing a biomass increase of a_2 units. In the steady-state approximation, the following expression for μ was obtained:

$$\mu = \frac{1}{A+B}\frac{d(A+B)}{dt}$$

$$= -\frac{1}{2}\left\{(r_1 s + r_2 + r_3) - [(r_1 s + r_2 + r_3)^2 + 4a_2 r_1 r_3 s]^{1/2}\right\}. \qquad [1.62]$$

This equation is more complex than Monod's but is similar to it in that it generates a saturation-type $\mu(s)$-curve which passes through the origin ($\mu = 0$ at $s = 0$) and approaches an upper asymptote $\mu = \mu_m a_2 r_3$, when $s \to \infty$.

3. **Account of diffusion effects**. Many models of this kind have been suggested. Here, we will present one example (Powell, 1967). The basic assumption of the model is that substrate consumption is performed by an enzyme which obeys Michaelis kinetics and is localized on the inner side of the cytoplasmic membrane. Actual substrate concentration around the enzyme active centres is smaller than in the solution, due to a limited diffusion rate. By applying a simplified Laplace equation, it was found eventually that

$$\mu = \frac{\mu_m(K_s+L+s)}{2L}\left(1-\sqrt{1-\frac{4Ls}{(K_s+L+s)^2}}\right) \approx \mu_m\frac{s}{K_s+L+s}, \qquad [1.63]$$

where L is a factor determined by membrane permeability and by the maximum rate of the enzymatic reaction.

4. **Derivations based on mechanistic considerations of protein synthesis** (Romanovsky *et al.*, 1984). The rate of protein synthesis, dp/dt, is supposed to be determined by rRNA concentration R, cell biomass, x, and by the size of the amino acids pool, A,

$$\frac{dp}{dt} = kxR\frac{A}{K_A + A}.$$

Other conditions were defined by the auxiliary relationships,

$$R = R_0 + \frac{(R_m - R_0)\mu}{\mu_m}, \quad A = \beta s, \quad \text{and} \quad P = \frac{p}{x},$$

where R_m and R_0 are the upper and lower limits of R variation respectively, and P and β are constants. For a steady-state chemostat culture,

$$\frac{dp}{dt} = 0 \quad \text{and} \quad \frac{dx}{dt} = \frac{1}{P}\,\frac{dp}{dt} = 0\,,$$

then

$$\mu = \frac{(kR_m/P)s}{(R_m K_A/\beta R_0 + s)} = \frac{\mu_m s}{K_s + s}\,. \qquad [1.64]$$

Here again, Monod's equation has been derived through a consideration of underlying intracellular processes.

Needless to say, none of these 'derivations' is free from criticism. In example 1, the degree of population heterogeneity is preposterously exaggerated by assuming that some cells completely avoid the capability to utilize substrate, while others operate at the maximum rate. In example 2, the suggested mechanism is clearly out of line. It might have been appropriate in simulating the holozoic mode of nutrition characteristic for animals, i.e. rapid swallowing of the food followed by its digestion; however, this mechanism has nothing to do with the osmotrophic nutrition of microorganisms. Next, it is hardly deniable that diffusion does accompany many reactions in microbial culture, but it does not mean that it controls their rates (as in example 3).

The most promising approach is taken in example 4, as soon as it aimed to reflect the real, although too simplified, growth mechanisms. However, more realistic structured growth models (see below) produce the relationships between μ and s which are no longer expressed analytically. Thus, in microbiological kinetics, in contrast to physics and chemistry, it is ultimately impossible to derive a single equation which would be adequate for all circumstances (as for instance in the case of Coulomb's law which is applicable to interaction of all charged particles). The main reason is that a microbial cell or a population is in fact a system including a number of subsystems which are sufficiently autonomous in their functioning. Therefore, the effects of s on μ are to a large extent dependent on the past history of the population, which is associated with the notion of the physiological state of microorganisms.

1.4.4 PHYSIOLOGICAL STATE OF CHEMOSTAT CULTURE

This term was first coined in 1958 by Malek (1958). It was not given, however, a clear definition. This seems to be intentional, because Malek anticipated the discussion-stimulated discovery of a suitable mathematical symbolism which would be able to quantify purely descriptive biological information. Subsequently, two trends emerged in considering the notion of the physiological state. Some microbiologists preferred to include in the term the entire data on cell composition and metabolism (Rabotnova, 1974). Others would confine its contents to a set of quantitative indices, which were established mainly in experiments with continuous cultures (Powell, 1967).

The impetus for the development of the concept of physiological state was Herbert's brilliant studies on the chemical composition of *A. aerogenes* at different D in chemostat culture (Herbert, 1961b). It has been found that some of the parameters studied remained constant (content of cell DNA and carbon), while others exhibited regular D-dependent variations, being either an increase (RNA content, cell sizes) or a decrease (the content of reserved polysaccharides). Since the 1960s such studies have become commonplace. Information was gathered for a large number of diverse microorganisms on how their chemical composition, activity, morphology, etc. were affected by chemostat dilution rate D (Khmel, 1970; Rabotnova, 1974). Those properties, which were D-dependent, were recognized as components of the **vector of the physiological state**.

The specific growth rate as an 'independent variable'

Reviewing the papers on chemostat studies, one would often locate references to the effects exerted by specific growth rate on some characteristics of the microbial culture. In literal terms, μ was no longer treated as a time-derivative of cell concentration $(1/x)\mathrm{d}x/\mathrm{d}t$ and became an 'independent variable' like temperature, medium tonicity, etc. First noted by Herbert, this jargon took deep-seated roots and until now has dominated microbiological mentality and lexicon. The point is illustrated also by the abundance of mathematical models incorporating equations of the type $y = F(\mu)$, where y is a dynamic variable (rate of product formation, maintenance energy, intracellular content of a compound), and $F(\mu)$ is a relatively simple function of μ (for an example, see Eqn. [1.36]).

The logic of Herbert's argument was as follows: first, a given characteristic of the culture, say, the intracellular RNA content, R, varies as dependent on D in chemostat in response to either s or μ changes; second, on transition from one growth-limiting factor to another (e.g. from the C- to N-limitation), the C-substrate concentration increases by several orders of magnitude, whereas the dependence of R upon μ remains the same. Thus, it follows that the variation in R is governed by μ.

Unfortunately, the inconsistency of this argument has never been exposed. To begin with, it would be wrong to regard μ and s as independent factors because, in chemostat, they are interrelated, i.e. μ is a function of s. The growth-limiting substrate concentration is then the only independent variable responsible for the variation of both μ and R in experiments involving changes in D. In addition, this point cannot be disproved by the second step in Herbert's deduction. Indeed, there are no grounds to exclude the possibility that metabolic reactions are slowed down by the deficiency of various substrates (the sources of energy, C and N) through the same molecular mechanisms, specifically through the mechanism of ribosomal synthesis.

For a long time these logical errors were not perceived because microbiologists quickly became accustomed to a corollary of chemostat theory which states that the steady-state μ value is equal to D. This is why it was tempting to identify the independent variable D with a dependent entity μ. In fact, careless interpretations of mathematical models may bring only 'immature sour fruit'! Moreover, the variables μ and D are equal only numerically, they are not equivalent in terms of their physical and biological meaning. D is set by the person running the experiment, whereas the μ-value belongs to the microbial population, whose growth rate in chemostat culture is self-adjusted to the predetermined D-value. This event does not occur at once and may involve a relatively long **transient process**. The signal causing both μ and R to change during transient periods is provided by the limiting substrate concentration in the medium. The foregoing discussion allows a new look to be taken at the problem of s effects on μ. The old and hackneyed approach based on a borrowing from enzymology should be considered as completely irrelevant. Indeed, in the case of a microbial culture, it is not only the rate of substrate consumption which is governed by s, but also certain qualitative properties of the cells expressed by different indices of the physiological state. This fact was first recognized by Powell, and paved the way for a quantitative interpretation of the physiological state concept.

Metabolic activity functional by E.O. Powell

On many occasions, the British mathematician Powell showed an exceptionally shrewd intuition of a genuine microbiologist. His benchmark paper of 1967 (Powell, 1967) grasped several key problems of microbial kinetics. In particular, he combined and put on a quantitative mathematical footing three notions, which were beforehand separated and cloudy:

- physiological state;
- past history; and
- non-steady-state growth kinetics of microbial culture.

The specific rate of substrate uptake, q_s, was presented by him as a product of two functions:

$$q_s(s) = q'_a(s)\,S(s)\,, \qquad [1.65]$$

where $S(s)$ is a simple function describing substrate saturation. For example, a Michaelis hyperbola, $s/(K_s + s)$, and $q'_a(s)$ is a function associated with the microbial physiological state. The instant value of $q'_a(s)$ is determined by the way in which s varies until the given moment (τ hours ago), and effects of later events which contribute more than earlier ones. This idea can be expressed conveniently in terms of a functional proposed by Volterra to describe the time delays in biological systems:

$$q_s(s) = \underset{\tau=0}{\overset{\tau=\infty}{Q}} [s(t - \tau)]\, S(s)\,. \qquad [1.66]$$

Transient processes are influenced by the past history of the culture in the following manner. Suppose that a steady-state growth of a chemostat culture is upset and the residual substrate concentration jumps from $s(0)$ to $s(1)$. Immediately, q_s will increase from $Q(0)S[s(0)]$ to $Q(0)S[s(1)]$. If no further changes in $s(1)$ occur, then Q will also eventually attain a new steady-state value equal to $q'_a[s(1)]$. The variable Q was termed the **metabolic activity functional** and proposed as a quantitative index of the physiological state of a microbial population. In essence, Q is the potential metabolic activity, that is the specific rate of a process at hand (e.g. the consumption of a growth-limiting substrate) measured just at the moment of relief from substrate limitation. Under steady-state conditions Q is related unambiguously to substrate concentration, $\bar{s}$. During transients, Q dynamics may be described by the following differential equation:

$$\dot{Q} = r(Q, s) - \mu(Q, s)\, Q\,. \qquad [1.67]$$

The first term in the right-hand side of this equation represents the production rate of the 'substance Q'', while the second term describes its dilution in the course of microbial growth.

In a later work, Powell (1969) associated the functional Q with a 'bottle-neck' in cell metabolism. The 'substance Q'' may, in reality, be represented by a single enzyme or polyenzymatic complexes, as well as by ribosomes or other cell components occupying the 'bottle-neck' position. Monod's chemostat model, supplemented by Eqn. [1.67], made possible at least a qualitative understanding of chemostat transient processes triggered by a D switch. For some specific cases, a fair quantitative agreement with experimental results could also be obtained by incorporation into Eqn. [1.67] of a term describing the turnover rate of the 'substance Q'' (Agrawal *et al.*, 1983).

Non-steady-state kinetics of microbial growth

A transient process in a chemostat culture induced by changes in D is an example of non-steady-state microbial growth. In such growth, μ and the metabolic activity of a microbial population exhibit continuous variation in time. The terms 'steady-state' and 'non-steady-state' stem from kinetics of enzymatic reactions. 'Steady-state kinetics' refers to the classic approach that assumes the constancy of ES complex concentrations during rate measurements. Non-steady-state kinetics of an enzymatic reaction is known to take place during the first milliseconds after reagent mixing or in response to some perturbations. The attractiveness of non-steady-state studies for microbiology is obvious. Firstly, there are a lot of encouraging examples showing the success of a similar approach in elucidating mechanisms of enzymatic catalysis (Berezin and Varfolomeyev, 1979). Non-steady-state experiments allow a much wider range of hypotheses to be tested and yield much more data on the dynamic behaviour of the studied objects. Secondly, in biotechnology steady-state operation is the exception rather than the common routine due to unavoidable disturbances in cultivation conditions. Thirdly, non-steady-state growth may potentially display greater efficiency and higher productivity for a whole range of metabolites (Cooney, 1979).

Intentionally, the development of non-steady-state microbial growth can be induced and maintained by an intermittent supply of the limiting substrate, and by deliberate fluctuations in D, s_0 and pH in the input medium, as well as incubation temperature. A consistent increase in Y, as well as in the rates of growth and particular metabolic reactions, have been reported for some non-steady-state chemostat operation modes as compared with steady-state ones (Pozmogova, 1983). Mechanisms underlying these effects have not yet been clarified.

We have already mentioned successful simulations of certain non-steady-state processes of microbial growth based on Powell's mathematical model. Powell's work has been advanced in two directions: first, as a formalistic approach based on the use of the so-called dummy state variables and on differential equations with time-delays of the type $\mathrm{d}x/\mathrm{d}t = f[x(t-\tau)]$ (Takamutsu *et al.*, 1981; Bazin, 1982), and second, as a mechanistic approach involving the use of structured models. Such models are discussed below in more detail.

Structured models; non-balanced microbial growth

By 'structured', we mean mathematical models describing growth-associated changes in microbial cell composition. Powell's model (Eqns [1.65–1.67]) is not yet a fully fledged structured model because 'substance Q' is, as it were, 'weightless', for its synthesis does not involve any substrate utilization. By contrast, a true structured model should include mass balance equations for all identified intracellular components. Their concentrations can be expressed either per unit volume of fermenter vessel ($c_1, c_2, \ldots, c_n$), or per unit cell mass ($C_1, C_2, \ldots, C_n$), and, hence, $C_i = c_i/x$. The mass balance equations can be written as follows:

$$\sum_{i=1}^{n} c_i = x\,, \quad \sum_{i=1}^{n} C_i = 1\,. \qquad [1.68]$$

For each variable C_i a differential equation is written which takes into account all sources, r_+, and sinks, r_-, as well as its dilution due to cell mass expansion,

$$\dot{C}_i = r_+(s, C_1, C_2, \ldots, C_n]) - r_-(s, C_1, C_2, \ldots, C_n) - \mu C_i. \qquad [1.69]$$

The first structured models contained the number of components, n, usually no more than two or three, and, accordingly, they were called two- and three-compartment models. Thus, a model developed by Ramkrishna *et al.* (1967) incorporated two compartments – nucleic acids and proteins combined with other active cell components. The model variables included also concentrations of the limiting substrate and the inhibitor. Compared with Monod's model, the proposed set of four equations was able to account for a much wider range of dynamic patterns. Specifically, it simulated *D*-dependent changes in the composition of cells in a chemostat culture as well as all known growth phases (including the lag and decline phases).

This work was the first successful simulation of the so-called **non-balanced growth,** which involves changes in the cell composition (Campbell, 1957). As a rule, the notions of steady-state and balanced growth are considered to be synonymous, but there are some exceptions. We will give one particular example. The growth in the second stage of a two-stage chemostat (Table 1.3) after some time generally tends to a steady-state, the biomass and residual substrate concentration remaining constant. However, such growth is not balanced, because cells delivered from the first stage differ in their properties and composition as compared with the bulk of cells in stage 2. This situation was termed the 'transient steady-state' (Powell *et al.*, 1967, pp.259–261).

The two-compartment model of Ramkrishna *et al.* (1967) was capable of simulating only some superficial manifestations of non-balanced growth. The choice of variables in this model was more or less arbitrary, and so should be regarded more as an illustration rather than a research tool.

Over the last decade much more realistic structured models based on biochemical data have been developed. A team of American biotechnologists headed by Shuler (Domach *et al.*, 1984) derived a simulation model of *E. coli* growth. That this model is quite comprehensive may be illustrated by listing the used dynamic state variables: glucose and NH_4^+ as exosubstrates, CO_2 and acetate as products excreted into the medium, amino acids, ribonucleotides, deoxyribonucleotides, monomeric precursors of cell wall components, rRNA and tRNA, non-protein polymeric components, glycogen, guanosine tetraphosphate, enzymes transforming ribonucleotides into deoxyribonucleotides, ATP, NAD(H) and protons.

Altogether, the dynamic model amounts to a system of 21 differential and 14 algebraic equations. An even more complicated model simulating growth of *Bacillus subtilis* (Jeong *et al.*, 1990) is the set of 39 non-linear and coupled differential equations containing nearly 200 parameters! These models are able to simulate particular growth features such as changes in cell sizes, shape and composition, as well as the D-dependent variations in replication time brought about by the shifts in glucose concentration. However, the predictive capability of such an intricate dynamic model should be still estimated as modest.

The two structured models discussed here are two extremes in the continuum of currently developed simulations. One is very rough and arbitrary but is readily analysed and can be run on a small personal computer. The second can be implemented only on mainframes. It gives a comprehensive picture of the prototype, authentic in details, and is based on information of actual molecular mechanisms involved in cell functioning. Needless to say, even this model is still nothing more than a 'caricature parody', since it by far oversimplifies the available present-day data on microbial biochemistry. However, despite this, it is too complex to be studied by conventional mathematical tools. For example, qualitative stability analysis for a set of 21 differential equations is desperately impossible. There is no hope of reliably identifying model parameters by fitting equations to experimental data. Therefore, such models are too cumbersome and awkward to be used efficiently in research or for microbial growth control. The optimal type of mathematical model lies, apparently, midway between the two extremes outlined. It is impossible to determine in advance either the best way to seek this optimum or the extent to which fine details of the reality could be ignored. Intuition and mathematical skill happen to be more important in this field than any quantitative criteria.

The development of new structured models remains a topical problem but one of high priority. Despite the few accomplishments discussed previously, no general theory (model) of non-steady-state growth of microorganisms has yet been proposed.

1.4.5 USE OF THE CHEMOSTAT IN STUDIES OF MICROBIAL GENETICS AND POPULATION DYNAMICS

A description of mutation and autoselection

The first studies of Novick and Szilard already contained the seeds of this further application of the chemostat theory. That is, the study of genetic heterogeneity of population cells in continuous culture and elucidation of microevolution regularities, including spontaneous mutations and adaptive selection. Pioneering studies in this area were carried out by the American researchers Novick, Moser, Atwood, Ryan, Marr and Painter. In the

1960s, similar work began in the USSR at the Institute of Physics, Krasnoyarsk (Pechurkin, Kovrov and Abrosov).

First of all, a chemostat was used for the quantitative evaluation of rates of mutations which do not affect the $\mu(s)$ dependence. Let N be the total cell concentration in a chemostat culture, m the concentration of mutants, μ the specific growth rate of the main non-mutated part of the cell population and η the specific growth rate of neutral mutants. Then we have

$$\dot{m} = \lambda\mu N + \eta m - Dm\,, \quad \dot{N} = \mu N - \lambda\mu N - DN\,, \qquad [1.70]$$

where λ is the mutation rate expressed as the ratio of the numbers of mutants to total number of cells formed. If $\lambda \ll 1$ and $\mu = \eta$, in the steady-state, we obtain $\mu = \eta = D$, and

$$\dot{m} = \lambda ND \quad \text{or} \quad m = m_0 + \lambda DNt\,. \qquad [1.71]$$

If $\eta > \mu$, then the original strain will be displaced by the mutant; otherwise, if $\eta < \mu$, m will tend to a lower limit $mL = \lambda N/(1 - \eta/D)$.

Experimental studies of phage-resistant mutants in a tryptophan-limited chemostat culture of *E. coli* showed that the period of linear m increase in accordance with Eqn. [1.71] was fairly short. Every 20–100 generations there was an abrupt fall in the number of mutants, after which the linear growth resumed at the same rate (Novick, 1959). The observed 'saw-tooth' dynamics in m was explained by Moser (1958) as a combined effect of mutation and selection. The original wild clone gives not a single but a whole array of mutations with subsequent reversions. Let us denote the total cell population in a chemostat culture as x, which is the sum of all subpopulations including original and emerging variants, $x = \Sigma x_i$. All possible transitions between variants are given by the matrix, $\lambda_{i \to j}$ ($j = 1, \ldots, n$; $i = 1, \ldots, n$; $j \neq i$). Then the chemostat model takes the following form:

$$\dot{x} = \sum_{j=1}^{n} \dot{x}_j = \bar{\mu}(s)x - Dx\,, \quad \bar{\mu} = \frac{1}{x}\sum_{j=1}^{n} \mu_j x_j\,,$$

$$\dot{x}_j = \mu_j(s)x_j - Dx_j - \sum_{i=j}^{n} \lambda_{j \to i} x_j + \sum_{j=i}^{n} \lambda_{i \to j} x_i\,, \qquad [1.72]$$

$$\dot{s} = D(s_0 - s) - \sum_{j=1}^{n} \frac{\mu_j(s)x_j}{Y_j}\,, \quad \mu_j(s) = \frac{\mu_m s}{K_{sj} + s}\,.$$

Every drop in the 'saw-tooth' dynamics of neutral mutants detected by their phage resistance can be interpreted as the appearance of another type of spontaneous mutant with a higher growth capability. In a chemostat culture, such mutants overcompete and displace all other cells by virtue of their higher affinity to limiting substrate (a decreased K_s value). Let such a mutant be denoted by the subscript k. The selection pressure for this mutant, σ, is given by

$$\frac{d(x_k/x)}{dt} = \sigma \frac{x_k}{x} - [\mu_k(s) - \bar{\mu}(s)] \cdot \frac{x_k}{x} . \qquad [1.73]$$

If $K_{sk} < K_{sj}$ ($j \neq k$), then $\sigma > 0$ until a new equilibrium is established. In this process, the original cells will be displaced by the mutants, the growth-limiting substrate concentration will decrease from $\bar{s}_1 = DK_{sj}/(\mu_m - d)$ to $\bar{s}_2 = DK_{sk}/(\mu_m - D)$, and the culture density will rise by $Y(\bar{s}_1 - \bar{s}_2)$.

Thus, Moser performed the first successful mathematical simulation of autoselection dynamics, which interpreted the selection pressure $\sigma = \Delta\mu$ in terms of a combined action of the environment conditions (limiting substrate concentration) and the kinetic features of individual cells comprising the population. As a matter of fact, the affinity to substrate was not always crucial for selection outcome. A number of instructive examples were reviewed by Pechurkin (1978):

1. In turbidostat and other types of fermenters with automatic cell density control, autoselection is in favour of mutants with higher specific growth rates, $\sigma = \mu_{mk} - \mu_{mj} > 0$. Since the total density of a heterogeneous population is kept constant by methods of engineering and the dilution rate is allowed to vary, then autoselection results in the increase in D from μ_{mj} to μ_{mk}.
2. Mutation towards a higher growth efficiency, $Y_k > Y_j$, will lead to the same result as an increase in μ_m: $\sigma = \mu_k - \mu_j = (\mu_{mk} - \mu_{mj})s/(K_{sj} + s)$, as soon as $\mu_k = q_s Y_k$ and $\mu_j = q_s Y_j$.
3. Growth of a mutant with a higher resistance to inhibitory metabolic products can be described by the Monod–Ierusalimsky equation with $K_{pk} > K_p$. Under selection pressure

   $$\sigma = \mu_k - \mu = \mu_m \frac{s}{K_s + s}\left(\frac{K_{pk}}{K_{pk} + p} - \frac{K_p}{K_p + p}\right)$$

   the original population will be completely displaced, and the product concentration will reach a higher steady-state level,

   $$\bar{p} = \frac{\mu_m K_{pk} \bar{s}}{(K_s + \bar{s}) D} - K_{pk} .$$

4. Growth of mutants that are resistant to an antibiotic will not be affected by competition if the respective antibiotic is continuously supplied: $\sigma = \mu_k(s)$, since $\mu = 0$ for all other forms. The dynamics of the total population will be governed by the initial density of the mutant.
5. A mutation resulting in elevated adhesion to fermenter walls will lead to accumulation of slow-growing cells, eventually (as the adhesion will be completed) we have $\sigma = \mu_k + D - \mu_j$.

It is noteworthy that the majority of autoselection models are based on Monod kinetics, which would normally ensure a fair fit to experimental data. For this reason, geneticists are not concerned with a refinement of the kinetic equations. Instead, they concentrate on the real underlying mechanisms responsible for changes in kinetic properties of cell populations, such as the affinity to substrate.

Let us review a few examples (for details, see Dykhuizen and Hartl, 1983).

1. Auxotrophic mutants displayed higher growth rates in a chemostat culture on a full medium as compared with isogenic prototrophs. Their advantage was explained as a 'saving' on those enzymes involved in the biosynthesis of the required metabolite. This explanation was put forward by Lwoff as early as 1944 and was not until recently experimentally tested. It was found (Dykhuizen and Davies, 1980), however, that the growth advantage of auxotrophs was not directly related to the reduction in the number of proteins synthesized and therefore was connected to other metabolic effects rather than energy supply.
2. A somewhat similar situation was found with the anaerobic growth of a chemostat culture of *Saccharomyces cereviseae*, petite mutants showing a faster growth than cells containing mitochondria. This effect arises from the lower energy cost of growth in a situation when for some reason mitochondrial biosynthesis is arrested and respiration is impossible (James, 1978).
3. The decrease of K_s in the course of autoselection in the chemostat is often explained by an augmented transport function. For example, with alanine-limited growth of *E. coli*, a gradual shift in proton simport stoichiometry was observed from a 1 : 1 (one proton per alanine molecule transported) ratio in the original cell population to 4 : 1 after 4 days and to 7 : 1 after 8 days of continuous cultivation (Collins *et al.*, 1976). As a result of autoselection, there was a 20-fold decrease in the K_s value.
4. The present-day 'hot spot' in microbial population genetics is the study of extrachromosomal elements (ECE) – plasmids, phages, transposons and insertion elements. Normally ECEs do not carry genes absolutely essential for growth, but are capable of fast replication, surpassing the chromosome DNA in the number of copies. R-plasmids are responsible for bacterial growth in the presence of antibiotics, but under normal conditions (with no antibiotics present) their synthesis becomes too heavy a burden for the host cell, which is manifested in a decreased growth rate. Among the more than 100 R-factors studied, about a quarter were found to increase the bacterial generation time by 15% (Adams *et al.*, 1979). For this reason, plasmid-bearing strains are unable to compete with plasmid-free populations, although there are a

few exceptions. Thus, colicin-positive cells carrying respective plasmids are able to withstand the competition with faster-growing plasmid-free strains by virtue of antagonistic inhibition.

In recent years, there have been proposed rather intricate and detailed dynamic models of autoselection which take into account the ECE-related effects, including the transfer of ECEs within the population, their segregation loss, changes in μ arising from the ECEs carriage etc. Some of these models have ecological applications and can be used to explain the behaviour of heterogeneous populations of enterobacteria in the animal intestine (Freter *et al.*, 1983).

Population and macrokinetic studies, stochastic and deterministic models

The theoretical analysis of autoselection in continuous culture which has been briefly discussed is related to the **population approach** in microbiological kinetics. It presumes that cells are inherently heterogeneous in many respects, including the kinetic properties of individual cells. Earlier (see Eqns [1.30–1.68]), such heterogeneity was completely neglected and it was neccessary to accept average kinetic parameters for the myriad of individuals in the population. This may be called the **macrokinetic approach**.

The two approaches were not always compatible, when the subject of research was not a special issue such as selection, but concerned other qualitative regularities of growth. This conflict was expressed by the different interpretation of the kinetic equations. Let us consider, as an example, a Monod-type relationship $\mu(s)$. A disciple of the macrokinetics approach might think about it in the following manner. A variation in s causes a smooth response in growth rates of **all cells** in the population. Even if kinetic properties of individual cells were not identical, the Monod equation will still be valid for single cells and for the entire population, except that in the latter case the parameters K_s and μ_m will be averaged over all cells in the population. An advocate of the population approach might take a different line of reasoning. He or she would first emphasize heterogeneity of the cell population, then consider the frequency distribution of cells in respect to μ and, as a first approximation, would distinguish two groups of cells, active and inactive. If the probability of a transition into the active state is proportional to s, then the population as a whole will obey the Monod equation.

Let us examine another authentic example from the recent history of microbiology. There was a fiery debate between the famous kineticist Hinshelwood and the geneticist Ryan about the mechanism of microbial adaptation. Transferred from one substrate to another, microbial cells display a latent period and then begin to produce enzymes which degrade the new substrate. Is this the result of a **regular and simultaneous** reorganization in the enzymatic apparatus of **every cell** in the population, or a consequence of a mutation giving a new property to a **fraction of cells** and thus causing vigorous growth of the clone on the new medium? A thrilling essay of this scientific controversy, as seen by contemporaries, can be read in *The Microbe's Contribution to Biology* (Kluyver and Van Niel, 1956). Today, when we know the molecular mechanisms underlying induction and derepression of enzyme synthesis (the operon concept by Jacob and Monod) and also the role of ECEs in high-frequency mutations, it becomes possible to tell where precisely Hinshelwood was right, and where Ryan was right, and also where both conflicting approaches need to be combined. These historical examples support the prosaic moral to seek the 'golden average' (in the meaning of compromise). Thus, it is necessary to take into account, on the one hand, the heterogeneity of real populations, autoselection, mutations and other stochastic events, and on the other hand, the deterministic regularities in the response to the environment of the majority of the individual cells in the population.

There are two types of model in mathematical biology, deterministic and stochastic. The former describes clear determined and regular processes. The latter deals with random, or stochastic processes. Needless to say, any real-life process has both deterministic and stochastic components, but professionals involved in modelling would usually recognize only one of them. As far as the chemostat theory is concerned, it was based on strictly deterministic models. All the equations derived by Monod, Herbert, Pirt, Powell and others belong to this class. Stochastic models of microbial growth in the chemostat (Kozheshnik, 1971) are not as effective in explanations and predictions of growth dynamics. The obvious reason for this is that a chemostat culture is a highly homogeneous and well-defined experimental system.

The stochastic modelling is more appropriate for population studies, because the fate of individuals in the population is not strictly determined and needs to be described by probabilities, frequency distributions, variance, etc. Of special importance for this type of modelling is its use in studies of the **microbial cell cycle**.

Using the frequency function of cell age distribution, the conservation of the number of cells can be written in the following differential form (similar to the continuity equation in hydrodynamics):

$$\frac{\partial n(t,\tau)}{\partial t} + \frac{\partial n(t,\tau)}{\partial \tau} = -\,[D(t) + w(t,\tau)]n(t,\tau)\ , \qquad [1.74]$$

where $n(t,\tau)d\tau$ is the number of cells which at time t have their age in the interval $[\tau, \tau+d\tau]$, $D(t)$ is the specific rate of cell elimination (dilution rate in the chemostat), and $w(t,\tau)$ is the specific rate of mother cells dividing into daughter cells. Equation [1.74] was first obtained by MacKendrick (1926) and was later rediscovered by von Foester (1959).

Collins and Richmond (1962) demonstrated that the kinetics of growth of individual cells determine the form of the size distribution of the cell population:

$$V(\nu) = \frac{\mu}{\lambda(\nu)}\left[2\int_0^{\nu}\psi(x)\,\mathrm{d}x - \int_0^{\nu}\phi(x)\,\mathrm{d}x - \int_0^{\nu}\lambda(x)\,\mathrm{d}x\right], \qquad [1.74a]$$

where $V(\nu)$ is the absolute growth rate of a cell of volume ν, $\lambda(\nu)$, $\phi(\nu)$ and $\psi(\nu)$ are frequency functions of the volume of surviving cells, dividing cells and newly formed cells respectively. The size distribution function can be obtained with required precision using a Coulter counter. Subsequent testing of different hypotheses on growth kinetics of individual cells during the cell cycle is performed by statistical comparison of experimental and calculated frequency curves (Grover *et al.*, 1987).

However, in practical terms this approach often fails, and differentiation of hypotheses is better attained by refinement of experimental technique rather than a mathematical approach. The well-known example is the 'growth law' of bacteria during the division cycle. To prove that it is closer to an exponential rather than a linear pattern, Cooper (1988) used an advanced membrane-elution technique and followed the dynamics of all major cell constituents throughout cell cycle. The results were much more convincing than statistical treatment of distribution curves.

Mixed cultures

Chemostat theory proved to be productive not only for the analysis of autoselection in single-species continuous culture, but also for the study of mixed cultures. The dynamic behaviour of polyculture may be described by the same set of equations as that applied to spontaneous mutation studies (Eqn. [1.72]). The only modification required is the replacement of the mutation rates $\lambda_{i\to j}$ by a matrix of mutual inhibition or stimulation for all species in polyculture. The main attention has been paid to the three following types of interaction in a polyculture: (a) competition for a common substrate; (b) metabiotic interactions (the product of one species is the only substrate for the other); and (c) the predator–prey and host–parasite interactions (Abrosov and Kovrov, 1977; Veldkamp, 1977; Pechurkin, 1978). Competition among different species in an open system, for a common limiting substrate, is basically similar to the previously discussed case of autoselection in a continuous culture. In both cases all initial forms are displaced by the most adapted one, a fact supporting the Gause rule on competitive exclusion. There was much effort to explain the violation of this rule observed in natural environments (the so-called 'paradox of plankton', that is, the coexistence of competing species in the same habitat).

Chemostat studies with mixed cultures (Pirt, 1975; Abrosov and Kovrov, 1977; Sommer, 1984) allowed the formulation of the necessary conditions to permit the sustainable coexistence of competing species:

1. Non-steady-state operation of chemostat culture caused by periodic switching of dilution rate between D_1, a value favourable for one competitor, and D_2, at which the second species grows faster.
2. The inhibition of the faster-growing species by its own metabolic products.
3. The stimulation of a slower-growing species by a metabolic product of a faster-growing population.
4. The presence in the open system with n potentially competing species of no less than n different substrates such that for each population there is a favoured substrate on which it can grow faster than any one of its competitors.

The components of syntrophic microbial associations cannot multiply separately in monocultures. This feature was simulated by the following model (Powell, 1985). The species X consumes the substrate S and excretes the product P into the medium. P inhibits the growth of X and serves at the same time as the only and unique substrate for the second species Y.

The classical predator–prey system, as developed by Volterra, was modified to incorporate the concentration of the growth-limiting substrate for the prey as the third variable. The logistic equation with its obscure parameters was also replaced by mechanistic equations based on Monod kinetics, which contain parameters having a clear-cut biological meaning (Curds and Cuckburn, 1968).

1.4.6 CONCLUDING REMARK ON THE CHEMOSTAT AS A RESEARCH TOOL

At last, we have completed our historical excursion into the development of chemostat theory. It is customary to conclude such a large amount of text with a general statement playing the role of an improving moral. Since 1975 the most comprehensive and consistent treatise on microbial cultivation theory is the famous 'yellow' book of John S. Pirt (1975). In this the author stressed the following advantages of using chemostat as a research instrument:

1. The chemostat permits the biomass growth rate to be varied with no change in environment other than the concentration, s, of limiting substrate.

2. The converse can also be realized, that is, D can be fixed while various environmental parameters, such as temperature, pH and medium tonicity, are changed.
3. The chemostat allows substrate-limited growth to be maintained for as long as is needed.
4. As a consequence, the chemostat method facilitates the investigation of growth and the reaction of the organism to its environment, as well as the control of microbial processes.

It should also be added that the extensive use of continuous culture improved the comprehension of the dynamic nature of microbial growth, made clear a distinction between apparent and true rates of processes, and disclosed all major components of the mass balance of microbial growth.

There are, however, a few widespread and persistent misconceptions related to chemostat theory. We have already mentioned 'the specific growth rate as an independent variable'. The other misleading point is the idea that chemostat cultures are more simple than batch ones.

To prove this point, it is often argued that it is possible, with a chemostat culture, to vary just one factor (D, which will affect $\bar{s}$), keeping all other parameters fixed. This is, however, wrong. Recall that steady-state chemostat culture at different D differs not only by substrate concentrations, $\bar{s}$, but also by physiological state of the cells (their composition and activity, sensitivity to stress, spectrum of exometabolites, etc.).

The complexity of batch culture is perceived first of all as variability, since cell properties and growth conditions continuously change with time. However, the dynamics of chemostat culture is in no way simpler during transients, when chemostat culture displays damped oscillations, overshoots, undershoots and other dynamic patterns which seldom could be seen in batch culture. During steady-state, growth biomass is constant, but this constancy is not equivalent to simplicity as there could be multistability, autoselection, wall growth, contamination and other factors which escape the experimenter. Lastly, a hidden complexity of steady-state chemostat culture stems from the fact that at each particular D, we have a cell population with unique physiological properties, this being the product of a long-term metabolic adjustment to constant growth conditions. This is why, at different D, we obtain somewhat different populations and hence comparison of kinetic measurements made over a wide D range must be performed with great care. Should then a constancy in time of a steady-state chemostat culture be viewed as a 'simplifying' factor?

A quite common misconception is that a simple mathematical model can adequately describe growth in chemostat but not in batch culture. True enough, by setting the derivatives in a dynamic model to zero, we significantly simplify mathematical analysis. For this reason, the chemostat provides a good starting point for model testing. At the same time, kinetic models which are adequate only for chemostat culture could not be accepted as an instance of certified knowledge. Just imagine what a physicist would think about a mathematical formula which produced a well-fitting projectile calculation for a shell fired from a mortar but gave large errors for rifle bullets. No doubt, such a formula would be proclaimed inappropriate and as being based on incorrect theoretical assumptions. The logic of the scientific method based on the elimination of alternatives is uncompromising. The existence of just one contradictory fact is sufficient to condemn an admittedly harmonious and consistent model as an inadequate alternative.

In reality, microorganisms remain as complex objects whatever cultivation device has been used for their growth. Also, the chemostat **model** may be regarded as the **theory** of microbial growth (i.e. a set of reliable and consistent knowledge) only after explanation of all well-established experimental findings. Today, however, many phenomena exist which are mysterious within the framework of even the most elaborate chemostat models. We will just list them:

- variations in the composition of cells and their metabolic activity as functions of D;
- yield variation on conserved substrates;
- the transient dynamics of chemostat culture following shift of D, or change in the growth-limiting factor;
- the behaviour of chemostat culture at extremely low D;
- variation in maintenance requirements as dependent on D and the type of limiting factor;
- dynamics of cell death and decay of biosynthetic activity in a starved culture (formally at $D \to 0$);
- quantitative regularities of secondary metabolites biosynthesis.

Even more 'blank spots' may be found in the quantitative aspects of microbial growth when we pass from laboratory cultures to microbial populations in natural habitats.

1.5 MICROBIAL GROWTH IN NATURAL HABITATS

The term 'natural' means any habitat where spontaneous microbial growth proceeds unaffected by human control. First of all, this category includes natural environments *sensu stricto*, i.e. soils, subsoils, sediments, rocks, waters, air, plants and animals, which harbour microbial populations and where they naturally grow, proliferate, undergo stress, and perish. In addition, the term 'natural habitats' should also include various man-made objects which are not designed specifically for microbial cultivation, i.e. soil substitutes, composts, industrial constructions affected by biocorrosion, etc.

In this section we will try to determine to what extent the kinetic principles established for laboratory cultures are valid for microbial growth in natural habitats. Are there any particular growth laws unique for natural populations which, in contrast to laboratory cultures, are involved in interactions with other micro- and macroorganisms as well as with a changeable and unpredictable environment? What are the true rates of microbial growth, multiplication and elimination in nature? What is the physiological state of microbial cells *in situ*? Let us consider how these questions were answered historically.

1.5.1 THE EARLY VIEWS

Originally, microbiology was not a differentiated science, and available information on microbial life was not only fragmentary but also not sufficiently specialized. As a result, microbiology was a somewhat 'syncretic' science, displaying a reconciliation of different approaches. In particular, microbiologists were as much interested in natural microbial growth as in its laboratory reproduction. It was typical of the best scientists of that time (Pasteur, Famintzin, de Bari, Winogradsky) to combine harmoniously simple and elegant laboratory experiments with *in situ* observations of visible microbial development, i.e. blooming of waters, cyanobacterial mats in lagoons and fungal mats on the forest floor, biofilms, etc.

At the turn of the 20th century the most exciting event in microbiology was the discovery and description of those microorganisms which carry out widespread geochemical processes, namely nitrification, oxidation of S, Fe, Mn and H_2, sulphate reduction, and methanogenesis (Winogradsky, Beijerinck, Omeliansky, Isachenko, Lebedev, Nadson, Perfil'ev). Most of these microorganisms are chemolithotrophic. Geologists became aware of such processes as early as the first half of the 19th century, and by the 1870s their biological nature was firmly established (Zavarzin, 1972). Although no rigorous *in situ* rate measurements were available at that time, there was still an agreement that lithotrophic growth on a laboratory media and in natural habitats were of comparable intensity. For example, NH_4^+ oxidation in a culture of nitrifying bacteria and in soil proceed at roughly the same rate, corresponding with rather slow bacterial growth (generation time about several days or weeks).

An entirely different situation occurred with heterotrophic bacteria. As early as in the middle of the 19th century it was found that such bacteria could grow on a nutrient broth at exceptionally high rates, dividing every 20–30 minutes. Cohn calculated that the progeny of just one bacterial cell could fill the entire world's ocean in only 5 days. This impressive calculation, originally made in 1866, is still safely passed from one textbook to another. It was, of course, no more than a spectacular learning demonstration of the 'explosive' nature of exponential growth, yet this straightforward extrapolation of laboratory growth data to reflect nature was characteristic at that time. Despite admitting that the real expansion of bacteria was somehow limited, microbiologists still took for granted the fast growth of microorganisms under natural conditions. As restraining factors they generally recognized the bactericidal action of sunlight, toxic substances, and the immune responses of higher organisms. Only the most perspicacious microbiologists identified the availability of organic matter as a limiting factor (Kostychev *et al.*, 1926).

Another early misconception stated that fast growth was a universal characteristic of all bacteria, being predetermined by their small cell sizes and simple life cycle (Rahn, 1932).

1.5.2 THE DEVELOPMENT OF SOIL MICROBIOLOGY

'The methods of soil microbiology'

This common title of a series of papers published by Winogradsky in 1925–1927 (Winogradsky, 1949, pp. 383–470) marked the birth of soil microbiology. The new branch of science acquired its specific methods and the guiding principle: 'to study microorganisms directly in soil or in soil-resembling media'. The implementation of this principle was based on two basic techniques: (a) the use of direct soil microscopy for the enumeration of microbial cells and assessing their morphological diversity; and (b) the study of so-called spontaneous cultures (*les cultures spontaneés*), which emerge on soil media or directly in soil amended by some particular substrate. In contrast to pure isolates, spontaneous cultures are, essentially, associations of many microbial species having selective advantage over others in the utilization of added substrate under specified growth conditions. As soon as these conditions are natural or very close to being natural, a spontaneous culture reveals those soil organisms which are responsible for the studied process *in situ*.

There are many reasons why pure cultures should be treated with care in soil studies but one particular argument was especially stressed by Winogradsky. He claimed that consecutive passages through artificial media might substantially alter the original physiological properties of the isolates. For example, as was observed by Müller at Van Niel's laboratory, long-term culturing of purple sulphur bacteria induces their ability to grow on organic compounds instead of hydrogen sulphide. On these grounds, Kluyver christened pure cultures as 'physiological artifacts', and Winogradsky stated that 'pure microbial cultures outside their natural habitats, i.e. grown on artificial media, can tell us nothing about the dynamics of biological processes in such habitats' (Winogradsky, 1949, p.387).

The first data, obtained using Winogradsky's technique, were quite shocking. To begin with, direct soil microscopy revealed a 100–1000-times larger bacterial population than conventional plating. Secondly, two groups of soil microbial populations were discovered which were characterized by different behaviour and called respectively **autochthonous** and **zymogenic** microflora.

Winogradsky's technique was enthusiastically taken up by followers. A significant advance in its development was made by Kostychev, a prominent Russian physiologist and biochemist. He augmented the direct microscopy with *in situ* measurements of soil microbial activity, such as respiration and N_2 fixation. Crimea tobacco plantations were chosen as the main site of field dynamic observations in combination with laboratory studies aimed at improving plant N nutrition through N_2 fixation (the acquisition of so-called 'biological' nitrogen). In fact, Kostychev introduced the **integrated (holistic) and dynamic** description of bacterial activity in an agroecosystem, coming close to what is known today as the system approach. Specifically, he

1. showed that the measured values of microbial biomass and soil nutrient content were of a dynamic nature being the function of the opposing processes of growth and elimination, production and consumption;
2. identified, at least qualitatively, the mass and energy fluxes between microorganisms and plants via root exudation, N_2 fixation, N-immobilization and mineralization;
3. proved that microbial biomass was a reservoir of plant mineral nutrition; and
4. posed the problem of controlling soil microbial activity, so that 'soil dynamics could be adjusted according to our needs' (Kostychev *et al.*, 1926).

The crisis in soil microbiology

Subsequent development of soil microbiology was prosaic and monotonous. More or less evident success was demonstrated in some particular fields related mainly to the descriptive biology of soil microbes, their spatial distribution and ecological interactions (Krassilnikov, Perfil'ev, Mishustin, Aristovskaya, Khudiakov, Novogrudski, Waxman, Kholodni, Rovira, Garret, Katznelson and others). At the same time, only minor progress was made in the development of the ecological and system approaches launched by Winogradsky and Kostychev. The implementation of quantitative methods and mathematical modelling in soil microbiology was extremely slow and sluggish.

The weakness of the theoretical background was spectacularly manifested in the failure of the campaign to boost the wretched Soviet agriculture by the implementation of 'bacterial fertilizers'. At this time in Russian history, dominated by Stalin and Lysenko, when showy immature initiatives were honoured higher than true and careful scientific researches, a large-scale campaign was instigated to amend soil with pure cultures of 'useful' microbes – N_2-fixing bacteria, phosphate-solubilizing microbes, producers of plant hormones, 'silicate' bacteria, etc. The introduced microorganisms, however, did not persist in soil and only occasionally fulfilled their intended function. At best, there was some positive effect on crop yield, brought about not by living cells, but by metabolites or nutrients originally present in bacterial preparations. Yet no activity, let alone multiplication of strains introduced into soil, was ever detected.

The failure of the bacterial fertilizer project, expected to bring quick and tangible return, was inevitable for the simple reason that microbial ecology and kinetics were completely ignored. At that time no attention was paid to issues related to population ecology, i.e. the interactions between introduced species and indigenous populations, competitive potential of released microbes, control of population dynamics, optimal inoculation rates and terms.

However, the main transgression was probably the neglect of such an important factor as the availability of nutrient resources (observed in the ecology of macroorganisms as early as the last century). A persistent idea expressed both explicitly and implicitly in papers published between 1940 and 1970 was that organotrophic microbial growth in soil might be limited by anything but the shortage of nutrients. In view of the high humus content of most soils, abundant growth was assumed to be guaranteed for all heterotrophic microorganisms in possession of the required degradative enzymes. For this very reason N_2 fixation was supposed to start in soil immediately upon *Azotobacter* introduction. At that time the important point was missed, that nitrogen fixation is an energetically expensive reaction, so that soil inoculation by diazotrophic populations needs to be combined with a supply of available organic substances.

The impact of the International Biology Program (IBP)

The issues of energetics and kinetics of microbial growth in natural habitats were considered as late as at the end of the 1960s, when the IBP was launched. Among the targets of this project was an estimation of the global natural resources of the Earth including soil biota. Specifically, there was an ambitious plan to measure 'microbial standing crop' as well as the microbial reproduction rates in the soils of all bioclimatic zones.

The first complex studies at IBP field stations by zoologists, botanists, phytocenologists, soil scientists and climatologists revealed the incompetence of microbiology in solving the stated problems because of the weakness of the theory and lack of pertinent methods. Summing up the preliminary IBP results, Aristovskaya (1972) had to admit bitterly that soil microbiology was a 'second-rate' science, lagging behind other branches of microbial ecology. (The most important progress between 1950 and 1970 was made in aquatic and geological microbiology, mainly due to the contribution made by Kuznetsov and his coworkers – Ivanov, Lalikova, Sorokin and others.)

As a result, the attention of soil microbiologists focussed on the development of methods and techniques. The highest priority was given to the following objectives:

- the accurate determination of soil microbial biomass;
- the measurement of microbial growth rates *in situ*;
- the evaluation of microbial contribution to the carbon budget of terrestrial ecosystems, the assessment of major fluxes of C-input to soil (plant litter formation, root exudation and sloughing), microbial respiration, and grazing by soil microfauna.

All the specified problems belong to the scope of the kinetics and energetics of microbial growth. With respect to pure homogeneous microbial cultures, they were, by this time, sufficiently well understood. As far as natural microbial populations were concerned, the main obstacle was identified as the absence of adequate techniques. Progress in this field was extremely slow, the first successes being in relatively simple environments, such as natural waters, waste treatment reactors, intestines (including the rumen) and the blood circulatory system of animals. All these environments are simpler than soil. We will term them provisionally as 'homogeneous', although it should be remembered that their actual homogeneity is often far from perfect.

1.5.3 ESTIMATING MICROBIAL GROWTH RATES *IN SITU* IN HOMOGENEOUS HABITATS

A splendid and comprehensive review of the early works (up to 1970) was presented by Brock (1971). An interested reader is directed to the original paper for details and references. Here, we shall restrict ourselves to comments on the most important approaches.

Microscopy *in situ*

Microscopic observations of this type are normally done only in aquatic habitats with the use of a submerged-slide technique. At regular intervals glass slides to which microbes are attached are removed from water for microscopy and afterwards returned. Microcapillaries may be used instead of standard glass slides (Perfil'ev and Gabe, 1969). An alternative, which has only been used by the most courageous ecologists, is to immerse a microscope directly into the pond and to carry out a diurnal observation of individual cells attached to glass surface (Staley, 1971). This approach revealed interesting observations; a single cell of *Chlorella* increased in size during the day, and at night this enlarged cell divided into four daughter cells. Occasionally this cycle was retarded, with no cell division until the second night.

Obviously, this technique requires discrimination between true growth of attached bacteria and their immigration from surrounding waters. The cell settlement or detachment could be accounted for by a microscopic count of ultraviolet-sterilized control slides. The generation time of aquatic bacteria was found to vary from 2 to 30 hours.

Methods based on the analysis of the cell-division cycle

In eukaryotes, the cell cycle consists of four phases, mitosis, G_1, S, and G_2. Mitosis can be recognized morphologically with, in many cell types, the time of mitosis (t_m) representing a constant fraction of the total cell-division cycle. If t_m is known, then the generation time, g, can be found from the relationship $t_m/g = 1.44\ R$, where R is the fraction of cells in mitosis. This method was used initially to measure the growth rate of the protozoa *Entodinium* in the rumen. The division frequency at night was higher than during the day, and the average generation time was about 15 hours.

Recently, Hagström *et al.* (1979) suggested that the growth rate of Gram-negative bacteria could be estimated from the frequency of occurrence of dividing cells. The division events (formation of septa and subsequent separation of the two daughter cells) is known to occupy a more or less constant time within the bacterial cell cycle. So, the higher the growth rate, the higher the probability of finding a cell at this stage. Calibration of the method was achieved using a mixed chemostat culture of marine bacteria. In the coastal region of the Baltic Sea, the frequency of dividing cells was below 5%, and the estimated mean generation time varied seasonally from 10 to 100 hours.

Genetic methods

An elegant method devised by Meynell (1959) is based on the use of bacteria with a non-replicating genetic marker. At each cell division the fraction of the population which contains the label will be halved. The dynamics of total and labelled populations is determined, after which the doubling time can be calculated from the rate of marker dilution. Meynell studied the growth of pathogenic enterobacteria in the gut and blood system of laboratory animals. The genetic markers were various superinfecting mutants of phages which enter the bacterial cells but do not replicate. One result obtained was as follows; after intravenous inoculation into a mouse, virulent *Salmonella typhimurium* cells became lodged in the spleen. Their viable count doubled every 24 hours, whereas the true doubling time as determined from the rate of marker dilution was 8–10 hours. The same strain grew on nutrient broth 20 times faster ($g = 0.5$ h).

Techniques stemming from chemostat theory

Many natural habitats are open systems, with a continuous supply of nutrients and the simultaneous elimination of cells. Under such conditions, the growth rate μ is eventually adjusted to the elimination rate, D (similar to the dilution rate in the chemostat). Now, the value of D is often easier to measure than μ. For example, in the case of bacteria growing in an animal's intestines, D is measured by feeding the animal food tagged with an inert label (lignin, silica-gel, dyes, etc.). The time of 50% reduction in the output label concentration is expressed as $t_{0.5} = \ln 2/D$. The steady-state (or quasi-steady-state) cell concentration, $\bar{x}$, is measured in the gut of sacrificed animals. By using this method, enterobacteria in laboratory rodents (mice, rats, hamsters) were shown to yield between 1 and 6 generations per day.

Another ingenious technique was developed by Brock for measuring the growth rate of thermophilic algae in hot-spring drainways (Brock and Brock, 1968). The technique involved the measurement of the algal wash-out rate after growth was prevented by darkening the system. The spring was sheltered from ambient light by a screen and the exponential decrease in cell concentration in the effluent stream was monitored. The value of D, which is equal numerically to μ under normal day–night cycles, was found to be 0.4 day^{-1} ($g = 40$ h).

In static aquatic environments such as lakes and ponds the main factor responsible for microbial cell elimination is no longer wash-out but their predation by protozoa and probably other small animals. The growth of cells and their grazing rate are about to be balanced. If, under *in situ* experiments, predation is completely suppressed (by passing the water sample through filters retaining large protozoan cells), then the μ value may be measured from the recorded increase in the bacterial population (Ivanov, 1955). However, this approach should be used with care. Suppose we are to measure the value of μ in a chemostat culture from the x dynamics after stopping the flow. It is obvious that by halting the pump operation we terminate not only cell wash-out but also the substrate input with fresh medium. Therefore the use of this method is confined to only non-limited microbial growth. These conditions are fulfilled in the chemostat only at $s \gg K_s$ (i.e. at high s_0 and subcritical D), so that reliable application of the method is limited to eutrophic natural habitats.

Isotope techniques

The high sensitivity of radioisotope techniques allowed the measurement of the rates of consumption of labelled substrates added to waters at nearly background concentration. The main problems were first, how to estimate the ratio between added labelled and natural non-labelled compounds and second, how to derive the rate of microbial growth from the measured rate of label consumption. Below we provide several examples.

Dark $^{14}CO_2$ fixation rate as a measure of total heterotrophic bacterial production was originally suggested by Romanenko (1964). Heterotrophic CO_2 fixation is an anaplerotic metabolic reaction, serving to regenerate those metabolic intermediates which are 'lost' from the tricarboxylic acid (TCA) cycle for the synthesis of macromolecules. Hence, the measured fixation rate is expected to be tightly coupled, through metabolic control, to overall cell growth. However, the experimentally observed ratio of carbon fixed from CO_2 to total carbon assimilated has been found to vary in a wide range, 0.01–0.12. In view of this fact, it was suggested (Overbeck and Daley, 1973) that measurements of CO_2 fixation should be accompanied by a determination of the activity of phosphoenolpyruvate (PEP)-carboxylase, the principal anaplerotic enzyme. This would allow more rigorous conclusions about the process stoichiometry to be made.

The primary productivity of phytoplankton is determined by the measurement of the rate of $^{14}CO_2$ photoassimilation. In recent modifications of the technique, the label incorporation was suggested to be determined in the fraction of chlorophyll *a* rather than in whole particulate matter. This allows the estimation of phytoplankton biomass and avoids possible underestimation caused by label transfer from algae to bacteria and zooplankton via excretion and grazing respectively (Laws, 1984).

Nowadays the most promising technique for the evaluation of secondary productivity (microbial growth rate) in waters is considered to be the measurement of the uptake of labelled precursors of nucleic acids biosynthesis, thymidine, uridine and adenine (Karl, 1979; Bell, 1986). The use of isotopes with high specific activity guarantees minimal alterations of *in situ* growth conditions. At the same time, the amount of added nucleoside should be large enough to suppress their synthesis *de novo* from endogenous cell compounds. The main deficiency of this method is the ambiguity of conversion factors from nucleoside uptake rate to microbial growth rate.

Until recently, the measured growth rates of natural populations could only be average values for an entire community. Today, it has become feasible to obtain frequency distributions of cells within natural populations with respect to their metabolic activity and, hence, their growth rates. For this purpose almost continuous rather than single-point dynamics of ^{3}H-adenine incorporation has been recorded. Subsequent statistical treatment generates the required frequency distribution of the cells' activity (Laws *et al.*, 1986).

We have briefly discussed the most important approaches for measuring microbial growth rates in homogeneous natural habitats and illustrated their application for a few specific cases. It is worth remembering that typical multiplication rates in such media are characterized by generation times of about 1 day, from 10 hours to 4 days.

1.5.4 ESTIMATION OF MICROBIAL PRODUCTIVITY IN SOIL

We now switch to a more complex natural object – the soil. Because it is opaque, spatially heterogeneous and has a large absorbing capacity, only a few of the above-mentioned techniques can be exploited with respect to soil microbes. As a result, soil microbiologists had to look for essentially different approaches.

Historically soil microbiology was developed mainly as a synecological discipline, the primary objective being no longer particular populations and species but the entire community, its total biomass and production per season. The specific growth rate, μ, and production, Δx, are closely related, since $\Delta x = \mu_{av} x_{av} \Delta t$, where μ_{av} is the mean specific growth rate and x_{av} is the mean soil microbial biomass over the time interval Δt (in our case, set equal to one season).

Assessment of productivity from fluctuation frequency of microbial biomass

The first systematic studies of bacterial production in soil were undertaken by a Leningrad team of microbiologists led by Aristovskaya (1972, 1975). The work involved daily measurements of the number and size of bacterial cells by direct microscopy of soil smears (a slightly modified Winogradsky procedure). Bacterial production was evaluated from the shape of the dynamic curve $x(t)$. This curve was always characterized by a seesaw pattern. Increases in x were observed every 3–8 days, followed by declines to background level. Fluctuations did not depend immediately on environmental factors and occurred even under stable hydrothermal conditions. This type of fluctuating dynamics was first observed as early as the beginning of the century and was explained by two mechanisms:

1. A predator–prey interaction of soil bacteria with microfauna (mainly with amoebae), which usually gives rise to oscillations in the population densities of both prey and predator (Catler *et al.*, 1923).
2. The accumulation in soil of self-inhibitory metabolic products (H_2, ethylene oxide, a hypothetical compound 'periodine', etc.), which

are susceptible to spontaneous autoxidation, decomposition or dispersion (Khudyakov, 1972; Nikitin and Nikitina, 1978).

For calculating productivity, Aristovskaya assumed that bacterial growth is periodically interrupted by toxin accumulation, while grazing of microbes was executed continuously. From this, the overall production of 'seesaw' bacterial growth was calculated as the following sum: apparent x increase (measured during intervals where $dx/dt > 0$) + bacterial biomass elimination (estimated as x decreases at time intervals when $dx/dt < 0$). This calculation algorithm may underestimate as well as overestimate the true bacterial productivity. The underestimation would be likely if bacterial growth is not totally arrested during 'intoxication' time. The overestimation would occur if the predator activity is not constant but instead oscillates in antiphase to the bacterial growth rate. Anyway let us discuss the obtained results. Very intensive rates of bacterial reproduction were observed in almost all the studied virgin and arable soils. The generation time was found to vary in the seasonal dynamics from 7 to 100 hours, with a seasonal bacterial production of 1–6 tonnes of dry weight per hectare. Compared with natural waters, the microbial growth rate in soils was roughly the same, whereas the seasonal productivity was higher by an order of magnitude. For example, the net bacterial production over one season in the Rybinsk water reservoir was as low as 200 kg ha^{-1} (Kuznetsov, 1981), while in podzolic soil of the same bioclimatic zone it was 1500 kg ha^{-1} (Aristovskaya, 1975).

The described approach was criticized mainly in respect to the suggested calculation algorithm involving dubious *a priori* assumptions. Moreover, scepticism was expressed even on the mere existence of the observed oscillations, since the spatial variation of x might have been erroneously accepted as the dynamic x variation (Zvyagintsev *et al.*, 1976). Besides, there was one ultimate theoretical contradiction. Oscillations in soil bacterial biomass imply an overall synchronization of growth of microorganisms occupying diverse and numerous microsites. However, this does not fit with the firmly established concept of soil microzonality (Krassilnikov, 1958; Zvyagintsev, 1987), which presumes a complete lack of interactive communication between various microsites. And, finally, the calculated bacteria production alone was close to the overall primary production of terrestrial ecosystems. It was doubted if such a high productivity of soil microorganisms (represented largely by heterotrophic species) could indeed be supported by the available energy resources.

Estimation of productivity from C-balance

In any event, these doubts have invoked a long series of studies where microbial production was estimated from the carbon balance of the entire terrestrial ecosystem, growth rate being related to either soil respiration or plant litter input to soil. The first approach was more reliable. It exploited the simple relationship between the respiration rate of aerobic chemoorganotrophs, v_{resp}, and their growth rate, $\mu_{av} : v_{resp} = Y_{p/x}\mu_{av}\, x_{av}$. Although the biomass yield, $Y_{p/x}$, actually depends on numerous factors, it can still be measured as the net average value in calibration experiments for the entire microbial community of particular soil. The main advantage offered by this method is the possibility for continuous and exact recording of *in situ* metabolic rates by CO_2 analysis, which is currently a well-advanced technique. Of course, one must be able to distinguish between the contributions of microbes and plant roots to the overall soil respiration, but this is basically feasible. A rough estimation based on soil respiration data (Gray and Williams, 1971) revealed lower productivity of soil microorganisms as compared with previous calculations, but this approach has not been implemented systematically.

The second type of mass balance evaluation of microbial production in soil is more common, with many such works having been performed over the last 20 years and new methods still appearing. The approach involves the estimation of C input to soil from plant litter and root deposition. These values are considered equal to C-substrate consumption by the heterotrophic microbial community. The accuracy of these estimates is inevitably low, but the method is interesting in two respects. Firstly, because it can provide an objective (although rough) estimate of average microbial growth rate, and secondly, because it clearly demonstrates the type of accepted theoretical framework. Production of microorganisms is calculated by the use of kinetic models similar to Eqns [1.30] and [1.32]. Such equations were originally derived for microbial growth in the chemostat. They describe the conversion of C-substrate into biomass, and allow for maintenance requirements. This raises the question of whether equations obtained for pure cultures can be applied to a soil microbial community. The answer will be negative, and to prove it we need to undertake a little theoretical investigation.

Calculation of productivity

Note that Eqns [1.30] and [1.32] describe microbial growth in the fermenter and not in the product vessel, which is the sink for outflowing cell suspension. However, for the soil community, both compartments should be taken into account, because soil retains those microbial cells which were produced during the primary growth on plant litter. The fate of these cells is resolved by the selection of the one of the following alternative pathways: (a) transition into the dormant state; (b) death and conservation of necromass as humus constituent; (c) death, autolysis and consumption of cell constituents by other microorganisms (reutilization); and (d) elimination of microbial cells by predators or parasites (exploiters), carbon from destroyed microbial cells being dissimilated to CO_2, incorporated into exploiter biomass with the remainder being excreted as waste products. Dead exploiters and excreted products are once again utilized by the surviving microorganisms.

For mature soils whose properties on a year-to-year basis are more or less constant, we may accept the steady-state approximation. Then some essential simplifications and conclusions can be made. As soon as a ratio between dormant and active cells is constant, then pathway (a) may be neglected, because the formation of dormant cells compensates their germination and both these processes do not affect the total productivity. We may also neglect pathway (b), because *de novo* humification (conservation of microbial necromass) is balanced by humus decomposition, and these processes are very slow. So, the only pathways to be considered are (c) and (d). In Table 1.4 the flow diagrams are shown for these two pathways along with the corresponding mass conservation equations. For comparison, similar derivations are presented for the unrealistic situation, which is persistently admired by soil microbiologists, that the biomass of cells grown on plant debris is assumed to be somewhere removed without reutilization or consumption, yet an allowance is made for maintenance requirements.

We are now ready to draw important conclusions. Incorporation into the model of reutilization terms (Eqn. [1.77]) leads to substantially higher microbial production as compared with conventional estimates by Eqn. [1.75]. Interestingly, with $Y \geqslant 0.5$, the secondary production may exceed the primary one, i.e. the newly formed microbial biomass during the summer season may be larger than the amount of utilized plant litter, and this is in no way a contradiction to the mass conservation law. Predation decreases microbial production as compared with simple microbial reutilization. Yet, even in this case, microbial production is much higher than the value calculated by a conventional chemostat model.

The next step of our analysis is to account for maintenance requirements, expressed by the negative term ax in Eqn. [1.76]. This is a very important issue, since the calculation of microbial productivity is exceptionally sensitive to the choice of a coefficient, and generally not supported by experimental data. As a matter of fact, there is no possibility for the direct determination of a characterizing a microbial community *in situ*. (The excellent experimental data reported by Anderson and Domsch (1985) need to be clarified with respect to terminology: they actually measured the rate of added glucose oxidation, which is in fact uncoupled wasteful respiration and is by several orders higher than maintenance ratio.) So, one must rely on data obtained with pure cultures, but this may often lead to absurd results. Even with the smallest value of

Table 1.4 Mathematical expressions used for calculation of microbial productivity in soils

Graph	Equation of mass balance	Calculation of microbial production	
Simple chemostat-type model			
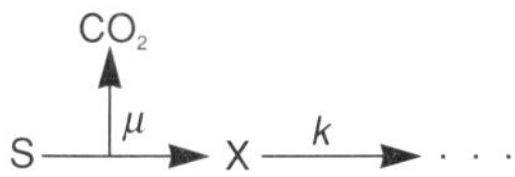	$\dot{s} = F - \frac{\mu x}{y}$ $\dot{x} = \mu x - kx$	$\mu x = YF$	[1.75]
Account of maintenance			
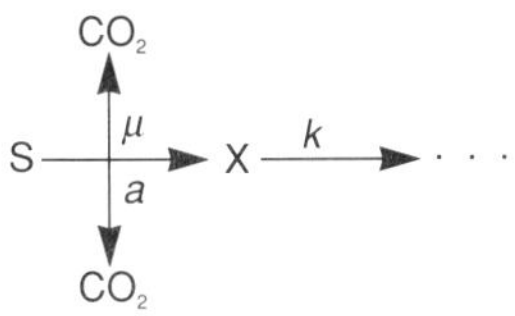	$\dot{s} = F - \frac{\mu x}{Y^{\max}} - \frac{ax}{Y^{\max}}$ $\dot{x} = \mu x - kx$	$\mu x = Y^{\max}F - ax$	[1.76]
Account of microbial biomass reutilization			
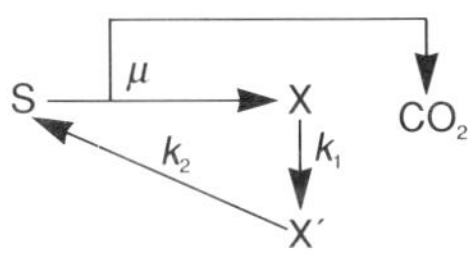	$\dot{s} = F - \frac{\mu x}{Y} + k_2 x'$ $\dot{x} = \mu x - k_1 x$ $\dot{x}' = k_1 x - k_2 x'$	$\mu x = \frac{FY}{1-Y}$	[1.77]
Account of grazing and reutilization			
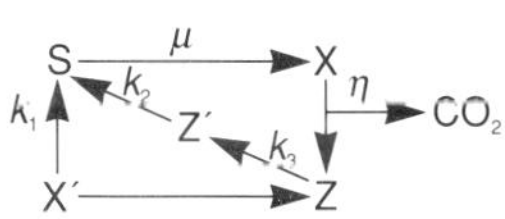	$\dot{s} = F - \frac{\mu x}{Y} + k_1 x' + k_2$ $\dot{x} = \mu x - \frac{\eta z}{\Psi}$ $\dot{z} = \eta z - k_3 z$ $\dot{x}' = \Psi' \frac{\eta z}{\Psi} - k_1 x'$	$\mu x = \frac{FY}{1 - Y(\Psi' - \Psi)}$	[1.78]

S, organic substrate; X, viable biomass; X′, necromass or microbial detritus; Z and Z′, respectively biomass and necromass of predators; *F*, input of OM; μ and η, specific growth rates of microbes and predators respectively; *Y*, $Y^{\max}$, Ψ, Ψ′, stoichiometric coefficients, k_1–k_3 and *a*, specific decay and death rates.

a ($0.001\ h^{-1} = 0.024\ d^{-1}$), the average growth rate in three of seven soils studied (Table 1.5) would be negative. Certainly, such a ridiculous result may be explained by overestimating x and a, or underestimating F, but there is also an obvious basic error in the use of Eqn. [1.76] which should by no means pass unnoticed. When the parameter a is introduced, the parameter Y acquires a new meaning. It is no longer just the apparent yield, but the 'true' maximal yield, Y^{max}, which stands for substrate-to-biomass conversion efficiency under the imaginary ideal condition $a = 0$. The apparent yield, Y, is variable, while the maximal yield, Y^{max}, is supposed to be a unique constant for a particular organism. The relationship between Y and Y^{max} is defined by the simplest of Pirt's equations (Eqn. [1.34]). Thus, the first widespread error is the mixing up of Y and Y^{max}; the second is that a is introduced as a constant which is not allowed to vary even with temperature (as shown with pure cultures, a rapidly decreases at low temperature (Tijhuis *et al.*, 1993)), not to mention any dependence of maintenance requirements on the cell's physiological state. However, since the accuracy of any corrections of this kind would always be questionable, the best choice is to exclude entirely the maintenance term from consideration and to use instead the apparent Y values measured *in situ*.

Table 1.5 presents the available experimental data on the C-budget of several terrestrial ecosystems, which were used to calculate the rates of microbial growth by different methods. Equations [1.75] and [1.77] give, respectively, an upper and lower boundary of the microbial growth rate. Calculated generation times vary within a range of 5 to 50 days, corresponding to 3–25 generations per season. (The only exception is the soil under continuous wheat, where microbial biomass is definitely overestimated; as a result average generation time increased here to 3 months.) The seasonal production of microbes is of the order of tons dry weight per hectare and exhibits a steady increase parallel to the primary production of an ecosystem from $1\ t\ ha^{-1}$ for tundra to $16\ t\ ha^{-1}$ for chernozem under virgin steppe. Compared with the 'seesaw' method, the mass balance calculation yields smaller microbial growth rates, but the results are closer when more realistic models (such as Eqn. [1.77]) are used. Thus, in four of six soils, the generation time is less than 1 week, which is close to the oscillation period found in most observations. Calculations with the use of incongruous kinetic models (such as Eqn. [1.76]) yield longer generation times, the difference being up to several orders of magnitude. Sometimes such models 'predict' that microbial communities in particular soils are not sufficiently supplied by energy even to maintain their viability. Ironically it was this type of incongruous kinetic model which shaped the dominating concepts on the physiological state of soil microorganisms.

Table 1.5 Estimation of soil microbial growth from mass balance of entire ecosystem

Ecosystem	x* (kg ha^{-1})	t† (days)	F‡ (kg ha^{-1} year^{-1})	Calculated mean generation time (days) according to equation			Maximum conceivable production (t ha^{-1} year^{-1})
				[1.75]	[1.76]	[1.77]	
Subarctic bog (Svensson and Rosswall, 1980)	310	100	1 560	34.4	$\mu<0$	20.7	1.0
Mixed forest (Chapman and Gray, 1986)	148.4	150	12 000	3.2	3.6	1.9	8.1
Coniferous forest (Parkinson *et al.*, 1978)	192	150	4 920	10.1	15.6	6.1	3.3
Arable soil under wheat (Lynch and Panting, 1980)	400	100	7 080	9.8	14.8	5.9	6.8
Soil under continuous wheat (Jenkinson and Ladd, 1981)	1 140	100	2 400	82.3	$\mu<0$	49.4	1.6
Typical chernozem under virgin steppe (Gilmanov and Bazilevitch, 1983)	800	190	23 900	11.0	17.8	6.6	16.0

* Mean microbial biomass; † warm period per year; ‡ annual input of plant litter to soil.

1.5.5 PHYSIOLOGICAL STATE OF MICROBIAL POPULATIONS *IN SITU*

Our current knowledge in this field is obviously quite approximate and incomplete. Nowadays, the misconceptions prevailing some 30 to 40 years ago no longer exist and there is little doubt that microbial growth *in situ* may by extremely slow due to a severe nutrient limitation.

Limiting factors

In the case of phototrophs, whose activity is the most adequately studied in aquatic ecosystems, the main limiting factor within the photic zone is supposed to be the availability of P and N. The significance of light limitation steeply increases with depth. For heterotrophic organisms, the primary limiting factor is the content of available organic compounds (rather than the total carbon content!). This was readily demonstrated in incubation studies with soil amended by various compounds (organic substances and mineral sources of N, P, K, Mg) in different combinations (Stotzky and Normann, 1961). The addition of mineral compounds had only minor effects on microbial activity assessed by monitoring respiration dynamics. Organic substrate was invariably found to cause a rapid increase in evolved CO_2. After the addition of glucose, however, soil microbes showed a greater demand for other nutrient elements, primarily N. It follows that in soil, as a patchy heterogenous environment, multiple limitation may occur.

Speculations

Mass balance calculations of microbial productivity (Eqns [1.75–1.77]) generate only averaged data on rather slow microbial growth, but how does such growth proceed in reality, in physiological rather than formal mathematical terms? The very slow microbial multiplication was thought to occur in two different patterns (Gray, 1976): first, short-term bursts of growth at the instants of substrate input, followed by a long-term dormancy state of spores, cysts or sclerotia; and second, continuous slow growth of vegetative cells interchanged with periods of complete starvation, when cells utilize their endogenous resources. According to Zvyagintsev (1987), there is always a pool of microbes in soil which are devoid of energy or some essential nutrient. These microorganisms do not multiply, but remain at a vegetative state of non-arrested metabolism without spore formation. (Note that the contribution of spore-forming bacteria to the microbial pool of most soils is rather small.) If so, then we must assume that soil microorganisms should be characterized by a peculiar physiological state which has been previously known only for multicellular organisms. That is, an active metabolism without cell division or cell death, the entire flux of gained energy being utilized for maintenance (resynthesis of decayed macromolecules, motility and osmoregulation). The existence of such a physiological state is purely conjectural. It was postulated in both ecological (Zvyagintsev, 1987) and physiological literature (Pirt, 1975), but *a priori,* without any experimental substantiation.

Facts

The declaration of the abnormality of the physiological state of soil microorganisms is fairly persuasive. It is in accord with the idea of Winogradsky and Kluyver that 'laboratory cultures are artifacts'. It can 'explain' why 99% of cells observed by direct microscopy are not plateable on nutrient media, why bacterial fertilizers introduced into soil so mysteriously disappear, and, to put it bluntly, 'why soil forms behave completely another way as compared with laboratory culture'. However, there are no sufficient direct experimental data confirming the reality of the declared abnormality in the physiological state of soil microbes. Rather, the counter-arguments are prevailing, when an attempt is made to use unbiased and quantitative approaches. For example, careful microscopic examination of soil demonstrates: (a) the absence of dead and damaged bacterial cells (Aristovskaya, 1972); (b) the occurrence of normal dividing cells which are especially abundant after rainfall (Clarholm and Rosswall, 1980); and (c) a rather high ratio of active motile bacterial cells and active fungal mycelium identified by means of vital staining with fluoroscein diacetate (FDA) or analogous techniques (Söderström, 1977). Less abundant but more informative and accurate are data relating to the chemical composition of *in situ* microbial cells. Most examples of this type come from aquatic microbiology. An advantageous technique is one based on phytoplankton-labelling *in situ* by $H^{14}CO_3^-$ under light conditions with subsequent fractionation of cellular constituents. By this method, the content of pigments, polysaccharides and proteins was determined in cyanobacteria from a fish-breeding pond (Rijn and Shilo, 1986). With the exception of phycocyan content, all other characteristics of natural populations were found to be similar to those of laboratory cultures. Another useful technique is the bioluminescent analysis of metabolites readily converted to ATP.

The physiological state of aquatic microorganisms is often characterized by biochemical indices such as the ATP content per cell, or the unit of chlorophyll per mass, *a* (their decline indicates the reduction of cell viability), or by the so-called adenylate energy charge (AEC), where

$$\mathrm{AEC} = \frac{\mathrm{ATP} + 0.5\,\mathrm{ADP}}{\mathrm{ATP} + \mathrm{ADP} + \mathrm{AMP}}.$$

The AEC index ranges from 0 to 1.0. For a totally viable and intensively growing population it is >0.8, for a damaged but still viable population it lays between 0.6 and 0.8, and for a starving declining population it is <0.5. In sea water, AEC is 0.8–0.9 near the photic surface, where density and activity of phytoplankton is maximal, after which it declines in regular fashion with depth. An especially sharp decrease of AEC occurs on the boundary between epilimnion and hypolimnion (Karl, 1980). Fresh soil samples were shown to have high ATP content per unit of microbial biomass and AEC values as high as 0.85 and over (Brookes *et al.*, 1983). A sharp reduction of AEC to 0.46 occurred only after soil drying, although the former AEC index was restored shortly after soil rehydration.

The physiological state of soil microorganisms was also characterized by nucleic acid analysis (Panikov, 1976). In upper organic horizons (A_0F and A_0L), the RNA/DNA ratio was the same as in active microbial cells obtained in laboratory culture. This ratio gradually decreased with depth from 2 to 0.6, which indicates a regular deceleration of growth and metabolism of the entire microbial community. A similar pattern of change in the RNA/DNA ratio is known to take place in a batch culture, except that it is displayed as a time series rather than as a spatial profile gradient.

Thus, the physiological state of *in situ* microbial populations does not appear to be basically different from that of laboratory cultures. However, we should not hurry to draw such a conclusion as it may be premature. Firstly, the properties of diverse laboratory cultures vary over a wide range. Using the currently available array of cultivation techniques, one can probably reproduce any conceivable physiological state (it is a different matter if the comparison is made with a simple batch culture). Secondly, only a few natural habitats have so far been studied. The most advanced analytical methods of biochemistry and molecular biology (such as measuring rates of DNA and RNA synthesis and determination of intracellular constituents) have been applied mostly to homogeneous media such as natural waters, activated sludge and rumen. Thirdly and finally, the physiological state has been portrayed too crudely and only in terms of averaged integral characteristics, and this might conceal the truly unusual features of some individual populations. Aggregation of dynamic variables as well as the use of mean and effective values is inevitable in ecological studies, but there are some limits beyond which it would be almost ridiculous to regard a complex and well-differentiated microbial community as an absolutely homogeneous non-differentiated mass. Thus, our next step is to look at how microbial diversity is organized into so-called microbial systems.

1.5.6 MICROBIAL SYSTEMS

The second half of the 20th century began with vigorous development of the system approach and with heroic attempts to quantify studies of complex natural systems (Bertalanffy, Lyapunov, Forrester, Ashby, Moiseev). In microbiology, this general trend has led to the development of extensive research of biodiversity, aimed at broadening rather than deepening our scientific knowledge of the microbial world. The main emphasis in 'extensive microbiology' is paid to systematization of the enormous amount of available data into various constructions (taxonomic, trophic, informational) based on the complete account of 'potentially available combinations' (Zavarzin, 1976, 1978). Here, we shall restrict our consideration to just one thin layer of this impressive stratum of knowledge, to the theory of the trophic and functional structure of microbial communities.

Zymogenic and autochthonous microflora

The first step from merely acknowledging the diversity of the soil microorganisms to an organized ordered knowledge was made by Winogradsky, who introduced these notions. Zymogenic microflora are characterized by explosive growth on fresh organic debris (allochthonous matter) and a quick loss of viability under starvation. The activity of autochthonous microflora is low but steady. 'This group includes humus-degrading bacteria specifically adapted to the environment they inhabit' (Winogradsky, 1949). Autochthonous microflora are found in fallow soils which are at the climax state, biological equilibrium. The notion of autochthonous microflora was extended by Winogradsky's disciples to include some new features such as poor growth on laboratory media, unusual morphology, and a relation to some taxonomic groups tentatively called 'new forms' (Conn, 1948; Aristovskaya, 1965; Nikitin *et al.*, 1966).

Microbial system of Zavarzin

The extended notion of autochthonous microflora turned out to be contradictory. The inconsistency was revealed by Zavarzin (1970) with the help of graph theory as a simple but quite effective mathematical technique. The entire operation of a microbial community was outlined by Zavarzin as follows. Plant debris composed of readily available soluble and less accessible insoluble components is degraded by microflora I and II, respectively. This process leads to the formation of humus substances and microbial biomass, the latter also consisting of readily available components (cytoplasm) and of less utilizable structural parts (cell walls). Both are degraded at the subsequent stage by microflora III and IV, respectively. Some products of hydrolysis of plant and microbial polymers are dissipated in the environment, escaping, therefore, the action of primary degraders. These low-molecular weight compounds are scavenged by **dissipation microflora** (dissipotrophic populations) V, which are characterized by high affinity to substrate (small values of saturation constant K_s in Eqn. [1.25]). Humus substances are attacked by microflora VI. Under anaerobic conditions some additional groups are suggested: microflora VII, degrading nitrogen-containing compounds (readily available components of plant litter and microbial necromass); microflora VIII, which ferment N-free organic compounds; microflora IX or 'secondary anaerobes' (sulphate-reducing and methanogenic bacteria); and microflora X oxidizing anaerobic metabolites as they enter the aerobic zone (hydrogen and methylotrophic bacteria).

This metabolic network is described by a graph and by a set of corresponding equations. It was only the mass balance analysis of fluxes between individual microbial groups that allowed the exposure of the aforementioned contradictory nature of the 'extended' interpretation of autochthonous microflora. Such microflora can be identified as group V or X, but not VI, which is responsible for humus consumption. Later on, Zavarzin (1976) highlighted the most important metabolic feature of primary degraders of plant debris, that is, their ability to produce extracellular hydrolytic enzymes. Dissipation microflora fall into three groups, aerobic, anaerobic and an intermediate. The outlined scheme is both complete and closed; it is not only able to describe the microbial community operation but also to fulfil an heuristic function by identifying gaps in our knowledge and suggesting experiments with which to fill them.

Oligotrophic and copiotropic organisms

Several other reduced schemes enjoying wide recognition are shown in Figure 1.1. Often aquatic, soil or subsoil microbial communities are considered to comprise just two components. One is **oligotrophic** microorganisms, while the second is **eutrophs** or **saprophytes** (Kuznetsov *et al.*, 1979), **heterotrophs** (Akagi *et al.*, 1977) or **copiotrophs** (Poindexter, 1981a,b). The term 'copiotroph' (from latin *copios* – abundant) appears to be more appropriate and we shall adhere to it hereafter.

There are no satisfactory definitions of oligotrophy and copiotrophy in the sense that no basic qualitative distinctions between these two groups are specified. Poindexter did not go beyond stressing a quantitative difference: oligotrophs are capable of growing under conditions of organic matter supply with a rate below 0.1 mg l^{-1} per day, whereas copiotrophs require a C-flux of 50 mg C l^{-1} per day and higher. Note, however, that these figures are meaningless if the microbial population density is not specified. The absence of a satisfactory theoretical criterion does not interfere with the differentiation of oligotrophs in practical terms. They can be successfully isolated from natural substrates on fairly diluted media (1–15 mg C l^{-1}) and maintained on such media as long as is required. A simple calculation shows that the amount of substrate introduced into the

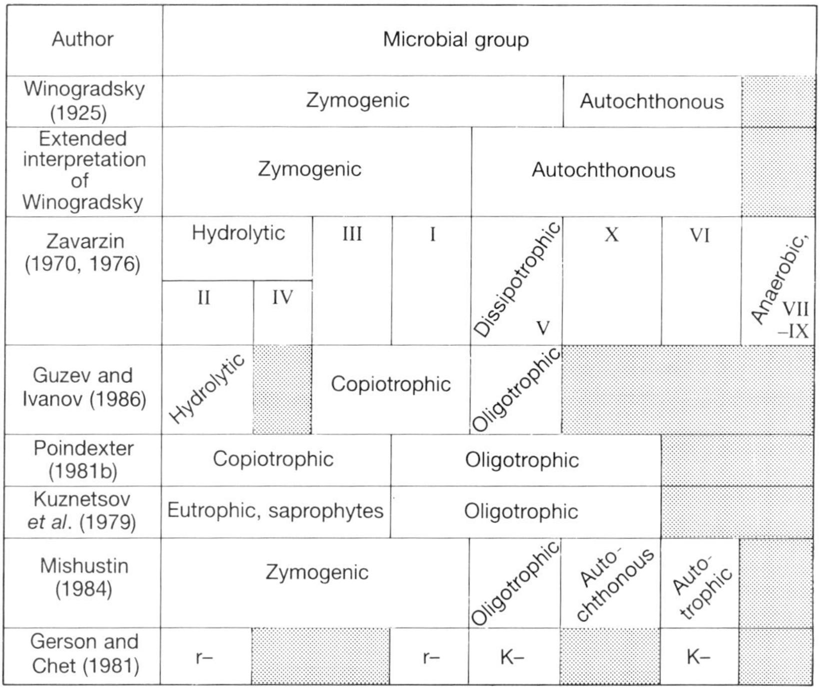

Figure 1.1 Functional structure of the soil microbial community: a comparative analysis. Shaded areas refer to organisms beyond the scope of the author's definition.

medium is not sufficient for even tiny colonies to be formed. An important component of mass balance for the growth of oligotrophs is, therefore, volatile organic substances from the laboratory air. This possibility was first demonstrated by Beijerinck as early as the beginning of the century (Beijerinck and Van Delden, 1903). Today, the ability of many oligotrophs to consume volatile compounds is well established. These may include C_1-compounds, volatile fatty acids, alcohols and hydrocarbons. Many oligotrophs have unusual morphology and belong to the so-called 'new forms': budding, prosthekate, stalked, toroid bacteria, etc. Oligotrophs also include *Arthrobacter globiformis* and some eukaryotic species (Mirchink and Babjeva, 1981; Poindexter, 1981a,b).

Distinctive features of oligotrophs in terms of their physiology and growth kinetics could by summarized as follows:

- a high-affinity transport system with a broad substrate specificity;
- an efficient metabolism (low maintenance requirements);
- the ability to survive under prolonged starvation;
- a strictly oxidative type of catabolism;

- the negligible expression or complete absence of hydrolase activity;
- slow growth;
- a susceptibility to substrate inhibition (Semenov and Vasiljeva, 1985; Aminov and Golovlev, 1987).

The last point is an especially important feature of growth kinetics because it draws a qualitative distinction between oligotrophs and copiotrophs. The mechanism underlying substrate inhibition is not clear. Sometimes it is explained as being due to the toxicity of H_2O_2, which is formed as an end respiratory product and cannot be deactivated since many oligotrophs are devoid of catalase (Kuznetsov, 1981).

The relationship between different systems is shown in Fig. 1.1. Oligotrophic species may often be identified as well as dissipotrophic and autochthonous microflora (Hattori, 1981). Copiotrophs *sensu* Poindexter correspond to groups I–IV of Zavarzin. According to Guzev and Ivanov (1986) the copiotrophic group does not include hydrolytics and comprises only organisms similar to microflora I and displaying an ephemeral type of behaviour.

The concept of life strategy

This is one of the most promising quantitative approaches which may be used to investigate the structure of microbial systems. This concept was shaped in general ecology and the definition of life strategy is rather general: 'a combination of adaptive reactions which provides the possibility for a given population to coexist with other organisms and occupy some part of niche hyperspace'. Two types of strategy are recognized most frequently, the r- and K-strategies (MacArthur and Wilson, 1967). This notation is derived from the logistic equation [1.14]. Climax communities are occupied by K-selected species, which are able to preserve a fairly stable population density, N, close to the medium capacity, K. They are well suited to severe competition with other organisms by means of an extreme differentiation of niches. On the other hand, the pioneer succession steps are dominated by r-strategists or opportunistic populations (for which the apparent specific growth rate determined by Eqn. [1.14] is close to its maximum value r). Such populations exhibit explosive growth and high colonization potential, but are displaced from climax communities by K-strategists as more powerful competitors.

Dynamic observations allow us to conclude that some species display fluctuations (r-pattern), while others remain stable (K-pattern). The similarity between these two patterns and the behavioural stereotypes of zymogenic and autochthonous microflora, respectively, was first acknowledged by Gerson and Chet (1981). Needless to say, a purely empirical logistic equation alone can hardly provide sufficient theoretical grounds to distinguish between strategies. It is more expedient to use a mechanistic approach based on equations of microbial kinetics (Andrews and Harris, 1986). Suppose a community of n potentially competing species is provided with a limiting substrate at a unsteady rate F. Then the following situations are possible.

1. If $F \geqslant \Sigma\mu_{m_i}x_i/Y_i$ ($i = 1, \ldots, n$), then, for some time, substrate limitation is relieved and species with a high μ_m and a short lag phase have the competitive advantage (the 'r-conditions'). The burst of growth leads to a depletion of excessive substrate and, as Σx_i increases or F declines, we come to the next case 2.
2. If $F \leqslant \Sigma m_i x_i$ ($i = 1, \ldots, n$), then the net available substrate is consumed exclusively to maintain the viability of the most adapted species, substrate concentration declines to the threshold level $s^* = \min [m_i Y_i K_{s_i}/(\mu_{m_i} - m_i Y_i)]$, $i = 1, \ldots, n$ (the 'K-conditions'). The limitation extent and the competition strength are now at their maximum, selection operating in favour of populations with high growth efficiency (parameters m and Y) as well as with high affinity to limiting substrate (parameter K_s).

We define r-strategists as those species which have an advantage under relieved competition. (Sometimes these species are erroneously referred to as strong competitors (Campbell, 1985)). K-strategists are those species which have an advantage in the second situation. It is not the absolute values of kinetic parameters that decide the outcome of competition, but rather how they interrelate for competing species. So there is no point in classifying all microbes into two groups of r- and K-strategists. It would be better to arrange the chosen populations along the r/K axis (the so-called ordination procedure). For example, within the group of dissipation microflora, we may position all known organisms along an axis of numerical values of the parameters K_s, m and μ_m. Then, on the r-side (corresponding to the maximal values of all parameters), we shall find species close to microflora I and to copiotrophs *sensu* Gusev and Ivanov, which have the highest probability to survive under changeable perturbed environmental conditions. On the K-side (minimal numeric values of parameters) there will be positioned ideal oligotrophs, similar to prosthecobacteria, which have the best competitive ability under low nutrient supply. Such an r/K continuum can be realized for any functional group of potentially competing populations.

The life strategies concept explains the diversity of growth characteristics of microorganisms by the fact that natural selection operates under different circumstances (r- and K-selection) and helps to predict the evolution of a microbial community. This concept is only just beginning to be recognized by microbiologists and is better developed for fungi (Rayner and

Boddy, 1988). Until recently, microorganisms were viewed as being exclusively the r-strategists. This point is well illustrated by a quotation from one of the best textbooks on microbiology (Stanier *et al.*, 1979):

> '. . . in the case of a unicellular microorganisms, the plan that directs cellular activity, and that is so strikingly revealed by the analysis of metabolic regulation, is a simple one: to use the nutrients immediately available to the cell in order to make two cells from one, with the greatest possible rapidity'.

Even today, this statement would meet no objections from biochemists and molecular biologists, but it will be vigorously challenged by ecologists who would argue that things are not that simple if one looks beyond *E. coli*. There is a basically different type of metabolic regulation which aims not at maximizing the growth rate, but at preserving the viability and stability of a microbial population under unpredictable environmental conditions.

In the 1960s and 1970s, the first steps were made in studying the molecular biology and biochemistry of those bacteria which are really dominating in nature and differ markedly from well-known enterobacteria in their dynamic behaviour and metabolic regulation. Today, the best studied are *Caulobacter* (Shapiro, 1976; Poindexter, 1981a) and coryneform bacteria (Golovlev, 1978, 1983).

The concept of microbial life strategies is very promising; however, we should admit several negative trends which might undermine and discredit the entire approach. Firstly, there are too many general statements and too few experimental works. Secondly, it is the use of oversimplified models (mathematical and conceptual) of microbial growth dynamics *in situ*. No advantage is taken of the possibility to describe the kinetics of non-steady-state and colony growth, microbial survival under starvation, cryptic growth, grazing, etc. To prove this point, we turn to the last subsection of our historical survey.

1.5.7 KINETICS OF MICROBIAL PROCESSES IN NATURAL HABITATS

The effect of substrate concentration

The first data of this type were obtained for sea water systems. The dependence between the initial rate of utilization of ^{14}C-compounds added to water samples and their concentration was described by a Michaelis–Menten equation (Parsons and Strickland, 1962). A deeper analysis of the problem led to the development of a method to estimate the turnover rate of the natural rather than the added substrate (Wright and Hobbie, 1966). The method was based on isotopic dilution, with the calculation following modified Michaelis–Menten equations:

$$v = V\frac{s_n + s^*}{K_m + s_n + s^*}, \qquad v^* = V\frac{s^*}{K_m + s_n + s^*}, \qquad [1.79]$$

where s_n and s^* are concentrations of the natural, non-labelled substrates and added labelled substrates, respectively; v is the overall rate of substrate utilization, and v^* is the rate of label consumption. By varying s^*, numerical values of the parameters V and K_m+s_n as well as of the turnover time of natural substrate, $T_{turn} = (K_m + s_n)/V$, may be found. The outlined approach was applied not only to aquatic media but also to lake sediments (Harrison *et al.*, 1971) and to soil (Ferroni *et al.*, 1985). The turnover time, T_{turn}, was found to range, in different natural environments, from several hours to several days, increasing in a regular manner with deterioration of ecotopic conditions (Nakas and Klein, 1981).

It is obvious now that Eqn. [1.79] holds only approximately for natural communities. The main source of deviation is thought to be the heterogeneity of the microbial community with respect to K_m (Williams, 1973; Tarapchak and Herche, 1986). If each of the n community members strictly conforms to the Michaelis kinetics, then the following equation is satisfied:

$$v^* = \sum_{i=1}^{n}\frac{V_i s^*}{K_{m_i} + s_n + s^*}. \qquad [1.80]$$

It is easily seen that Eqn. [1.80] no longer produces an hyperbolic curve because K_m is a non-additive parameter. Also it is difficult to determine uniquely from experimental data the distribution of microorganisms with respect to K_m (at least within the framework of stationary kinetics). A good fit to experimental points is feasible even at $n = 2$, but the obtained values of parameters will be entirely *ad hoc*. There are other reasons for the deviation from the simple Michaelis kinetics, e.g. diffusion control of reaction rate (Robinson and Tiedje, 1982). However, these sources of deviation can be readily accounted for and corrected.

Degradation kinetics of individual compounds added to soil or natural waters

Some examples are given in Table 1.6. The kinetic equations which are used are, as a rule, very simple. The typical features of kinetic analysis as it stands today are as follows:

1. The analysis is not intended to produce new information about the natural mechanisms of microbial activity regulation. It is fairly superficial and purely formal, the only purpose being to fit experimental points to a kinetic equation.

Table 1.6 Kinetic models of the decomposition of organic compounds in amended soil and water samples

Added substances/ sample source	Equations for dynamics of:		References
	Substrate	Biomass	
Glucose benzoate/ activated sludge	$\dot{s} = k_1 \frac{s}{K_m + s}$	$\dot{x} = 0$	Simkins and Alexander (1984)
Xenobiotics/ lake waters	$\dot{s} = -ksx$	$\dot{x} = 0$	Paris *et al.* (1981)
Xenobiotics/ lake waters	$\dot{s} = -akst$	$\dot{x} = a$	Larson (1980)
Fungal metabolites/ soil	$\dot{s} = -(k_1 + k_2 x)s$	$k_2 x = \begin{cases} k_3 t \\ x_0 e^{\mu t} \end{cases}$	Brunner and Focht (1984)
Glucose benzoate/ activated sludge	$\dot{s} = -\dot{x}/Y$	$\dot{x} = \mu \frac{x}{1 - x/x_m}$	Simkins and Alexander (1984)
Glucose benzoate/ activated sludge	$\dot{s} = -\dot{x}/Y$	$\dot{x} = \mu_m \frac{sx}{K_s + s}$	Simkins and Alexander (1984)
Glucose, amino acids/ sea waters	$\dot{s} = -\dot{x}/Y$	$\dot{x} = \mu_m x$	Li (1984)
Phenol, aniline/ soil	$\dot{s}_1 = -as_1 + bs_2$ $\dot{s}_2 = cs_1 - ds_2$	$\dot{x} = 0$	Scow *et al.* (1986)

2. As a result of feature (1), mathematical modelling is restricted to a *post factum* description of obtained dynamical curves; no prediction is possible.
3. The only criterion for agreement is the residual error; no attempts are made to reproduce qualitative features of microbial behaviour *in situ*.
4. The scope of used dynamic variables is restricted to the concentration of decomposed compound or CO_2 as the end degradation product, while the microbial biomass, and parameters of physiological state, are completely neglected.

Kinetics of microbial decomposition of natural organic matter

The first studies on decay of plant litter and soil humus were carried out during the late 1940s (Jenny *et al.*, 1949) and since then have become more popular. Degradation dynamics are traditionally described by a first-order equation or by a sum of such equations, supposedly being able to account for the heterogeneity of detritus (Hanson *et al.*, 1984; Marayama, 1984):

$$\dot{s} = -(k_1 s_1 + \ldots + k_n s_n) = -\sum_{i=1}^{n} k_i s_i \,. \qquad [1.81]$$

Even though Eqn. [1.81] does enable a reasonably good fit to the observed dynamic curves (usually at $n = 2$ or 3), it has, in fact, got nothing to do with the actual operating mechanism. Moreover, it is definitely misleading in assuming a parallel and independent degradation of different individual compounds. If it were so, humus in soil would have consisted of the most resistant plant litter components (lignin, waxes and resins), while no readily available compounds (polypeptide, amino acids, hydrocarbons, etc.) would have been detected in decaying plant litter in late degradation stages. This, however, is in direct contradiction with the evidence on the chemical composition of plant litter at different degradation stages (Wessen and Berg, 1986). In fact, it is only the water-soluble fraction that is rapidly eliminated, whereas the structural polymeric litter components (cellulose, hemicellulose, lignin, polypeptide) are present at a more or less constant rate during the entire dynamics. Why then do we observe a gradual deceleration of the decomposition? So far there are no satisfactory explanations of this problem.

In order to account for the fact that microbial degradation is not equivalent to the simple destruction of organic substances but also includes transformation reactions (the formation of microbial cells and exometabolites, interaction of metabolic products with soil solid phase or humic substances, etc.), Carpenter (1981) devised the following dynamic equation, to replace Eqn. [1.81]:

$$\frac{\partial f(z,t)}{\partial t} + \frac{\partial V(z,t)}{\partial z} = - U(z,t)\, f(z,t)\,. \qquad [1.82]$$

Here $f(z,t)$ is the distribution of detritus components with respect to their susceptibility to microbial degradation, z, where z ranges between 0 and the maximum value z_m; $V(z,t)$ is the amount of detritus converted into more or less available compounds; and $U(z,t)$ is the degradation rate. However, application of Eqn. [1.82] is impossible because neither $V(z,t)$ nor $U(z,t)$ can be represented in a simple analytical form. It seems more natural to describe the process by equations of microbial kinetics which take into account the diversity of individual substrate within plant litter. Such equations should include as dynamic state variables the biomass of selected groups of the soil microbes (e.g. Zavarzin's microflora) and their exploiters (predators and parasites), concentrations of metabolic products, residual amounts of the polymeric compounds of litter available for enzymatic hydrolysis and concentrations of monomeric products. In addition, ecotopic conditions (temperature, moisture, salinity, etc.) have to be incorporated into a realistic simulation model. Unfortunately, no such models have so far been formulated.

Simulation of soil as environment for microbial growth

This is a traditional objective of quantitative soil microbiology. The first studies of microbial interaction with the solid phase of soil had already been carried out at the start of the century by Aizenberg, Catler, Khudiakov and Novogrudsky. In the 1950s and later such studies greatly benefitted from advances made in neighbouring disciplines – physical chemistry, soil chemistry and biochemistry, physics and mineralogy. Investigations carried out by McLaren, Esterman and Marshall revealed some of the molecular mechanisms underlying the interaction of microbial cells, their metabolites and growth substrates, with clay minerals, organic ion exchangers and other sorbents used for the physical simulation of the soil solid phase. A conceptual model was developed of microbial growth on the solid–liquid interface. A physicochemical interpretation was given to the soil microsite structure and to the microscale distribution of bacteria and fungi with different properties, i.e. aerobic–anaerobic, heterotrophic–autotrophic, prey–predator, antagonistic species, etc. (Szabo, 1974).

The accumulated evidence indicated that the spatial factor may be crucial for microbial performance *in situ*. To account for the spatial factor we have to use so-called distributed models in which dynamic state variables are functions of not only time but also of spatial coordinates. Such models could be relatively straightforward when one deals with macroscale spatial gradients, e.g. in the case of continuous-flow soil columns, similar to the tubular reactors used in chemical engineering. The mathematical models of the performance of these reactors are traditionally based on partial differential equations. Consumption of substrate in a continuous-flow soil column can be described by the following equation (Saunders and Bazin, 1973; McLaren, 1980):

$$\frac{\partial s}{\partial t} + f\frac{\partial s}{\partial z} = -\mu_m x \frac{s}{(K_s + s)Y} \approx -\mu_m x \frac{s}{K_s Y}, \qquad [1.83]$$

where f is the feed rate of the nutrient solution and z is the linear coordinate (identifying a location along the column). Equation [1.83] was used to describe the nitrification of NH_4^+ continuously pumped into soil. The results obtained were in reasonable agreement with the dynamics observed under steady-state, but not under transient conditions.

A possible approach to the modelling of soil heterogeneity at microscopic level is exemplified by another study. To estimate the amount of nitrogen loss via denitrification, Smith (1980) proposed a model of oxygen diffusion from the air through aeration pores into soil aggregates. The size distribution of aggregates was found to be log-normal with variance, σ, and mean, M. The author also estimated the effective diffusion coefficient, D, and the aggregate respiration rate, Q. Given these values, the fraction of anaerobic zones, W, was estimated from the concentration of O_2 in aeration pores, C, as

$$W = \frac{1}{\sigma\sqrt{2\pi}} \int_{r_c}^{\infty} \frac{r_0{}^3}{r^4} \exp\left[-\frac{(\log r - \log M)^2}{2\sigma^2}\right] dr, \quad r_c = \sqrt{\frac{6DC}{Q}}, \qquad [1.84]$$

where r is the radius of aggregates, r_0 is the radius of their anaerobic part, and r_c is the maximum radius of totally aerobic aggregates.

Modelling of microbial growth in rhizosphere

Among all ecotopic factors, we are always most interested in the trophic factor. The spatial pattern of microbial growth is to a large extent determined by the substrate distribution. The main source of organic matter in the soil is higher plants and especially their roots. In a model by Newman and Watson (1977), this problem is examined at microscopic level. The mass balance equation taking into account the spatial variations of microbial activity around a cylindrical root, r, is written as

total change	diffusion from root	uptake for growth	uptake for maintenance	input from soil

$$\frac{\partial s}{\partial t} = \frac{D}{r}\frac{\partial}{\partial r}\left\{r\frac{\partial s}{\partial r}\right\} - \frac{\mu_m x s}{Y(K_s\theta + s)} - mx + \lambda, \qquad [1.85]$$

where D is the diffusion coefficient and θ is the volumetric water content in the soil. The boundary conditions for Eqn. [1.85] are given by

$$-D\frac{\partial s}{\partial r} = \begin{cases} Q(t), & \text{at } r = a \\ 0\,, & \text{at } r = b\,, \end{cases}$$

where $Q(t)$ is the rate of root exudation, a is the root radius and b is the outer boundary of substrate diffusion (distance between the neighbouring roots).

Results of calculations using Eqn. [1.85] are in reasonable quantitative agreement with experimental data on the density of bacterial cells occupying a root surface. One unexpected conclusion which was difficult to test by conventional microbiological technique was that the spatial extent of the rhizosphere could not be large, and that the domain of immediate root influence was actually confined to the surface of the plant's tissue. At the same time, this model should not be expected to yield a realistic holistic picture of the dynamics of microbial growth in the rhizosphere, since it does not take into account some important links existing within the system. For example, no allowance was made for bacterial elimination by predators, for the dependence of root exudation upon ontogenetic stages of plant growth and on ecotopic conditions, and for the positive and negative bacterial effects on plant growth. We may conclude that, although it is very difficult, soil microbiologists should not restrict themselves to the studies of microbes alone. To gain insight into and understand the nature of the observed phenomena, there is a need to invade neighbouring scientific fields, in particular plant physiology.

The last term, λ, in Eqn. [1.85] is the flux of monomeric compounds released during the slow depolymerization of soil organic matter. This flux is related to one enigmatic phenomenon, which is known as the **priming effect**. Any burst of microbial activity induced by the amendment of soil with a readily available substrate leads to an increase in λ, i.e. to the acceleration of the decomposition of stable soil organic matter (Broadbent, 1947). So far all explanations that have been suggested are purely conjectural and include co-metabolism, additional synthesis of hydrolytic enzymes during explosive growth, the positive effect of amendment on the transport of endogenous compounds from soil into cells, and changes in the physical and chemical properties of soil (Kunc, 1975). One hypothesis, based on the analysis of a mathematical model, explains the priming effect as an optimization of the C/N ratio by added soil substrates (Parnas, 1976).

1.6 CONCLUSION

The results of historical survey

Progress in quantitative microbiology has been determined largely by the exploitation of approaches borrowed from exact sciences. There has been an active mathematicization of microbiology and development of novel cultivation techniques over the last decade. This, of course, should by no means prevent us from paying tribute to the founders of microbial growth theory – not only the honoured patriarchs of microbiology but also those brilliant scientists whose names are now almost forgotten. Modern literature on microbial cultivation hardly ever mentions the names of Egunov, Rubner, Buchanan, Müller, Terroine and Tauson. Their contribution certainly deserves recognition, but this is not the only conclusion to be drawn from our historical survey. The chronological order of facts in our presentation has allowed us to follow the evolution of ideas on microbial growth. At different times, different proximate disciplines had notable influences on the way in which microbial processes were analysed quantitatively. At the beginning of the century it was mathematical demography and thermodynamics; in the middle of the century, it was chemical and enzymatic kinetics; and today, it is molecular biology, cybernetics and system analysis. Some ideas and concepts were correctly guessed from the very beginning (at least as we see it today), but not always recognized by contemporaries. This applies to the Rubner–Terroine–Tauson concept of maintenance metabolism, Egunov's colony growth theory and the first attempts to perform controlled continuous cultivation. Other ideas turned out to be erroneous, although for a long time they dominated the scientific community. Among the examples one finds Henrici's concept of cytomorphosis, Droop's cell-quote theory, and Herbert's treatment of the growth rate as an independent variable. True enough, it is often impossible to segregate ideas into absolutely valid and entirely incorrect ones. It seems more sensible to judge such ideas in terms of how productive they have been as guidelines in the quest for new knowledge and in developing new concepts of microbial growth.

What actually is 'microbial growth theory?'

Quantitative investigations in microbiology have been given different names: mathematical theory (Buchanan), growth laws (Egunov), growth kinetics (Hinshelwood, Van Niel). In our view, all modern mathematical microbiology can be identified as being related to two research areas, the **stoichiometry** and the **kinetics** of microbial growth. The first is concerned with static quantitative characteristics of growth; the second with their development in time. Both stoichiometry and kinetics can be analysed using macroscopic and microscopic approaches, at cell and population level. These areas also include the cell cycle theory, statistical analysis of cell distribution in age, size, etc. Growth principles expressed in the form of mathematical equations used to be called 'laws of growth'. Today, a more discreet stand is usually taken and such equations are merely known as mathematical models, stressing the approximate nature of the suggested mathematical descriptions. As we have seen with $\mu(s)$, any mathematical formula, however complex, is no more than an approximation to the actual dependence between μ and s. Being convinced by this fact, the professional researcher is no longer attempting to produce a universal and fundamental growth model. A more important objective is to have a model that would allow cultivation conditions to be optimized, population dynamics to be at least approximately predicted, and a selection to be made between a number of alternative hypotheses.

At the same time it is hardly reasonable to be too pragmatic and regard mathematical modelling as having only an auxiliary and operational role, devoid of any self-reliant theoretical importance and meaning. Just a good fit could be accomplished using more or less arbitrary mathematical expressions (high order polynomials, for example), but it is an altogether different matter to speculate on those functions that would reflect the mechanism of the studied phenomenon. In this survey, we have been mainly interested in the second type of model, and there are good reasons to believe that it does indeed portray links and interrelations existing objectively within a microbial cell or population. Such relations are often intrinsic, fundamental, stable and reproducible, as required by the category of the law of nature. The laws and regularities of microbial growth are in fact consequences of more general laws and can be split into the three following groups:

1. **Mass and energy conservation laws**. This group encompasses all quantitative relations of macro- and microstoichiometry of microbial growth. It includes expressions for mass and energy yields of growth, the mass-and-energy balance theory, characteristic tables of microbial metabolism and the concept of maintenance requirements. This group also includes the sets of mass balance differential equations describing sources and sinks for dynamic state variables (e.g. equation set [1.28] for the chemostat or equations [1.75–1.78] for a soil microbial community).

2. **Intermediary dynamic complexes**. Microbial growth is based on the activity of enzymes and of supermolecular structures such as ribosomes and polyenzymatic assemblage organized on cytoplasmic membranes. One fundamental feature of their functioning is the reversible formation of dynamic complexes with substrates, products, inhibitors and activators. This explains why metabolic reactions are of mixed kinetic order (revealing concentration–activity curves which exhibit saturation) and also such

effects as reversible inhibition and activation, substrate inhibition, and so on. This group of microbial growth laws embraces all the relations derived from the kinetics of enzymatic reactions.

3. **Cybernetic laws**. Live systems are capable of expedient self-regulation and adaptive self-control by feedback. The corresponding cell mechanisms have been developed and refined by evolution. This group combines laws of microbial growth which describe adaptation of microbial populations to new environmental conditions with autoselection and selection in a microbial community, the cell cycle and differentiation. One implication of the cybernetic laws is the possibility to exploit optimal control theory in analysing the behaviour of microbial populations.

The outlined laws and principles apply to microorganisms as well as to macroorganisms but, in the case of the simplest living things, the general fundamental growth principles can, obviously, be revealed more readily and accurately.

Ecological aspects of microbial kinetics

The growth of microbial populations, no matter whether it takes place in a laboratory culture, in an industrial fermenter or in a natural habitat, must be governed by the same basic laws. So far we have only a very limited knowledge of these laws because laboratory or industrial fermenters do not reproduce all the diversity of growth conditions found in natural habitats. The greatest disappointments in ecological studies stem from abusing the kinetic models derived for very restrictive laboratory conditions (one example is the calculation of microbial productivity in soil using the simple chemostat model; see section 1.5.4). Studies of the ecological aspects of microbial kinetics involves, therefore, not only field research, but also considerable extension of laboratory work (elaboration of new cultivation techniques, new growth regimes, a larger number of research objects) as well as the further development of 'general' microbial growth theory to enable quantitative description, understanding and prediction of a wider range of microbial behaviours.

Diversity of patterns of microbial growth *in situ* and *ex situ* 2

2.1 NATURAL MICROBIAL POPULATIONS AND 'LABORATORY ARTIFACTS'

In this book, we intend to have more liberty than is customary for classical ecologists, in the extrapolation of laboratory data to reflect natural habitats. This is because many subtle features of microbial behaviour can only be disclosed in accurate laboratory studies with pure cultures. Is such extrapolation really justified? Are axenic laboratory cultures sufficiently representative of their natural progenitors? Different opinions have been expressed in respect of this eternal problem. On the one hand,

> 'pure cultures could certainly be regarded as a physiological artifact' (Kluyver)

and so

> 'a clear demarcation line should be drawn between data obtained under abnormal experimental conditions, which *invoke* microorganisms to reveal some new features, and data from observations on ecological factors in nature' (Winogradsky, 1949, pp.25–47).

On the other hand,

> 'many properties of pure laboratory cultures are also exhibited by microbial populations under natural conditions' (Ierusalimsky, 1963, p.189)

and

> '*unless there are indications to the contrary,* it is justifiable, and operationally necessary, to assume that in most characteristics pertinent to the habitat, pure cultures do resemble their progenitors in nature' (Hungate, 1962, pp.100–101).

In order to resolve this contradiction, it would be better to turn away from prejudices and instead analyse the problem in quantitative terms.

Properties of microbes in pure laboratory cultures may differ from those of their ancestors in natural habitats because of the following factors:

- the lack of interaction among populations;
- autoselection in the course of culture isolation, purification and maintenance;
- phenotypic changes in the physiological state of microbial cells in response to a changed environment (different with respect to the availability and spectrum of substrates and modifiers, temperature, humidity, etc.).

It was primarily the third factor that Winogradsky was referring to when he wrote of 'invoking' laboratory forms to grow abnormally. We also regard this as a fundamental factor. The first point is not decisive since, in soil, microbial growth is confined to microsites where practically pure cell clones develop. In homogeneous natural habitats, such as waters, there are negative and positive interactions via metabolites. However, antagonists are not able to sustain coexistence and, in the case of positive cooperative effects, it is in fact microbial associations that are isolated from the natural habitat rather than pure cultures. The second and third points above are almost indistinguishable in practical terms. They are also fairly similar in principle, because both the selection and phenotypic variations are not random but tend towards a better adjustment of the population to the given growth conditions.

In terms of quantitative microbiology, the third point may be interpreted as a difference in the vector of physiological state of a laboratory culture as compared with a population *in situ*. According to Winogradsky and Kluyver, this difference is great and important, whereas, in the view of Ierusalimsky and Hungate, it is not absolutely crucial. However, the physiological state of microorganisms is not static and constant, but rather a dynamic function of previous and instant growth conditions. It makes, therefore, no sense to compare laboratory and native forms without specifying the cultivation method and the microenvironment inhabited by the given population. At Winogradsky's time, two major cultivation techniques were available, namely plating on solid agar media and liquid batch culture with nutrient broth. Microbial cells grown under such conditions do have a peculiar physiological state which is indeed dissimilar from that of *in situ* soil microbes.

Today we have a much wider assortment of cultivation techniques. Consequently, we have increased our ability to control the physiological state of a laboratory culture and may intentionally shape it by cultivation conditions. Moreover, we are 'in principle' able to imitate the physiological state of native populations. The only problem is the lack of exact information on the physiological state of natural populations and those values of the vector components we are aiming to reproduce (i.e. the true specific growth rate, biomass chemical composition, metabolic activity, etc.). Soil microbial communities are conspicuously heterogeneous with respect to these parameters. It was suggested by Hattori (1985) that this heterogeneity could be estimated from the time distribution of colony emergence during microbial enumeration by plating technique.

We propose another approach. First of all we have to estimate the entire diversity of growth patterns for native microbial populations *in situ*, and then relate each growth pattern to a particular laboratory cultivation method. This provides a firmer ground for our extrapolation to nature. As a measure of congruence, one may use such integral characteristics as the true specific growth rate, μ_i, of the given population or group of soil microorganisms. If the values of μ_i for *ex situ* and *in situ* populations coincide, then the physical model may be regarded as relevant and may be used as a basis for conclusions about other less general microbial parameters and properties. Here, we shall explain what we mean by growth patterns and how to embrace their diversity.

2.2 PATTERNS OF MICROBIAL GROWTH IN SOIL

Soil as a microbial habitat is a system of microenvironments drastically dissimilar in the nature and quantity of their growth substrates, aeration and moisture regimes, physical and chemical properties. The dynamic behaviour of microbial populations is determined by the type of occupied microsite. But what is their actual diversity, and can any regularity be discerned in this kaleidoscope? There are several classifications of soil microsite based on micromorphology data (Szabo, 1974), providing an extensive although static description of soil conditions. Without claiming an overall solution to the problem of classification, we will suggest a method of categorizing natural habitats with respect to those few features which determine **microbial population dynamics** (Panikov and Zvyagintsev, 1983).

The most important features are the processes of mass and energy flow, and in particular the **dynamic pattern of nutrients input to the microsite.** As a first approximation, two extremes may be identified: (a) a continuous and (b) a discontinuous or single-term substrate input. A continuous input pattern means that the limiting substrate is either always supplied to the microsite at a non-zero rate or that a temporary pause is negligible as compared with the cell generation time. A single-term input pattern denotes an instant local rise in substrate concentration with no further replenishment.

Another key factor is the presence (α) or absence (β) of **exogenous elimination** of propagating microbial cells or mycelium. By exogenous elimination we mean all the factors, external to the microbial population, which either induce the removal of microorganisms from the microsite or cause their death. Well-known examples of elimination are cells leached with soil solution or animals' excrements, grazing by predators or lysis by parasites, and active migration in exploring favourable microsites, including not only the taxis of motile cells but also the tropism of mycelial organisms via directed apical growth.

Finally, the third essential factor is the **spatial organization** of the microbial population within a habitat, which can be either (*a*) homogeneous or (*b*) heterogeneous. Spatial homogeneity implies the absence of spatial gradients in concentrations of substrate, biomass or metabolic products. (It should be noted that the requirement of an absolutely uniform distribution of the aforementioned dynamic variables along a spatial coordinate is too strong with regard to natural conditions. It would suffice to require their **random** distribution over the microsite volume.) The homogenization of virgin soils is mainly performed by animals, whereas in arable soils tillage will be much more important. Homogeneous microenvironments are also formed when the principal growth substrates are mobile, such as gases and volatile organic compounds.

The three independent characteristics of ecotopic conditions produce eight possible limiting case combinations (Table 2.1). Each entry in Table 2.1 denotes one of the vertices of a three-dimensional cube with axes 1–2, *a*–*b*, and α–β. Every specific situation in the natural environment can be represented by some point inside the cube. We prefer to consider a continuum of microsites rather than several distinctly defined types because in nature there are no ideal cases: perfectly steady supply of substrate, entire cell elimination, or complete homogenization. Nevertheless, particular habitats often tend to be closer to one or another of the cube's vertices. Below we present several examples of microbial habitats where different growth patterns occur.

- 1*a*α – the growth of microorganisms in the gut of soil invertebrates fed on plant litter. The microbial substrates are supplied continuously as a result of the digestion of the plant material swallowed by animals. Elimination of microorganisms is also continuous, since they are 'washed-out' from the intestines with excrements. Medium homogeneity is provided by the movement of the intestines resulting from the muscular contractions of the animal.
- 1*a*β – the growth of dissipotrophic microorganisms on monomeric volatile or readily soluble compounds which continuously diffuse from the sites of anaerobic digestion or depolymerization of organic matter. The homogeneity is ensured by high substrate mobility. Elimination is either absent or rather weak due to the immobility of the particular microbial species and low density of the population, which renders predation ineffective.
- 1*b*α – the growth of mycelial organisms (mostly fungi) on plant debris,

Table 2.1 Diversity of microbial growth patterns in soil habitats

Spatial organization of the microbial population	Substrate input			
	Continuous (1)		Single-term (2)	
	Cell eliminated (α)	No elimination (β)	Cell eliminated (α)	No elimination (β)
Homogeneous (*a*)	1 *a* α	1 *a* β	2 *a* α	2 *a* β
Heterogeneous (*b*)	1 *b* α	1 *b* β	2 *b* α	2 *b* β

i.e. forest litter, sloughing roots, tree stumps, etc. The continuity of the substrate input is provided not only by fresh litter but also by the continuous action of extracellular hydrolases, which attack the insoluble polymeric substances from the surface and produce monomeric substrates. Elimination involves grazing and apical hyphae expansion. The spatial heterogeneity of the mycelium is manifested in its differentiation into a small growing peripheral zone and the non-growing compartments which are responsible for transport and formation of reproductive structures.

- $2a\alpha$ – the explosive growth of microorganisms on available organic substrate instantly supplied into soil as a result of some perturbation, i.e. the application of manure, input of plant foliage exudations after heavy rains, drying–rewetting or freezing–thawing cycles, temporary rises in hydrolase activity, etc. Growth is usually accompanied by elimination in the form of grazing, myco- and bacteriolytic activity, as well as the active migration of starving microbial cells after the depletion of the available substrate. The extent of spatial heterogeneity of major dynamic variables is not generally high, since substrate amendment combines formerly separated microsites into one reactive zone. However, sometimes the particular research objective dictates the necessity to take into account the establishment of spatial gradients.
- $1a\alpha \rightarrow 2a\alpha$ – microbial growth in the rhizosphere on root deposition products (exudations and sloughing). Continuous excretion by roots of monomeric substances occurs only over short periods of time, due to the diurnal dynamics of plant photosynthesis, and also because the ‘address’ of the excretion loci may frequently change. This habitat is, therefore, an intermediate one, between the extreme types 1 and 2. Most rhizosphere bacteria are motile (clearly, they benefit greatly from fast chemotactic motility to elevated substrate concentrations and from loci of toxin formation) and perpetual water films around the roots provide free access to different microsites. Thus the rhizosphere and rhizoplane should be regarded as spatially homogeneous systems (unless some specific problems are studied, see above (Newman and Watson, 1977)). Elimination is performed via grazing by nematodes and protozoa and, possibly, via plant immunity factors.

The dynamics of microbial growth in any type of habitat may be described by the following mass balance equations:

$$\dot{s} = F(t) - G(s) - \frac{\mu(s)\,x}{Y} - mx\,, \qquad [2.1]$$

$$\dot{x} = V(t) - H(x)\,x + \mu(s)\,x\,,$$

where $F(t)$ is the input of the limiting substrate into the microsite, $G(s)$ is its dissipation (leaching) from the habitat, $V(t)$ is the microbial immigration rate, and $H(x)$ is the rate of microbial elimination; the rest of the notation is conventional. To simulate microbial growth dynamics in a particular homogeneous system one has to make the following selection: $F(t) = 0$, $s_0 > 0$ for an instant single-term input of substrate; $F(t) > 0$ for continuous supply of substrate; $H(t) = 0$ for systems with preserving proliferating microbial cells; and $H(t) > 0$ in the case of their elimination. The contribution of allochtonous microorganisms indicated by term $V(t) > 0$.

Spatially heterogeneous systems can be simulated either by distributed models employing partial derivatives, similar to Eqns [1.82–1.85], or implicitly, by an increase in the number of state variables. For example, the total biomass x of a microbial colony may be represented as a sum of the peripheral and central components.

Having delineated the diversity of microbial growth patterns in natural environments, we shall attempt to find, for each of them, the corresponding counterpart within laboratory cultivations.

2.3 ARRAY OF LABORATORY CULTIVATION METHODS

All known laboratory cultivation techniques can be positioned in the same three-dimensional space defined by the axes 1–2, *a*–*b* and α–β (Table 2.1). Obviously, this ordination may be performed in a straightforward manner without the ambiguity associated with natural habitats.

1*a*α – Homogeneous continuous culture (continuous-flow fermenters with complete mixing)

There are two subgroups within this type of cultivation technique. In the first, steady-state growth is maintained naturally by the microbial culture itself. Self-regulation is performed through negative feedback which originates from the dependence of growth rate on substrate concentration (chemostat) or on temperature (caloristat).

In the second group, electronic devices are used for the automatic adjustment of dilution rate to the instantaneous growth rate of the microbial culture. Electronic control is based on the sensing of cell density or growth-linked products, i.e. optical density (turbidostat), culture liquid viscosity (viscostat), CO_2 concentration in output air, culture pH (pH-auxostat), etc.

In the first type of fermenter, in theory steady-state growth may be established between 0 and μ_m, but in practical terms neither very low nor very high values are attainable because of the long time needed to reach the steady-state in the first case and the risk of culture washout in the second. The second type of technique is capable of maintaining steady growth at high s when either $\mu \rightarrow \mu_m$ or, under substrate inhibition, when $d\mu/ds < 0$. There is also a 'potpourri' mixed technique known as the 'bistat' (Minkevich *et al.*, 1987), which combines a chemostat and a pH-auxostat.

The mass balance equations for the chemostat and its modifications have already been given above (Eqn. [1.28]; Table 1.3). In a simple complete mixing cultivator, all cells have an equal probability of being washed-out and hence $\mu = D$. If there is substantial wall growth, or biomass retention or feedback, then $\mu < D$, this difference increasing with the extent of biomass retention in the fermentation vessel. In terms of our scheme, such cultivation systems correspond to points on the edge $1a\alpha$–$1a\beta$ of our cube.

1*a*β – Continuous cultivation without cell wash-out

Examples of such fermenters are listed in Table 2.2 along with the corresponding mass balance equations. This group embraces cultures with batch or continuous dialysis, fed-batch culture (FBC) and batch culture with a supply of limiting substrate via the gas phase (gases and volatile compounds). It also includes the chemostat with complete biomass feedback by means of filtration (Chesbro *et al.*, 1979; Markvichev and Manakov, 1985). The latter fermenters are less reliable in practical terms as compared with dialysis culture, because the membrane filters are quickly plugged with cells. The limiting substrate is supplied into the dialysis culture through a semipermeable membrane and, in the case of a gaseous nutrient, through the gas–liquid interface. In both cases mass transfer is reasonably well described by Fick's Law. The culture volume remains fixed in all systems, with the exception of a FBC. In the latter a constant nutrient feed, F, provides a linear increase of culture volume, V, during the cultivation span, dilution rate, $D = F/V$, being decreased hyperbolically. The value of these cultivation techniques in the context of this work is that they provide the possibility of realizing very slow continuous growth. With a constant limiting substrate flux, Fs_0, the absence of cell washout means that at each subsequent moment an equal ration is shared by an increasing microbial biomass, and, as a result, μ eventually falls to negligible values or even to zero (the maintenance state). Unlike the chemostat, no true steady-state is established in this case, but when the substrate is virtually depleted, we have $\dot{s} \approx 0$ and the system reaches a quasi-steady-state (Pirt, 1975). If a quasi-steady-state approximation is found to be sufficient, then extremely slow continuous growth can be obtained after a reasonable period of time, say in a couple of weeks, as compared to the several months needed in a chemostat.

Table 2.2 Continuous cultivations without cell wash-out

Cultivation method	Design	Equations of mass balance	
Batch dialysis	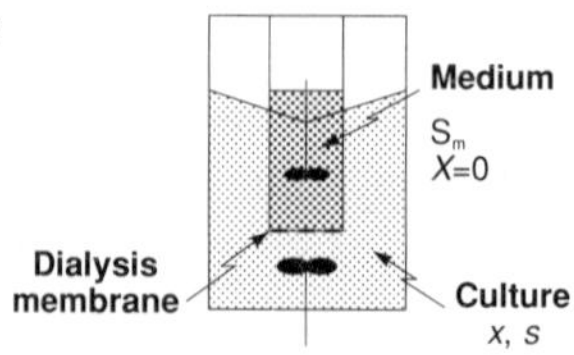	$\dot{s} = \Psi(s_m - s) - q(s)x$ $\dot{x} = Y[q(s) - m]x = \mu(s)x$	[2.2]
Continuous dialysis	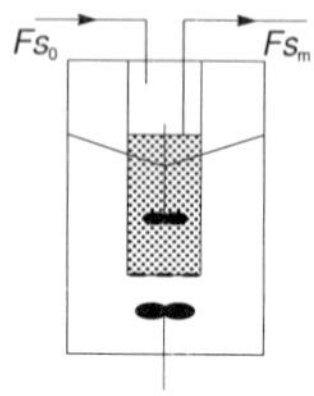	$\dot{s} = \frac{F}{V_m}(s_0 - s_m) - q(s)x$ $\dot{x} = Y[q(s) - m]x = \mu(s)x$	[2.3]
Fed-batch culture (FBC)	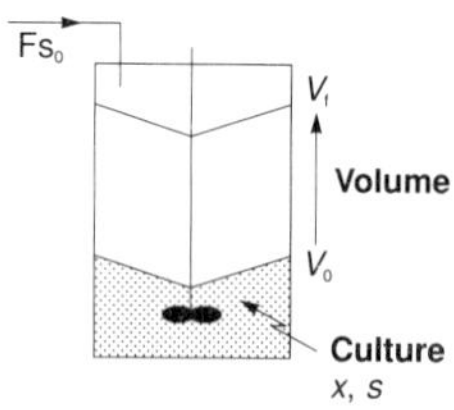	$\dot{s} = D(s_0 - s) - q(s)x$ $\dot{x} = Y[q(s) - m]x = [\mu(s) - D]x$ $D = \frac{D_0}{1 + D_0 t}$	[2.4]
		$\dot{S} = Fs_0 - q(s)X$ $\dot{X} = Y[q(s) - m]X = \mu(s)X$	[2.5]
Batch culture with substrate supply via gas phase	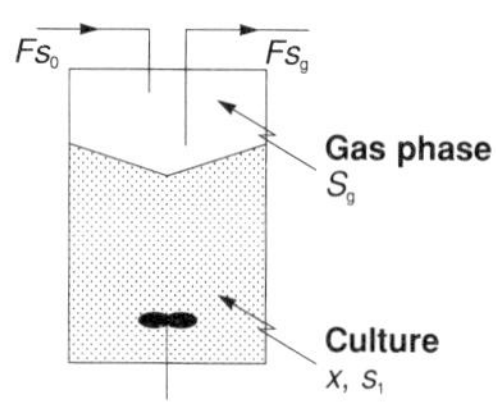	$\dot{s}_g = \frac{F}{V_g}(s_0 - s_g) - \Psi(p_g - p_l)$ $\dot{s}_l = \Psi(p_g - p_l) - q(s_l)x$ $\dot{x} = Y[q(s_l) - m]x = \mu(s_l)x$	[2.6]

s, x, limiting substrate and biomass concentrations per volume unit of culture; S, X, respective total quantities per culture vessel; p, gas partial pressure; V, reactor volume; V_0, V_f, volumes of FBC at the start and at the end of the experiment respectively; subscripts m, g, and l refer to dialysis compartment, gas and liquid phases respectively; Ψ is mass transfer coefficient.

2$a\alpha$ – Continuous cultivation with a discontinuous supply of limiting substrate

Suppose we have a simple chemostat culture fed by nutrient medium lacking just one essential component. This is added as a small volume of concentrated solution at regular and sufficiently large time intervals, Δt. Then

$$\dot{s} = D[s_0(t) - s] - q(s)x, \quad \dot{x} = \mu(s)x - Dx\,, \qquad [2.7]$$

$$s_0(t) = \begin{cases} A > 0, \text{ when } t = i\Delta t, i = 0, \ldots, n\,, \\ 0, \text{ otherwise.} \end{cases}$$

This cultivation method was originally employed to obtain synchronized cell division (Goodwin, 1969). It is still occasionally used to study chemostat transients (Litvinenko *et al.*, 1980).

2$a\beta$ – Simple batch culture

Cultivation begins at the initial limiting substrate concentration, s_0, and inoculum size, x_0. The biomass reaches its maximum, x_m, when the limiting substrate is depleted, $s = 0$, and then declines even in the absence of exogenous elimination, so that $x \to 0$ as $t \to \infty$. A continuous-phased culture (Dawson, 1965) is no more than a repeated simple batch culture, i.e. at regular intervals, Δt, half of the culture volume is withdrawn and the fermenter is refilled with an equal volume of fresh medium. Under such conditions, repeated batchwise growth proceeds with biomass increasing cyclically from x_0 to $2x_0$. Obviously the growth dynamics are governed by the time interval between consecutive substrate additions, Δt. If $\Delta t = \ln 2/\mu_m$, a saw-tooth non-limited growth takes place as in a turbidostat. With $\Delta t < \ln 2/\mu_m$ the culture should be washed out and, when $\Delta t > \ln 2/\mu_m$, the maximal attainable biomass decreases with increasing Δt due to endogenous biomass decomposition and waste respiration during the lag phase.

Spatially heterogenous microbial growth is realized in many versions, sometimes unintentionally as a result of non-perfect mixing. We start our consideration with continuous flow methods.

1$b\alpha$ – Plug-flow (tubular) culture

The inoculum and the medium are mixed on entry into a long reactor tube and the culture flows in the tube at a constant velocity without mixing. In each small element of culture liquid, moving along the spatial coordinate z at a linear velocity $f = F/A$ (where F is flow rate, $cm^3\,h^{-1}$, and A is the cross-sectional area, cm^3), growth of biomass proceeds as in a simple batch culture. To account for the culture movement *per se*, we have to pass from ordinary to partial derivatives and replace dx/dt by $\partial x/\partial t + f\partial x/\partial z$ ('along-the-flow growth rate'). Allowing in addition for some dispersion of the moving front by diffusion, we can write

$$\frac{\partial s}{\partial t} + f\frac{\partial s}{\partial z} = D_s\frac{\partial^2 s}{\partial z^2} - q(s)\,x\,, \qquad [2.8]$$

$$\frac{\partial x}{\partial t} + f\frac{\partial x}{\partial z} = D_x\frac{\partial^2 x}{\partial z^2} + \mu(s)\,x\,,$$

where D_s and D_x are the diffusion coefficients of the substrate and microbial cells, respectively.

1*b*β – Continuous-flow reactors with microbes attached

The nutrient solution is pumped through a column filled with adsorbent material and is utilized as it moves by growing immobilized cells. The mass balance equations for a packed column are obtained from Eqn. [2.8] by simplifying the equation for x,

$$\frac{\partial s}{\partial t} + f\frac{\partial s}{\partial z} = D_s\frac{\partial^2 s}{\partial z^2} - q(s)\,x(z)\,, \qquad [2.9]$$

$$\frac{\partial x}{\partial t} = \mu(s)\,x = Y[q(s) - m]x\,.$$

Bacterial cells accumulate faster on the top of the column due to larger $q(s)$ values and as a result a distinctive spatial biomass distribution develops in the form of an hyperbolic decrease of x with column depth. Growth does not reach steady-state with respect to x until cell elimination becomes well expressed (the effects of inhibitory products, endogenous cell decomposition and leaching).

Colonies

Besides continuous-flow columns, other heterogeneous systems are widely used. Especially popular is plating on **solid media** made from natural or synthetic gels (agar, polyacrylamide gel (PAAG), silica gel, synthetic alumosilicates, etc.) as well as some porous materials (sand, glass beads, glass fibre). Impregnated with nutrient solution, such materials are used to grow microorganisms in the form of colonies or lawns.

At first glance, it is tempting to consider these techniques as analogies of a simple batch culture with single-term substrate input (type 2*b*β according to our scheme). But . . . no rush! A closer look at the mechanism of colony growth reveals a greater resemblance to chemostat culture! Below we will outline our considerations.

The spread of a colony over a solid substrate, for example a layer of agar, proceeds by the growth of only a peripheral zone with biomass, x_p. Then

$$\dot{s} = -q(s)x, \quad \dot{x} = \mu_w x_p - ax, \qquad [2.10]$$

where a is the specific decay rate of cell (mycelium) components. Straightforward geometric analysis shows that the linear spread rate for a regularly shaped colony (a cylinder, a sphere or a strip) of size R is given by the following relation (Pirt, 1975):

$$\dot{R} = \mu_w w = K_r, \quad \text{or} \quad R = R_0 + K_r t, \qquad [2.17]$$

where K_r is the colony linear expansion rate in mm h^{-1}; μ_w is the microbial specific growth rate within the peripheral zone; and w is the zone width.

In the case of unicellular organisms (e.g. bacteria, yeasts) which are incapable of penetrating into the gel's matrix, the substrate is available through passive diffusion to the peripheral zone from the underlying gel layer. As the colony grows this flux diminishes, K_r decreases, and eventually growth stops altogether. Mycelial organisms (fungi and actinomycetes) are able to propagate both on the surface and in the depth of gel. As a result their growth is not completely dominated by diffusion effects and the colony front advances at a faster rate than substrate is depleted in the 'frontier zone'. The colony spreads at a constant radial rate, K_r, until the Petri dish is filled or the agar deteriorates from dryness. (In the case of rich nutrient media, there are also effects of self-inhibition by metabolic products, but this will not be discussed at this point.)

We have seen that colony growth is: (a) continuous; (b) substrate-limited; (c) directed; and (d) spatially ordered. Properties (a) and (b) suggest a strong similarity between growth of a colony (especially a fungal one) and that of a chemostat culture. If so, a colony's peripheral zone is analogous to the cell culture in the fermenter, while its central part is comparable with the waste cell suspension which is discharged into the product bottle. A steady running of a pump, delivering medium at a rate F ($cm^3\ h^{-1}$), corresponds to the active and uniform substrate utilization by the growing mycelium at a rate $K_r A$ ($cm^3\ h^{-1}$, A is the area of colony–medium interface). Finally, both cultivation systems are characterized by the elimination of propagating biomass. That is, by the expulsion of grown cells (mycelium) from further active growth (either by washing-out or by the motion of the colony front). In both cases, the elimination rate is equal to biomass growth in the active compartment. The essential deviation from the chemostat lies in properties (c) and (d). Spatial differentiation of hyphae and direction of colony expansion are governed by spatial gradients of limiting substrate and, possibly, some metabolic products. Steepness of gradients is partly diminished by the effects of metabolic translocation along hyphae over distances of the order of w.

It can be concluded from the previous discussion that colony growth belongs to class 1*b*α, and not to 2*b*β. In general, it is very likely that the 2*b* combination is an empty, logically forbidden combination, because a single-term momentary input of substrate may only be realized in an homogeneous system. Any spatial segregation whatever will actually prolong the consumption of substrate and, therefore, transform a batch process into a continuous one. For this reason, growth on any insoluble substrate (lignocellulose and other plant polymers, oil droplets, grains of sulphur, etc.) should always be treated as a continuous process. The **solid-phase fermentation** can also not be anything but continuous, whether new portions of substrate are added to the reactor or not. (Obviously, this applies to the growth mechanism itself and not to the engineering operation.)

2.4 CONCLUSION

In this chapter we have put forward a unified ordering scheme embracing the entire diversity of patterns of microbial growth in laboratory cultures and in natural media. Microbial growth patterns are distinguished by three features: the dynamics of substrate supply (1, continuous supply, and 2, single-term momentary addition); the elimination of growing microorganisms (α, significant and β, absent); and the magnitude of spatial gradients (*a*, homogeneous systems and *b*, heterogenous systems). Each specific situation can be represented by a point inside a cube with the axes 1–2, α–β and *a*–*b*. The 2*b* combination is logically forbidden because any spatial segregation results in protracted substrate utilization and so transforms a batch process into a continuous one.

The equivalent types of growth pattern *in situ* and *ex situ* are described by the same or closely related sets of kinetic equations. Firstly, this puts the extrapolation to natural environments of the principles established in laboratory experiments onto a sound basis. Secondly, our approach suggests how the physiological state of native microbial forms could be imitated in laboratory examinations. There are three or four homogeneous cultivation techniques well suited for this purpose: the batch culture, the chemostat, and the dialysis or the fed batch culture. Taken together, they can cover the entire range of the physiological state variation, from feast (substrate surplus) to famine (total substrate absence). In addition, there is the possibility of adequately simulating different dynamic patterns, i.e. (a) a sudden transition of starving cells to nutrient abundance and *vice versa*; and (b) a gradual and predictable, almost steady-state transfer of microbial populations from one state to another. Anticipating possible objections, we would still argue that the specified homogeneous cultivation systems are much more useful for solving numerous ecological problems than physical models such as sterilized soil or soil agar media, clay–sand mixtures, etc.

The latter models resemble soil only in its superficial attributes and give a distorted perspective of the real diversity of microbial growth patterns.

If a researcher is to study non-trophic interactions of soil microbes within the environment, then there is a whole range of other laboratory techniques, including heterogeneous ones. Even so, we believe, one should not be content simply with an external and cosmetic similarity between the model and soil. There remains a great deal to be learned from analysing quantitative laws of microbial colony formation, even on the familiar agar plates.

Microbial growth under homogeneous conditions

3

This and following chapters will deal with our original experimental data and model developments. We shall begin with the simplest case, homogeneous culture, which provides the best opportunity to gain insight into the mechanisms of microbial growth. Firstly, experimental data will be considered, paying special attention to those aspects where conventional growth theory fails. Then, towards the end of the chapter, we will suggest a new type of kinetic model which provides an adequate description and explanation of all available experimental data.

3.1 MATERIALS AND METHODS

Microbial cultures

Our choice was determined by intention to embrace as fully as possible the potential diversity of kinetic properties of microbial populations with different life strategies. This is why we used isolates of microbial species with known ecological properties and a wide natural distribution. At the same time we restricted ourselves mainly to aerobic and prototrophic organisms which were able to grow on relatively simple and chemically defined media. Brief information on the strains which were used is presented in Table 3.1.

Media

The composition of the defined nutrient media which were used is specified below (the associated microbial cultures are shown in Table 3.1).

1. Base glucose–mineral medium (mg l^{-1}): glucose 200, $(NH_4)_2SO_4$ 100, KH_2PO_4 100, K_2HPO_4 1000, $MgSO_4{\cdot}7H_2O$ 200, $CaCl_2{\cdot}2H_2O$ 40, $CuSO_4{\cdot}5H_2O$ 0.04, KI 0.1, $FeCl_3.6H_2O$ 0.4, $MnSO_4{\cdot}4H_2O$ 0.4,

$Na_2MoO_4 \cdot 2H_2O$ 0.2, $ZnSO_4 \cdot 7H_2O$ 0.4, H_3BO_3 0.5, Na-EDTA 5.0, pH 7.5. Some components were varied depending on the aims of the particular experiment and this will be indicated where neccessary.

2. Base medium with substitution of $(NH_4)_2SO_4$ by asparagine, 800 mg l^{-1}, and yeast extract, 1 mg l^{-1}.
3. Base medium with substitution of $(NH_4)_2SO_4$ by peptone, 800 mg l^{-1}.
4. Base medium with addition of yeast extract, 10 mg l^{-1}.
5. Base medium with addition of yeast extract, 50 mg l^{-1}, and standard mixture of vitamins (μg l^{-1}): biotin 20, folate 20, thiamine 50, pantothenate 50, cyanocobalamine 1, riboflavine 50.
6. Base medium with the following modifications (mg l^{-1}): KH_2PO_4 0, K_2HPO_4 850, biotin 0.02, pH 5.0.
7. Base mineral medium without glucose.
8. Base medium with 1.7–10.0 mg l^{-1} urea instead of $(NH_4)_2SO_4$ and KH_2PO_4 + K_2HPO_4 mixture at a concentration of 0.08–4.4 mg P l^{-1}, 5% CO_2 in gas phase, pH 6.5.

To avoid precipitations and chemical interactions, individual components were sterilized separately with aseptic mixing after cooling.

Cultivation conditions

Continuous chemostat cultivation was performed with simple apparatus consisting of interchangeable components (Figure 3.1). The difference between particular devices was only in the use of different fermentation vessels.

Semicontinuous dialysis was carried out in a modified standard shaking flask (Figure 3.2a). Glucose diffusion via the semi-permeable membrane was reimbursed with a concentrated solution from a syringe, every 3–4 days. The dense cellophane membrane provided a rather low permeability (parameter Ψ in Eqn. [2.2] was about 0.5 d^{-1}) and consequently provided an almost constant supply of glucose to the culture. Thus, batch dialysis used in this manner yielded almost continuous microbial growth.

Batch cultivation was performed either with one of the continuous apparatus after stopping the flow, or with modified shaking flasks (Figure 3.2b) and rotary shakers at 180 r.p.m. These flasks were equipped with a port for sterile air sampling and a cylindrical cuvette for on line determination of culture turbidity.

The cultivation temperature was 25°C (except *Micrococcus* and *Lipomyces* which were grown at 20°C). Acidity was generally kept constant by the presence of phosphate buffer without any special devices (pH 6.5–7.5 for bacteria and algae and pH 4.5–5.0 for fungi including yeasts). At elevated glucose concentration (s_0 > 500 mg l^{-1}) 5–10 times more

Table 3.1 Microbial isolates used in kinetic studies

Species, strain	Isolation site	Location	Nutrient medium for		Strain source*
			Maintenance	Experiments	
Bacteria					
Arthrobacter globiformis C-4	Typical serozem, field under cotton	Chimkent region	A	1	a
A. globiformis VCM-658	Soils	–	A	1	b
Micrococcus agilis 63H	Lichen mat	Pamirs	A	3	a
Rhodococcus erythropolis 61X	Mountain brown soil	Tien Shan	A	1	a
Pseudomonas fluorescens 312	Sod-podzolic soil, lysimeter water	Botanical gardens of Moscow University	A	1	a
P. fluorescens VCM-1472	Plant rhizosphere	–	A	1	b
Labrys monachos 42	Lake sediments	Estonia	J	5	c
Prosthecomicrobium hirschii 22	Takyr	Alma-Ata	E	5	c
P. polyspheroidum 1313	Chernozem	Krasnodar	J	5	c
Caulobacter vibrioides C-12	Fresh water and neutral soils	–	E	5	c
Rhizobium meliloti A3	Sod-podzolic soil under alfa-alfa	Latvia	C	2	c
Bacillus subtilis VCM-428	Soils	–	A	1	b
Yeasts					
Debaryomyces formicarius VCM-1555	Nests of grey forest ants	Moscow region	B	6	a
Lipomyces tetrasporus 130	Ordinary chernozem	Voronezh region	B	6	a

Table 3.1 Continued

Species, strain	Isolation site	Location	Nutrient medium for		Strain source*
			Main-tenance	Experi-ments	
Fungi					
Penicillium funiculosum	Soil under corn	Moscow region	B	7	a
P. spinulosum	"	"	*B*	*7*	*a*
Muscor plumbeus	Podzolic soil under spruce	Moscow region	B	7	a
Mortierella ramanniana	"	"	B	7	a
Trichoderma koningii	"	"	B	7	a
Zygorynchus heterogamus	"	"	B	7	a
Ulocladium consortiale	Mountain brown forest soil	Carpathian Mountains	B	7	a
Algae					
Chlorella vulgaris 211/8K	Soils, fresh water	–	E	9	e
C. vulgaris 234	"	–	E	9	d
Bractococcus minor 222	"	–	E	9	d

* a, Culture collection of Soil Biology Department of Moscow State University; b, All-Russian Collection of Microorganisms, VCM; c, Institute of Microbiology, Moscow; d, Botany Department of Kirov Agricultural Institute; e, Culture Collection of Algae and Protozoa, Cambridge

† A, Meat-peptone agar; B, malt agar; C, bean-extract agar; D, starch-ammonium agar; E, defined medium 8 with 2% agar; F, defined medium 4 with agar; G, potato agar

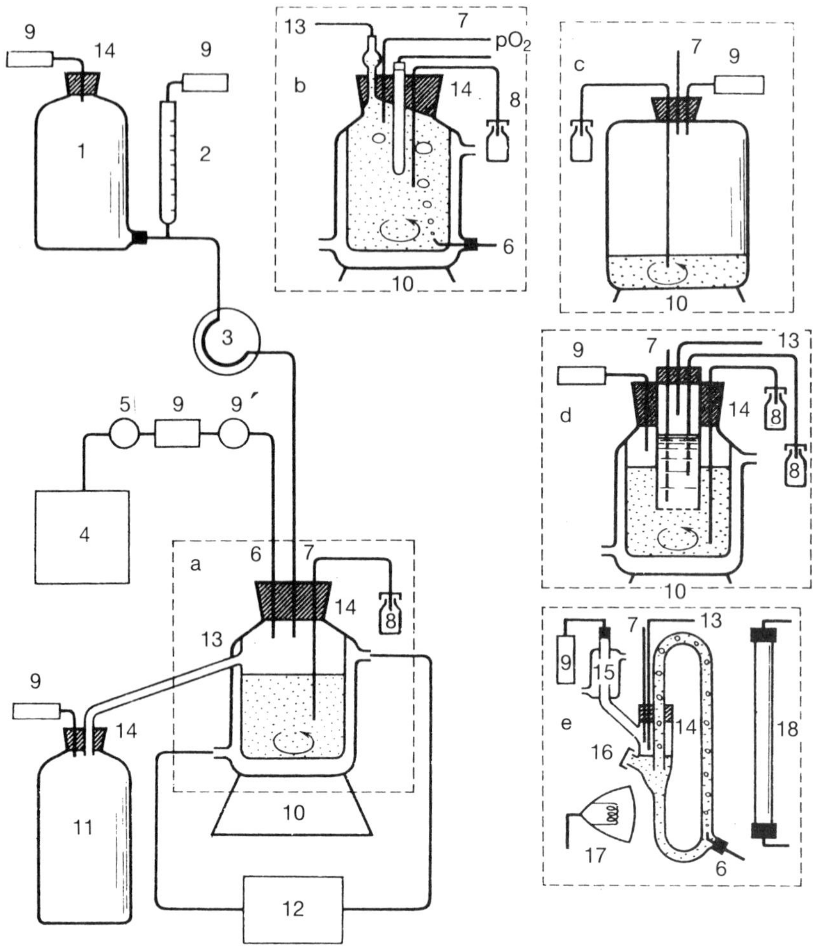

Figure 3.1 Continuous culture apparatuses used in the present studies. (a) Basic set-up; modifications used for: (b) instant measurements of dissolved oxygen uptake (Panikov and Bondarenko, 1981); (c) for fed batch cultivation; (d) for continuous dialysis (Panikov *et al.*, 1982); and (e) for obtaining dense culture of phototrophs (Pirt *et al.*, 1979). (1) Medium reservoir; (2) measuring burette; (3) peristaltic pump; (4) air flow inducer; (5) rotameter; (6) inlet for sterile air; (7) medium feed; (8) sampling port; (9) glass-fibre filter; (10) magnetic stirrer; (11) product bottle; (12) circulation thermostat; (13) cultural liquid outflow; (14) non-toxic rubber bung; (15) condenser; (16) sampling port for syringe; (17) infra-red heater; (18) fluorescent light.

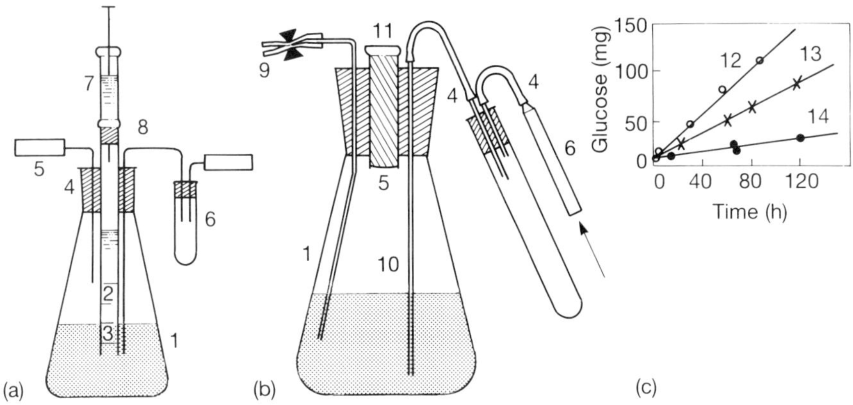

Figure 3.2 Modified shaking flasks for (a) batch dialysis and (b) simple batch cultivation; (c) data on kinetics of glucose diffusion through cellophane membrane. (1) Regular shaking flask, 750 ml; (2) dialysis compartment, 25 ml; (3) cellophane membrane; (4) rubber bung; (5) glass-fibre filter; (6) sampling device; (7) syringe with concentrated glucose solution; (8) rubber ring; (9) tubing; (10) capillary tube from stainless steel (to sample liquid for determination of dissolved CO_2 + HCO_3^-, when flame use is forbidden); (11) septum. Glucose concentration inside dialysis compartment: (12) 1.0%; (13) 0.5%; (14) 0.1%.

concentrated phosphate buffer was used. Aeration was controlled by a pO_2-sensor, with pO_2 never falling below 50% of air saturation.

Microbial count and biomass determination

Cell counts were performed either manually with phase-contrast microscopy or with a Coulter counter (Coulter Electronics, model Z_B1). For viable counts we refined the traditional plating technique by: (a) using an automatic pipette with hydrophobic plastic tips instead of ordinary glass ones; (b) making a few 100–1000-fold dilutions instead of many 10-fold serial dilutions; and (c) employing a pour-plate modification instead of a surface colony count. As a result reproducibility was significantly improved (standard error $<5\%$). Viable biomass was calculated as

$$x_{via} = N\rho V(1 - W), \qquad [3.1]$$

where N is colony-forming unit (CFU) number per culture volume, V is average CFU size, ρ is moist cell density, and W is moisture content. W and ρ values were accepted from Fichman (1967) as 1.08 g cm^{-3} and 0.75 respectively. *V*-values were estimated from the direct microscopy of cells and cell aggregates.

The total microbial biomass (concentration of solids in the cultural liquid) was determined from (a) dry-weight measurements; and (or) (b) dichromate oxidation. For the assessment, cells were separated from liquid by filtration (Synpor, Czechoslovakia, 0.24 and 0.45 μm pore size for bacteria and yeasts respectively), washed, dried at 80°C and weighed. In the dichromate oxidation method 1.5 ml of analysed liquid was blended with 2.4 ml of oxidant (0.2 M $K_2Cr_2O_7$: concentrated H_2SO_4 = 1 : 5), incubated for 20 min at 140°C and cooled. The yield of reduced chromium Cr^{2+} was estimated from absorbancy at 590 nm. Determinations were made from the liquid culture before and after the removal of cells by centrifugation, the biomass being represented by the difference in the two measurements.

Dichromate oxidation displays a close correlation with dry-weight determinations ($r = 0.99$) and has an obvious technical advantage (rapidness and higher sensitivity). However, it measures biomass in somewhat awkward units of chemical oxygen demand (COD), rather than in conventional mass units. For conversion we need to know the value of the stoichiometric coefficient *b* in Eqn. [1.41]. To determine its value we used two approaches: (a) the regression of COD data against dry-weight determinations; and (b) the calculation according to Eqn. [1.41] using experimental data on cell composition (CHN-analyser, Carlo Erba 1106, Italy). The first and second approaches give for the used microbial species B-values of 23.3–26.5 and 36.5–43.2 mmole O_2 g^{-1} dry biomass respectively. This was an unambiguous indication that dichromate provides an incomplete (55–64%) oxidation of cell biomass. Nevertheless, the application of an empiric conversion factor gave quite appropriate results from a practical point of view.

Dry weight and dichromate oxidation characterize the total concentration of solids which is the sum of viable biomass, necromass and mass of cell debris (cell fragments or their secondary associations):

$$x = x_{via} + x_{necro} + x_{deb} . \qquad [3.2]$$

To determine the ratio x_{via}/x_{necro} an agar film technique (Postgate, 1967) was used: bacteria and yeasts were grown on meat–peptone and pure agar respectively, those cells which produced microcolonies being counted as viable. To express x_{via} in terms of mass concentration the total biomass x and optical density at 600nm (OD_{600}) were measured simultaneously. The relationship between OD_{600} and x depends on many factors (especially on medium tonicity and cell aggregation); however, under constant environmental conditions and absence of flocculation we may put

$$\mathrm{OD}_{600} = E_1 x_{\mathrm{via}} + E_2 x_{\mathrm{necro}} + E_3 x_{\mathrm{deb}}, \qquad [3.3]$$

where E_1, E_2 and E_3 are extinction coefficients for viable biomass, necromass and cell debris respectively. Since an E-value is determined by the difference in refractive index of cells and medium and the maximum difference is attributed to viable cells, we may write $E_1 > E_2 \approx E_3$ and

$$x_{\mathrm{via}} = \frac{\mathrm{OD}_{600} - E_2 x}{E_1 - E_2}. \qquad [3.4]$$

Numerical values of E_1 and E_2 were measured in suspensions of exponentially growing and heat-killed cells respectively.

Determination of microbial respiration activity

The rate of oxygen uptake was recorded in a special fermentation vessel (Figure 3.1b), which allowed an instantaneous isolation of culture from air after stopping the air flow. The dynamics of dissolved O_2 (c_0) were recorded with a quick-response Clark-type electrode, where during aeration

$$\frac{\mathrm{d}c_0}{\mathrm{d}t} = K_\mathrm{L} a H (p_\mathrm{g} - p_\mathrm{l}) - q_{\mathrm{resp}} x \qquad [3.5]$$

and after stopping aeration

$$\frac{\mathrm{d}c_0}{\mathrm{d}t} = - q_{\mathrm{resp}} x, \qquad [3.6]$$

where p_g and p_l are oxygen partial pressure in gas phase and liquid phase respectively, H is Henry's constant, and q_{resp} is the specific respiration rate. The mass transfer coefficient, $K_\mathrm{L}a$, was found from the dynamics of p_l after the restoration of aeration. A small Clark-cell (about 1 ml) was used for the respiratory measurements in culture samples.

The rate of CO_2 production was estimated by both dynamic and static methods. In the dynamic method, sterile air was passed continuously through the fermenter at rate F, CO_2 concentration was continuously measured in the outflowing and inflowing air (c and c_0 respectively) using an infra-red gas analyzer ('Infralyt4', Junkalor Dessau, Germany) or in discrete samples by gas chromatography with a catharometer. Respiration was calculated using

$$q_{\mathrm{resp}} x = \frac{F(c - c_0)}{V_\mathrm{l}}, \qquad [3.7]$$

where V_l is the culture volume.

The static technique was applied only for batch culture (Figure 3.2b). In this approach, the aeration port containing a glass-fibre filter (5) was closed by a rubber septum and the dynamics of CO_2 in the gas phase (c_g) as well as CO_2 + carbonates in liquid phase (c_l) were recorded during a period of 1–24 hours. The values of c_l were determined as CO_2 after sample acidification and the value of $q_{resp}x$ was calculated from the total rate of $c_g + c_l$ accumulation. Aeration of the flask was resumed after completion of the respirometry by removing the septum and flushing the culture with sterile air via (6) and (5) (indicated by an arrow).

Chemical assays

Glucose was determined with a hexokinase (kid 'Orion') or a glucose oxidase–peroxidase enzyme assay. Phosphate was determined with ammonium molybdate, ammonium with indophenol reagent, nitrate with disulphophenolic acid, and starch by iodometry. Urea was determined as ammonium after urease treatment. The radioactivity in solutions was measured in samples transferred to a scintillation cocktail GC-8, the count being made using the scintillation counter Ultrabeta ('LKB Vallac', Sweden). The content of ^{14}C in bacterial biomass was estimated by calculating the difference between the total radioactivity of suspension and its filtrate. $^{14}CO_2$ was absorbed by an α-phenylethylamine : ethanol : scintillation liquid 'cocktail' GC-106 (1 : 1 : 3).

3.2 MICROBIAL GROWTH LIMITED BY THE SOURCE OF CARBON AND ENERGY

The availability of organic substances most frequently limits the growth of chemoorganotrophic microorganisms both in nature and in laboratory cultures. Here, our objective is to reveal quantitative principles which would be adequate not only for a specific culture type, but for all the known systems of homogeneous cultivation. We tested four such systems: chemostat (1*a*α), dialysis culture and fed-batch culture (1*a*β), as well as simple batch culture from the lag phase to the phase of decline (2*a*β).

3.2.1 CHEMOSTAT

Dynamics of culture stabilization: bistability phenomenon

We observed three patterns of microbial dynamics at the onset of the continuous culture (Figure 3.3):

- rapid and smooth achievement of the first steady-state and subsequent ones following every consecutive change in D (*Debaryomyces*, *Pseudomonas*);

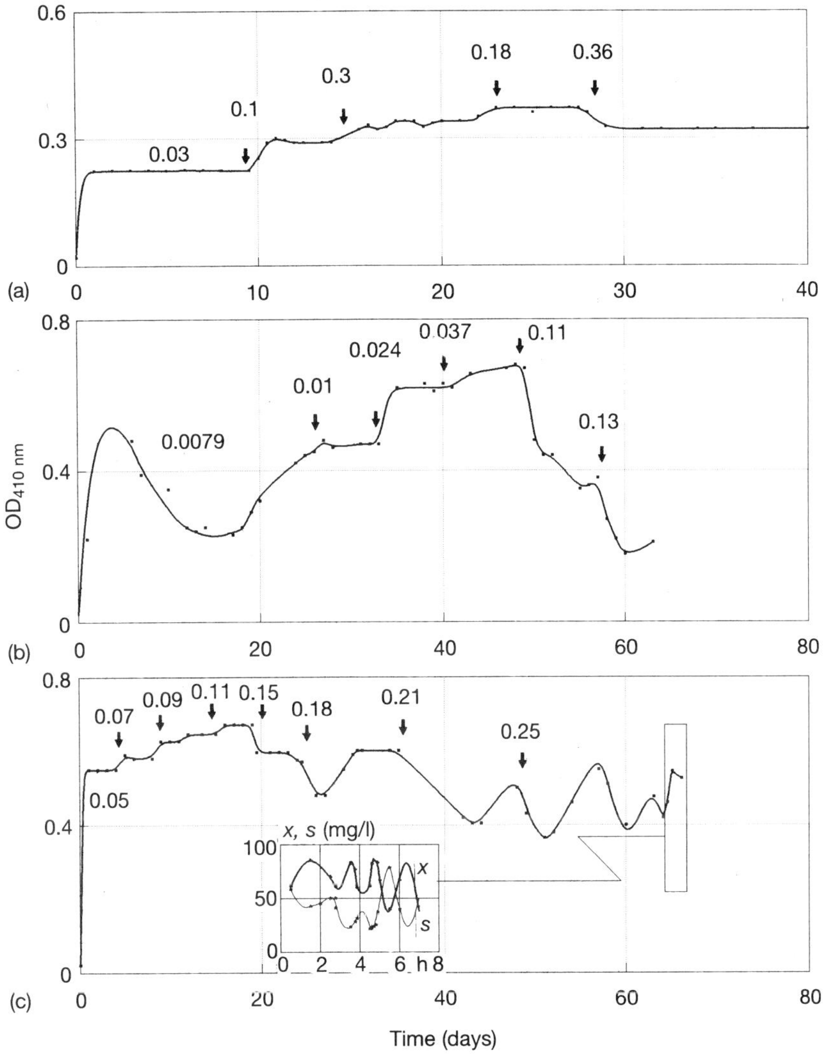

Figure 3.3 Biomass dynamics in continuous culture exemplified by three microbial species after consecutive changes in dilution rate, D (shift to the new D value, h^{-1}, is indicated by number and arrows). (a) *P. fluorescens* 312; (b) *L. tetrasporus*; (c) *A. globiformis* C-4. The insert in (c) shows sustained oscillations of biomass (x) and residual glucose (s) in *Arthrobacter* continuous culture at $D = 0.25\ h^{-1}$. (From Panikov, 1991a.)

- extended non-stable growth at the start of the chemostat culture before the first steady-state was attained (*Bacillus*, *Lipomyces*, *Rhizobia*); and
- retarded stabilization after any change in D, the high probability of sustained oscillations and multistability (*Arthrobacter*, *Rhodococcus*, *Micrococcus*).

The most intriguing is the third type of dynamic behaviour, which has been studied more carefully with *A. globiformis* as a model organism. At low D, stabilization settled fairly quickly (Figure 3.3c). At D about 0.5 μ_m it was delayed, e.g. the change in D from 0.15 to 0.18 h^{-1} at first caused the slow wash-out of culture during one week (corresponding to 15–20 generations); the growth rate then increased and a new steady-state was smoothly attained. At high D, steady-state was entirely impossible and instead sustained oscillations of x took place with an amplitude ±20–30 mg l^{-1}. The instantaneous growth rate, calculated as $\mu = (dx/dt)(1/x) + D$ ($D = 0.25\ h^{-1}$), varied in the range 0.12–0.47 h^{-1}. The oscillation period (1–2 h) was less than the average generation time ($g = (\ln 2)/D \approx 2.8$ h) and hence oscillations could not be explained by the synchronization of cell division. Residual glucose $s(t)$ varied in the opposite direction to biomass, as should be expected from Eqn. [1.29] for a simple chemostat model. However, correlation between s and x was rather poor, indicating a significant variation of the stoichiometric parameter $Y = -dx/ds$ or a time delay in biomass growth as compared with substrate uptake.

Now we consider what happens during movement in the opposite direction. Following consecutive decreases in D from the highest values causing oscillations to low ones, the new steady-states were attained rather quickly, but deviated significantly from those observed at the same D in the previous experiment (compare curves (ii) and (ii′) in Figure 3.4b). This is evidence for the so-called **multistability phenomenon**: when at one particular dilution rate, two or more non-identical steady-states may be achieved depending on their past history. The first series of steady-states (curve (ii)) corresponds to normal substrate-limited growth (high $\bar{x}$, low $\bar{s}$). The second series (curve (ii′)) is characterized by 1.5–2 times lower values of $\bar{x}$, high residual glucose concentration ($\bar{s}$ about 100 mg l^{-1}), accumulation of exometabolites, and an unusual cell morphology. That is, instead of regular spherical or rod-shaped cells there was a domination of intensively branched and swollen cells. We will elaborate on the nature of this second steady-state a bit later but for now we will define it tentatively as an **inhibitory** state.

We did not make systematic tests for multistability with respect to other organisms of our collection. However, it may be stated that *Debaryomyces*, *Pseudomonas* and *Bacillus* are not susceptible to this phenomenon in that their values of $\bar{x}$ and $\bar{s}$ generally did not depend on the previous steady-state or the direction of D changes. On the other hand, all examined

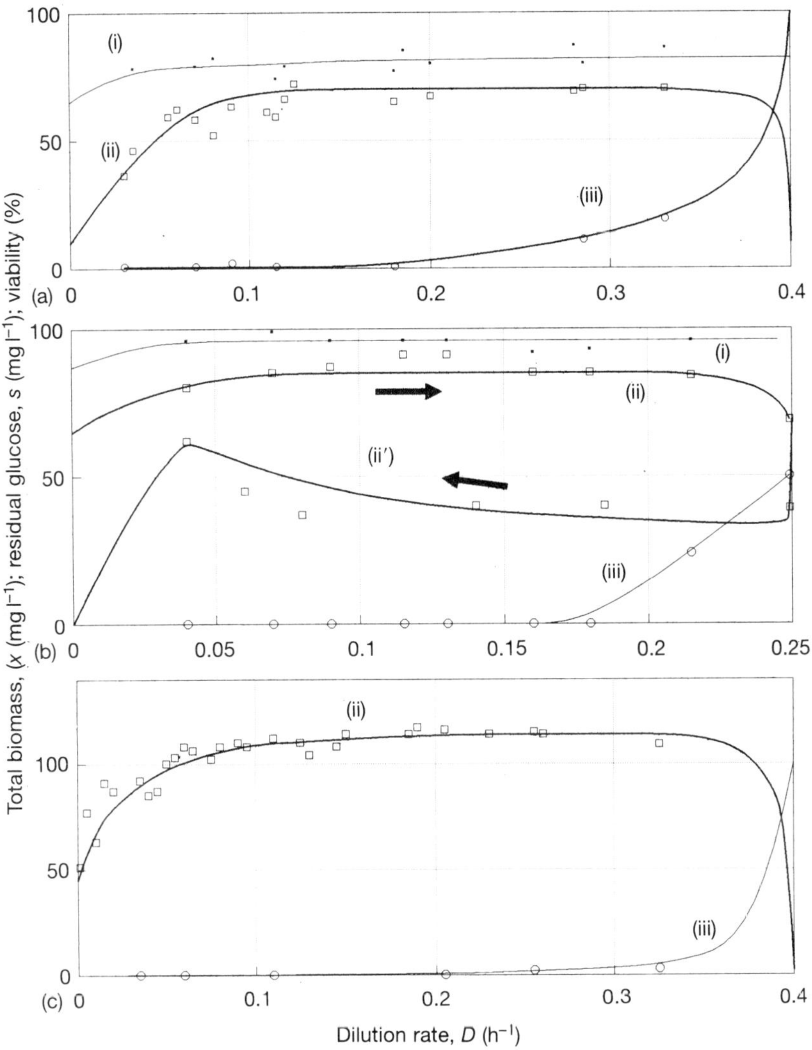

Figure 3.4 Glucose-limited steady-state growth of (a) *P. fluorescens* 312; (b) *A. globiformis* C-4; and (c) *D. formicarius*. Fraction of viable cells (i), total biomass (ii), and residual glucose (iii) are plotted *versus* dilution rate, *D*, of chemostat culture. The curves are calculated using the synthetic chemostat model (SCM). Note two sets of steady-state values of biomass, ii and ii′, for *A. globiformis*, illustrating the phenomenon of multistability. Arrows indicate the direction of *D* consecutive changes. (From Dorofeyev *et al.*, 1992.)

coryneform bacteria (*Arthrobacter, Rhodococcus, Micrococcus*) did display multistability. Most frequently it was revealed as poor reproducibility of growth parameters in chemostat cultures, unexpected abrupt changes of kinetic properties like loss of affinity to substrate, decline in growth rate, distortion of cell morphology, etc.

Multistability was reported for such well-studied microorganisms as enterobacteria and fodder yeasts grown on unusual substrates like phenol or in the presence of inhibitory compounds (Muller *et al.*, 1979; Gurevich, 1984). What we declare here is that the multistability of coryneform bacteria is manifested on quite usual substrates such as glucose. The trigger from a limitation-type of steady-state to an inhibition one was shown to be most probable at medium and high *D*. Spontaneous transition in the reverse order, from inhibition to normal limitation, has been never observed. Recovery from inhibition was achieved only through a temporary starvation at zero or very low *D*.

Note that coryneform bacteria belong to those organisms for which growth kinetics have not yet been carefully studied. In our experiments they demonstrated the following surprising features: (a) multistability of continuous growth; (b) ambiguity of the term 'critical dilution rate', since wash-out may take place at a relatively low *D*, and then end abruptly; and (c) principal inability of the microbial population to attain a true steady-state at some *D* range, where growth is possible only in the form of sustained oscillations.

Endogenous respiration and turnover of cell components

The stoichiometry of chemostat culture limited by energy source should be strongly dependent on the maintenance requirements. Maintenance coefficient *m* in Eqn. [1.32] is assumed to be a constant. We suspect that this is not true, but in order to prove it, we need to perform a **direct** estimation of the energy source consumption for maintenance. Unfortunately the complete solution of this problem remains, to date, impossible. However, we may determine directly the so-called endogenous respiration, that is the rate of O_2 uptake or CO_2 production resulting from the oxidation of cellular constituents rather than substrates immediately taken up from the medium. The specific rate of endogenous respiration, $q^e{}_{resp}$, is related to the average specific turnover rate of cell polymers, *a* (see Eqns [1.30] and [1.31]), and maintenance coefficient, $m = a/Y^{max}$.

Accurate measurements of $q^e{}_{resp}$ need to be made under normal growing conditions. It is known that the simple deprivation of the culture of nutrients (by removal through centrifugation or washing on filters with subsequent incubation in buffer) produces extreme stress and may alter the normal turnover rate (Marr *et al.*, 1963). To avoid the danger of artifacts we used a label substitution technique (Figure 3.5). The chemostat culture

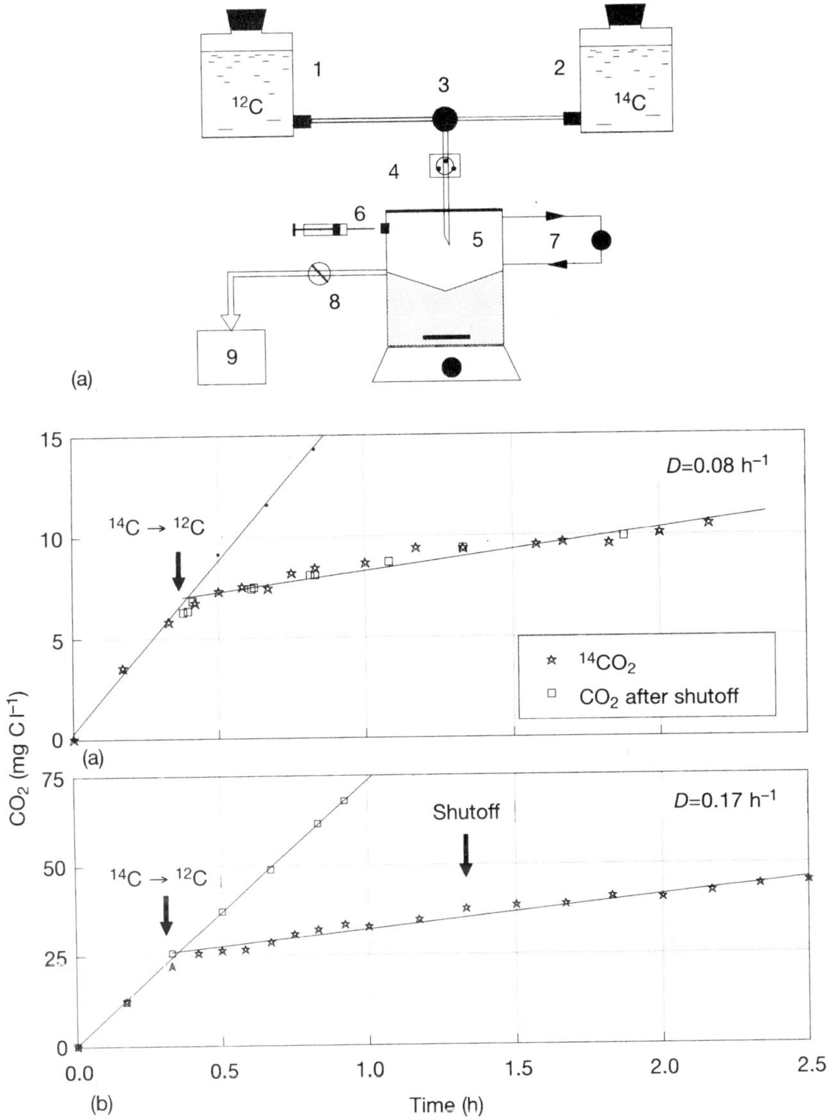

Figure 3.5 Label substitution technique for determination of turnover rate of cell macromolecular constituents (Dorofeyev and Panikov, 1994). (a) Experimental set-up: (1) medium reservoir with unlabelled glucose, s_r = 5.55 mM; (2) medium reservoir with labelled glucose, s_r = 5.55 mM; (3) two-way valve; (4) peristaltic pump; (5) fermenter; (6) gas sampling port; (7) air circulation by flow inducer; (8) stopcock; (9) product bottle. (b) Dynamics of ^{14}C evolution before and after switching of the medium feed from ^{14}C- to ^{12}C-glucose in chemostat culture of *P. fluorescens* 1472 at two dilution rates. The curves are fitted by Eqn. [3.9].

Table 3.2 Modified chemostat model of microbial growth on the mixture of labelled (^{14}C) and unlabelled (^{12}C) limiting substrate

Variable	Equation
Unlabelled substrate, (mol ^{12}C) l^{-1}	$\dot{s} = D(s_0 - s) - q_s x + a(1 - \theta)x$
Labelled substrate (mol ^{14}C) l^{-1}	$\dot{s}^* = D(s_0^* - s^*) - q_s^* x + a\theta x$
Unlabelled biomass, (mol ^{12}C) l^{-1}	$\dot{x}' = Yq_s x - a(1 - \theta)x - D(1 - \theta)x$
Labelled biomass, (mol ^{14}C) l^{-1}	$\dot{x}^* = Yq_s^* x - a\theta x - D\theta x$
Total biomass, (mol C) l^{-1}	$\dot{x} = Y(q_s + q_s^*)x - (a + D)x$
Molar fraction of ^{14}C in biomass, dimensionless	$\dot{\theta} = Y[q_s^* - \theta(q_s + q_s^*)], \quad \theta = \frac{x^*}{x^* + x'}$
$^{14}CO_2$, (mol ^{14}C) l^{-1}	$\dot{p}^* = (1 - Y)q_s^* x$
Total CO_2, (mol C) l^{-1}	$\dot{p} = (1 - Y)(q_s^* + q_s)x$
Specific rates of limiting substrate uptake, mol C (mol biomass C)$^{-1}$ (h)$^{-1}$	$q_s = Q_s \frac{s}{K_s + s^* + s}$, $q_s'' = Q_s \frac{s^*}{K_s + s + s^*}$

is fed alternately from two bottles containing unlabelled and labelled ^{14}C(U)-glucose, respectively. Initially the culture was continuously fed with ^{14}C-substrate until steady-state was obtained. The feed source was then quickly switched to the alternate line. Just before the $^{14}C \rightarrow {}^{12}C$ transition, the headspace of the fermenter was isolated (by stopping liquid and air outflow while running the liquid flow inducer) and respiration was recorded as the total of CO_2 and $^{14}CO_2$ increase in the gas phase. During this short-term experiment (less than 2 h) the culture volume was increased by no more than by 5%, and the dilution rate was kept almost constant. Supplementary determination of label in the culture liquid showed that only during the first 5–10 min a notable amount of $^{14}CO_2$ was produced from the oxidation of extracelluar substrate, so that the predominant part of the label originated from cell constituents.

The dynamic data were fitted by the modified chemostat model, which took into account microbial growth on the mixture of labelled and unlabelled substrates (Table 3.2). The main assumptions were as follows: (a) that there is no significant isotope fractionation (due to the short duration of experiment); (b) that ^{12}C and ^{14}C uptake obeys Michaelis kinetics with competitive inhibition terms (competition between ^{12}C- and ^{14}C-substrate for common transporters); and (c) that the products of cell decay enter the substrate pool and are reutilized with the same efficiency as exosubstrate.

To describe the $^{14}C \rightarrow {}^{12}C$ transition we may simplify the complete equation set (Table 3.2) by putting $s_0^* = 0$, $q_s^* \ll q_s$ (no label input), and $\dot{s}^* \approx 0$ (quasi-steady-state approximation justified by direct s^* determinations), then

$$\dot{p}^* = (1 - Y)\, ax\theta\,, \qquad [3.8]$$

$$\dot{\theta} \approx -\, Y\theta q_s = (\mu + a)\theta\,,$$

or after integration

$$p^* = p_0^* + \frac{(1 - Y)\, ax}{\mu + a}\,[1 - e^{-(\mu + a)t}] \approx p_0^* + (1 - Y)axt\,. \qquad [3.9]$$

For the first few hours after the $^{14}C \rightarrow {}^{12}C$ switch (because changes in label fraction θ are not significant), turnover rate a is proportional (but not equivalent!) to the $^{14}CO_2$ evolution rate:

$$a \approx \frac{\dot{p}^*}{(1 - Y)\bar{x}}\,. \qquad [3.10]$$

The calculated a-value was found to be rather high at about $0.02\ h^{-1}$, which accounts for 20% of the actual growth rate at $D = 0.08\ h^{-1}$.

Endogenous respiration was the same following the cessation of the flow of nutrients to the chemostat vessel, as in the growing continuous culture (Figure 3.5). These results were surprising, since the transition to starvation should accelerate the breakdown of cell proteins (Marr *et al.*, 1963). Thus, our results indicate that the turnover of cell polymers can be activated merely by substrate limitation, achieved in chemostat culture at moderate D, and not only by starvation.

Nevertheless, the constancy of turnover rate under limitation and starvation conditions did allow the simplification of the entire technique for the assessment of a. Instead of label substitution we may safely use the measurements of endogenous respiration in chemostat culture after shut-off. For this purpose we employed fermenter type (b) (Figure 3.1) and recorded amperometrically the consumption of dissolved oxygen before and after shutoff (Figure 3.6). The dynamics of the respiration rate v_{resp} was approximated by equation:

$$\begin{aligned} v_{resp} &= k_1 s_1 + k_2 s_2 \\ &= k_1 s_1(0) e^{-k_1 t} + k_2 s_2(0) e^{-k_2 t} = v_1(0) e^{-k_1 t} + v_2(0) e^{-k_2 t}\,, \end{aligned} \qquad [3.11]$$

where s_1 and s_2 represent concentrations of exosubstrate and endogenous oxidizable substrates respectively, and k_1 and k_2 are corresponding rate constants. Extrapolation to zero time (the instant of culture shutoff, when $s_1 = s_1(0) = \bar{s}$ and $s_2 = s_2(0)$) gives the initial respiration rates $v_1(0)$ and $v_2(0)$ which correspond to the oxidation of exosubstrate and cell

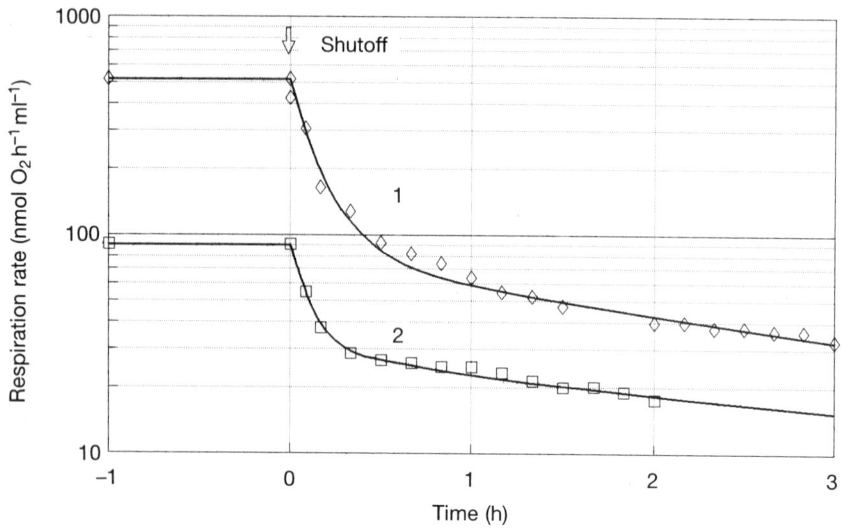

Figure 3.6 Dynamics of oxygen consumption rate in chemostat culture of *D. formicarius* before and after shutoff (↓). Input glucose concentration, $s_0 = 200$ mg glucose l^{-1}. Dilution rates D before shut-off (h^{-1}): (1) 0.252; (2) 0.033. The dynamics of respiration rate was approximated by Eqn. [3.10]. Endogenous respiration was found by extrapolation of the linear part of the curve to $t = 0$ (Panikov, 1984b).

constituents in steady-state culture. The specific rate of endogenous respiration then may be calculated as $q^e_{resp} = v_2(0)/\bar{x}$. An example of q^e_{resp} determination is given in Figure 3.6. The values of q^e_{resp} increased from 0.49 to 0.92 mmol O_2 h^{-1} (g biomass)$^{-1}$ with an increase of $\mu = D$ from 0.033 to 0.252 h^{-1}. The maintenance coefficient, measured by Pirt's method ($m = 0.56$ mmol O_2 h^{-1} (g biomass)$^{-1}$), was close to the low end of the observed range of q^e_{resp} variations. This general principle was confirmed by the other organisms studied (Figure 3.7), i.e. a monotonous but non-linear decrease of q^e_{resp} at low D was always observed.

The effects of limiting substrate concentration on growth and respiration rates

In Chapter 1 we stated that the influence of substrate concentration on microbial growth is complicated because of its dependence on culture prehistory. This is why we studied this central problem of microbial kinetics using a combination of two approaches: (a) the instantaneous response of cells to different glucose amendments was recorded (**short-**

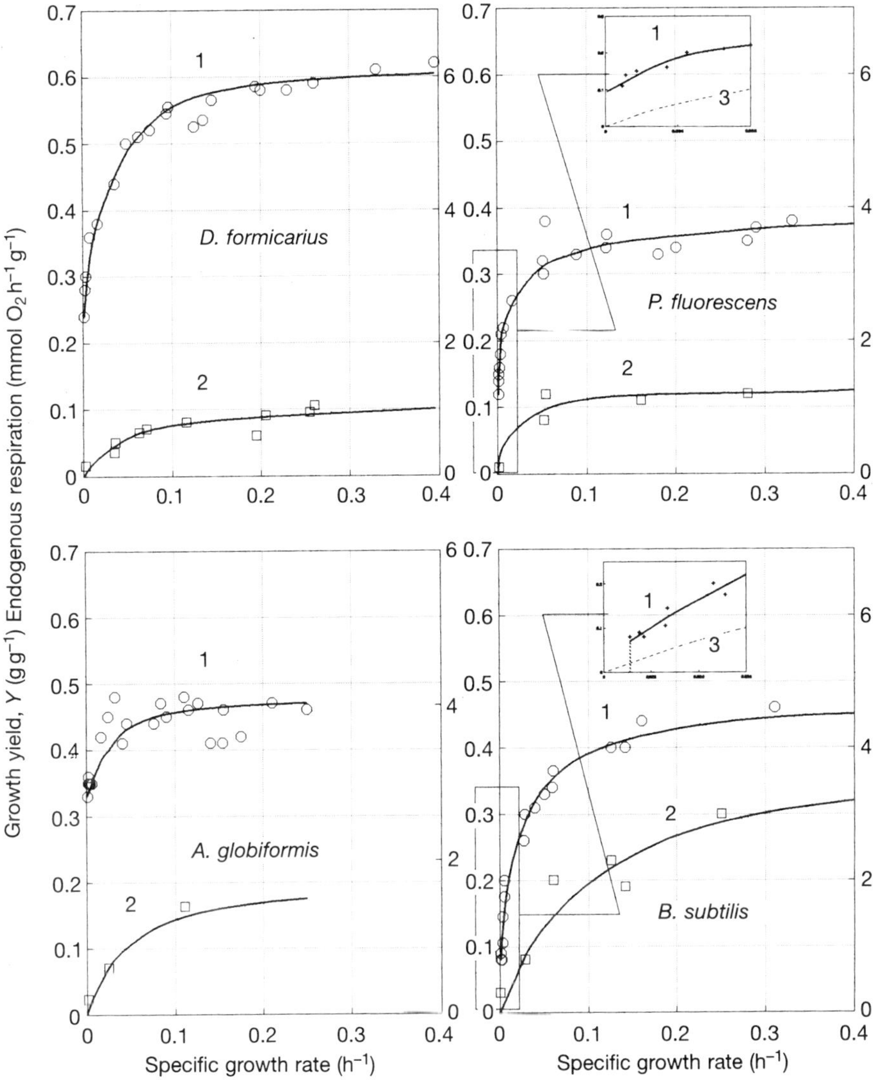

Figure 3.7 Variation of growth yield, Y (curve 1), and endogenous respiration (curve 2) as dependent on dilution rate in chemostat cultures. The curves were calculated using Eqns [3.17]; the dotted line (curve 3) is a prediction of Eqns [1.33–1.35], which is clearly inadequate (Dorofeyev and Panikov, 1988).

term effects); and (b) the residual substrate concentrations and uptake parameters at various D of steady-states chemostat culture was measured **(long-term effects)**.

The performance of the short-term experiments is illustrated by Figure

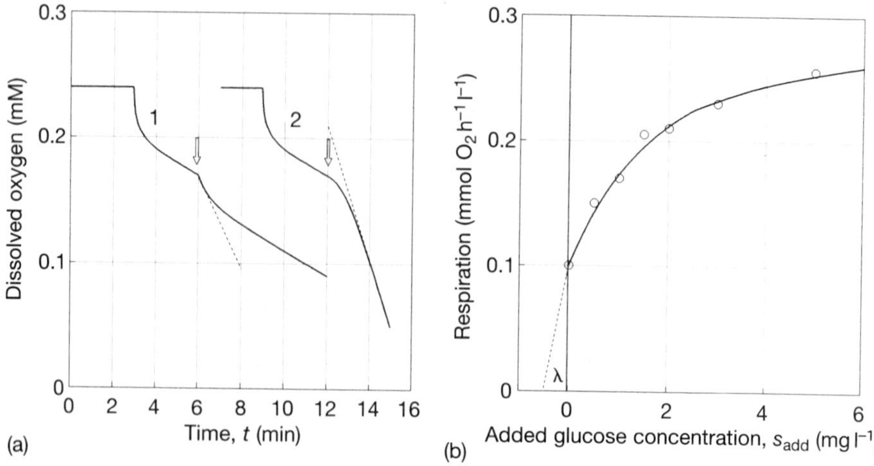

Figure 3.8 Short-term experiments to examine instant response of chemostat culture to amendment with limiting substrate. (a) Illustration of technique: 1 ml of culture liquid was transferred to the Clark cell and dissolved oxygen concentration recorded before and after glucose addition (↓). Amounts of added glucose: (1) 1.0 and (2) 20 μg ml^{-1}. (b) The initial respiration rates, v_{resp}, were estimated from slopes of dotted lines and plotted *versus* concentration of added glucose, s_{add}. The concentration of oxidizable substrate in the culture before amendment (λ) was found by extrapolation using Eqn. [3.11] (Panikov, 1984a).

3.8. In each case, cells sampled from chemostat culture responded to an amendment with accelerated respiration (it was immediate at low amendment rates (s_{add} 0.25–4 mg l^{-1}) or delayed by 1–2 min lag phase at high s_{add} (20–50 mg l^{-1})).

If we take into account the total substrate concentration s (added glucose and residual substrate, $s_{add} + \lambda$), then the response of cells sampled from steady-state culture at one particular D may be adequately approximated by the Michaelis–Menten equation:

$$q_{resp} = Q_{resp}\frac{s_{add} + \lambda}{K_s + s_{add} + \lambda} = Q_{resp}\frac{s}{k_s + s}. \qquad [3.12]$$

However, the parameters of this equation varied with D: the higher the dilution rate the higher were both Q_{resp} and K_s, their ratio Q_{resp}/K_s being more or less constant (Figure 3.9).

The quantitative description and explanation of the D-dependent variation of growth parameters is the main objective of our kinetic analysis. Obviously, the underlying physiological mechanism of this phenomenon is the long-term adaptive adjustment of microbial metabolic machinery to the size of the limiting substrate flux. This adaptation takes place during the

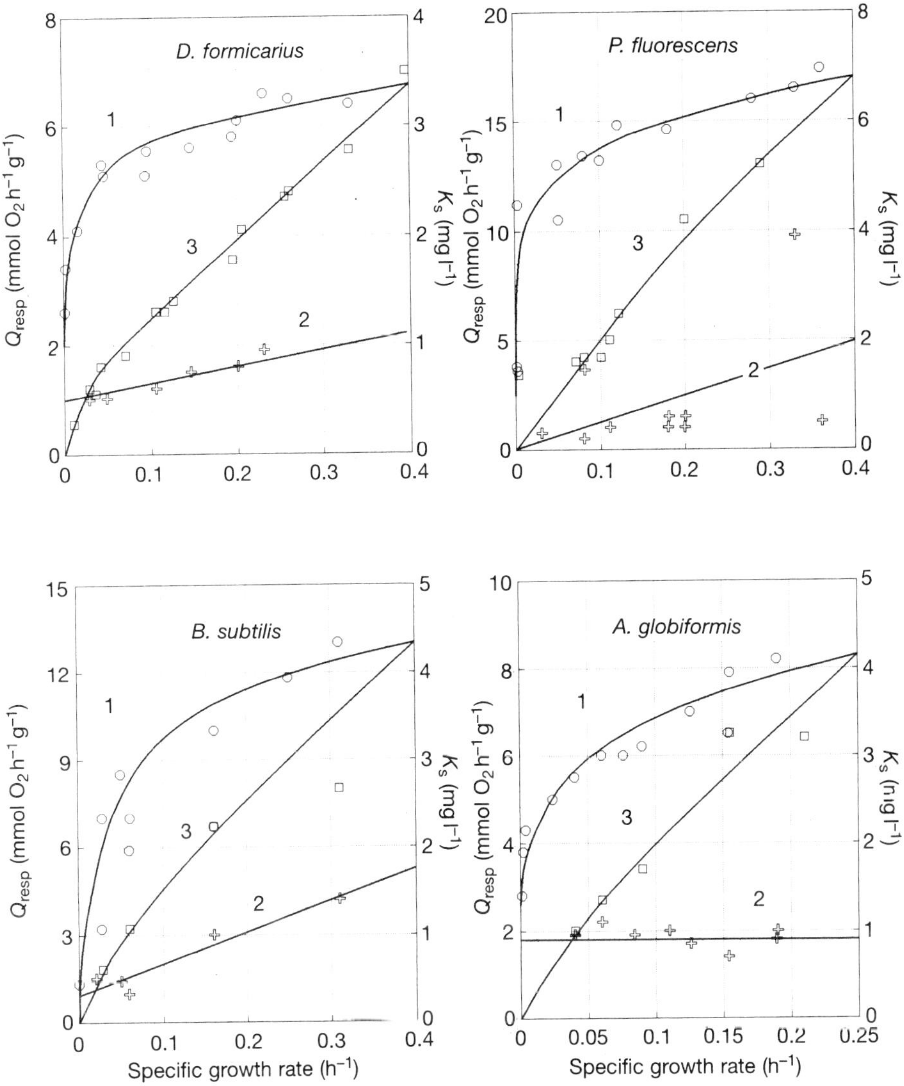

Figure 3.9 Variation of parameters Q_{resp} (curve 1) and K_s (curve 2) estimated in short-term experiments (see Figure 3.8) as dependent on steady-state (quasi-steady-state) specific growth rate. The curves are calculated using Eqns [3.17]. Curve 3 is actual respiration rate measured in the fermenter.

establishment of steady-state in continuous culture and should in fact be related not to $\mu = D$, but to the substrate concentration in the medium $\bar{s}$ (see discussion of this point in section 1.4).

The values of $\bar{s}$ at different D were estimated in the present study by two

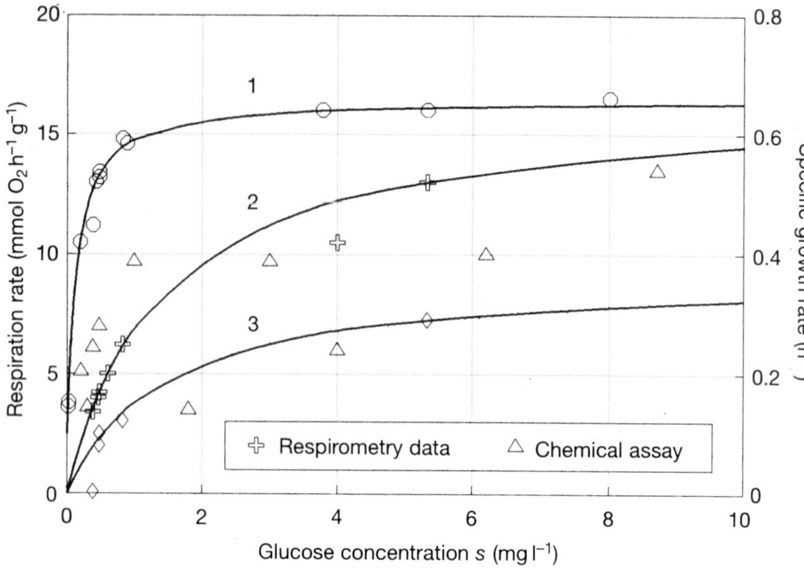

Figure 3.10 Effect of limiting steady-state glucose concentration on the maximal respiration rate exhibited in short-term experiments (curve 1), actual respiration rate (curve 2), and specific growth rate (curve 3). The data for the range of dilution rate in chemostat culture of *P. fluorescens* residual glucose were determined with glucose oxidase or from respirometric assays using Eqn. [3.13]. Note the higher scattering for chemical determinations.

methods: (a) by the chemical determination of residual glucose; and (b) by respirometry and subsequent calculation according to Eqn. [3.12]:

$$\bar{s} = \frac{\bar{q}_{\text{resp}} K_{\text{s}}}{Q_{\text{resp}} - \bar{q}_{\text{resp}}}, \qquad [3.13]$$

where $\bar{q}_{\text{resp}}$ is the specific respiration rate measured by the dynamic method directly in a steady-state chemostat culture at some D-value, while Q_{resp} and K_{s} are parameters estimated in short-term experiments for the same D. This second method is much more reliable than the first, because chemical determinations suffer from many interferences (Figure 3.10). Nonetheless, generally both methods were in fairly good agreement.

The dependence of Q_{resp} on $\bar{s}$ may be approximated by

$$Q_{\text{resp}} = Q^{0}_{\text{resp}} + (Q^{\text{m}}_{\text{resp}} - Q^{0}_{\text{resp}}) \frac{\bar{s}}{K_{\text{Q}} + \bar{s}} = \frac{Q^{0}_{\text{resp}} K_{\text{Q}} + Q^{\text{m}}_{\text{resp}} \bar{s}}{K_{\text{Q}} + \bar{s}}, \qquad [3.14]$$

where Q^{0}_{resp} and $Q^{\text{m}}_{\text{resp}}$ are lower and upper limits of potential respiration activity repectively, and K_{Q} is the saturation constant.

Two other parameters, which are traditionally assumed to be constants, may be expressed through $Q_{\text{resp}}(\bar{s})$:

$$K_s = \frac{Q_{\text{resp}}(\bar{s})}{G}, \quad m = m_0[Q_{\text{resp}}(\bar{s}) - Q^0_{\text{resp}}],$$

$$G = \text{constant}, \quad m_0 = \text{constant}. \qquad [3.15]$$

The biological interpretation of these semiempirical relationships is quite transparent. The first implies the well-known fact that an increase of a cell's metabolic capacity after the partial relief of substrate limitation is accompanied by a decrease of affinity to the substrate (as a result of the *de novo* synthesis of enzymes with a higher K_s). The last relationship suggests that maintenance requirements are proportional to the amount of active cell components rather than to the total biomass.

Thus, in order to describe both long-term and short-term effects of substrate concentration on microbial growth, we need to append the classic Monod–Herbert's chemostat model with algebraic equations [3.14–3.15]. In addition we have to convert respiration rates to rates of substrate uptake. This could be done from a mass balance for oxygen (or CO_2):

microbial respiration q_{resp}	=	combustion of consumed substrate $A_{\text{glu}} q_s$	−	combustion of biomass produced μB	, [3.16]

where A_{glu} and B are chemical oxygen demands for the complete oxidation of 1 g glucose and 1 g of cells respectively. Then we have

$$\dot{s} = D(s_0 - s) - q_s x, \quad q_s = Q_s \frac{s}{K_s + s},$$

$$\dot{x} = (\mu - D)x, \quad \mu = Y^{\max}(q_s - m), \quad m = m_0(Q_s - Q^0_s), \qquad [3.17]$$

$$Q_s = \frac{Q^0_s K_Q + Q^m_s S}{K_Q + s}.$$

Equation set [3.17] is already able to reproduce D-dependent variations of all measured values ($\bar{x}$, $\bar{s}$, Q_{resp}, K_s, m) in chemostat culture under steady-state conditions. A fairly good agreement of calculations with experimental data is demonstrated in Figures 3.7, 3.9 and 3.10. New growth parameters (Table 3.3) are kept constant within the entire range of tested D. However, we are not quite satisfied with the present stage of model development because of its *ad hoc* nature and failure to predict transient phenomena.

Table 3.3 Kinetic parameters of Eqn. [3.17], relevant to entire range of D variation in a chemostat culture

Culture	Q^0_{resp} (mmol O_2 h^{-1} per g biomass)	Q^m_{resp} (mmol O_2 h^{-1} per g biomass)	K_Q (mg l^{-1})	G mmol O_2 h^{-1} (mg glucose)$^{-1}$ per g biomass	m_0 (dimensionless)
D. formicarius	2.0	6.8	0.14	8.3	0.21
L. tetrasporus	0.6	2.9	0.45	3.3	0.38
P. fluorescens	2.5	16.5	0.14	13.0	0.09
B. subtilis	0	17.0	0.35	11.0	0.28
R. erythropolis	0.8	4.2	0.25	5.8	–
A. globiformis	2.6	8.3	0.33	(K_s = 0.7 mg l^{-1})*	0.29

* For *Arthrobacter*, K_s values measured in short-term experiments were found to be constant at all D-values.

Table 3.4 The simplest structured model describing microbial growth limited by source of carbon and energy

Variable	Equation
Flow chart:	
Transport (q_S), respiration (q_{resp}), biosynthesis (q_L), and turnover (a)	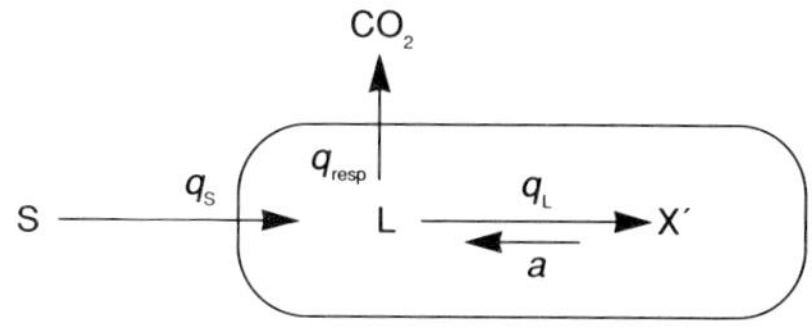
Exosubstrate (mg l^{-1})	$\dot{s} = D(s_0 - s) - q_s x$
Intracellular intermediate (mg l^{-1})	$\dot{l} = q_s x - q_{resp} x - q_L x + ax' - Dl$
Intracellular intermediate (g (g of biomass)$^{-1}$)	$\dot{L} = q_s - q_{resp} - q_L + a(1 - L) - \mu L$ $= q_s - \theta q_{resp} + a(1 - L) - \mu L$ $\theta = \dfrac{q_L + q_{resp}}{q_{resp}}$
Cell polymers (mg l^{-1})	$\dot{x}' = q_L x - ax' - Dx'$
Total biomass (mg l^{-1})	$\dot{x} = l + \dot{x}' = (\mu - D)x, \quad \mu = q_s - q_{resp}$

The dependence of respiration rates on intracellular substrate concentration

Up to this point we have discussed the effects produced by the concentration of limiting substrate in the medium. In reality almost all metabolic processes depend on **intracellular concentrations** of respective substrates, these effects being delineated in so-called **structured** kinetic models. Our main objective is to resolve the problem of the experimental testing of these models.

Let us consider the simplest case of a structured model, which describes cell growth as a conversion of exosubstrate, S, to biopolymers, X′, via intracellular intermediates L (Table 3.4). Intracellular fluxes q_L and q_{resp} are supposed to be immediate functions of L. The chemical nature of L metabolites is well defined and includes amino acids, nucleotides, polycarbonic acids and other low-molecular weight compounds, which are often collectively termed the **intracellular metabolic pool**. In our study, L concentration was estimated (a) chemically as ^{14}C-compounds extracted by cold trichloroacetic acid (TCA) from ^{14}C-labelled cells, and (b) by the novel kinetic technique based on the execution of short-term starvation–enrichment experiments and subsequent calculation of L from a carbon balance (Dorofeyev and Panikov, 1994).

In the starvation–enrichment experiment samples from steady-state chemostat culture were transferred to a 1-ml Clark cell. During the first 20 min of endogenous respiration, the uptake of dissolved O_2 was recorded. Then, 10 μl of glucose was added to give a final concentration of 5 mg l^{-1} and measurements were continued until the respiration rate approached the former endogenous level (usually 15–30 min). In identical experiments ^{14}C(U)-glucose was used to follow the dynamics of substrate consumption. Every 1–2 min after ^{14}C-glucose addition, 0.1 ml subsamples were analysed for the residual radioactivity of the filtrate (glucose + exometabolites). In some experiments respiration was measured also from the rates of CO_2 and $^{14}CO_2$ evolution by an analysis of the gas phase.

For these particular experimental conditions the complete structured model (Table 3.4) may be significantly simplified. During the first **starvation** phase we have $s \approx 0$, $q_s \approx 0$, $dL/dt \approx 0$, and $L = \hat{L} \approx$ constant, thus $a \approx \theta q^{e}_{resp}$. The addition of a small pulse of glucose leads to an instantaneous increase in limiting substrate concentration from $s \sim 0$ to some positive value $s(0)$. Since $s \ll x/Y$ and $L \ll 1$, we may put $\mu L \approx 0$ (neglecting microbial biomass increase, and L dilution due to cell expansion) and $a(1 - L) \approx a$. Thus, we have for the second *enrichment* phase

$$\dot{L} \approx q_s - \theta q_{resp} + a \qquad [3.18]$$

or after integration

$$\Delta L(t) = L(t) - \hat{L} = \frac{\Delta s(t) - \theta[\Delta p(t) - \Delta p^{e}(t)]}{x}, \qquad [3.19]$$

where ΔL is the intracellular substrate concentration in excess of the quasi-steady-state starvation value $\hat{L}$, $\Delta s(t)$ is the exosubstrate consumed, $\Delta p(t)$ is the total amount of CO_2 produced during the transient, and $\Delta p^{e}(t)$ is CO_2 produced from endogenous respiration.

Conversion of CO_2 production to oxygen consumption is straightforward: $\Delta o(t) = \Delta p(t)/RQ$ and $\Delta o^{e}(t) = \Delta p^{e}(t)/RQ$, where RQ is a respiratory quotient, supposed to be constant. The stoichiometric parameter θ can be determined at an end-point time $t = T$, when both the added substrate and the intracellular pool are practically exhausted:

$$\theta = \frac{s(0)}{\Delta p(T) - \Delta p^{e}(T)} = \frac{s(0)}{RQ[\Delta o(T) - \Delta o^{e}(T)]}. \qquad [3.20]$$

Combining Eqns [3.19–3.20] results in the final expression for ΔL:

$$\Delta L = \frac{s(0)}{x}\left(1 - \frac{\Delta o(t) - \Delta o^{e}(t)}{\Delta o(T) - \Delta o^{e}(T)} - \frac{s(t)}{s(0)}\right) \qquad [3.21]$$

Figure 3.11 illustrates the procedure and shows how ΔL values were found from respirometry and substrate uptake data during starvation–enrichment experiments. Stopping the supply of medium resulted in starvation with an almost constant respiration rate, q^{e}_{resp}. A glucose pulse initiated an instantaneous burst of bacterial activity. However, respiration did not exhibit a strong parallel with substrate uptake. Rather, these two reactions developed in succession, i.e. maximal rates of respiration lagged maximal rates of glucose consumption by ~ 1 min. This was the delay which gave rise to the increase in the concentration of internal substrate.

The values of the stoichiometric parameter θ varied in different individual trials within the range 2.07–2.23 g C/g CO_2-C, which is close to the related parameter $Y_{s/p} \approx 2.1$ g glucose C/g CO_2-C (note that $Y_{s/p} = q_s/q_{resp} = \theta - a(1 - L)q^{e}_{resp}$, so $Y_{s/p} \approx \theta$, since the second term $a(1 - L)q^{e}_{resp} \ll 1$). Thus, the stoichiometry of *P. fluorescens* growth was principally the same under steady-state chemostat conditions as in the short-term starvation–enrichment experiments.

In Figure 3.12, specific respiration rates are plotted versus increments ΔL for several independent starvation–enrichment experiments. The functional relationship was of a hyperbolic type shifted to the left by the value $\hat{L}$, the pool size of starving bacteria. Formally it could be expressed by the modification of the Michaelis–Menten equation analogous to Eqn. [3.11]:

$$q_{resp} = \frac{Q_{resp} L}{K_L + L} = Q_{resp} \frac{\Delta L + \hat{L}}{K_L + \Delta L + \hat{L}}, \qquad [3.22]$$

where L is the total intracellular substrate concentration ($\Delta L + L$), Q_{resp} is the maximal respiration rate, and K_L is the saturation constant. Non-linear regression gave the following numerical values for one particular D ($0.08\ h^{-1}$): $\hat{L} = 0.54$ mg C (g biomass C)$^{-1}$, $K_L = 12.1$ mg C (g biomass C)$^{-1}$, and $Q_{resp} = 29.1$ mmol $O_2\ h^{-1}$ (g biomass C)$^{-1}$.

Equation [3.22] may be used also for resolving the inverse problem, i.e. how to calculate L from respiration rates:

$$L = \frac{q_{resp} K_L}{Q_{resp} - q_{resp}}. \qquad [3.23]$$

This expression enables the instantaneous determination of pool sizes, provided the parameters Q_{resp} and K_L are known. In our experiments, pool sizes varied in the range 0.3–40 mg C (g biomass C)$^{-1}$, i.e. the pool contained only minor cell constituents with an extremely short turnover time.

The comparison of kinetic and direct chemical methods of L determination in the cold TCA-extract (TCA-pool) was made during the transition from steady-state growth to starvation (Figure 3.13). A steady-state TCA-pool was established at the level 23.2 mg C (g biomass C)$^{-1}$. Amino acids

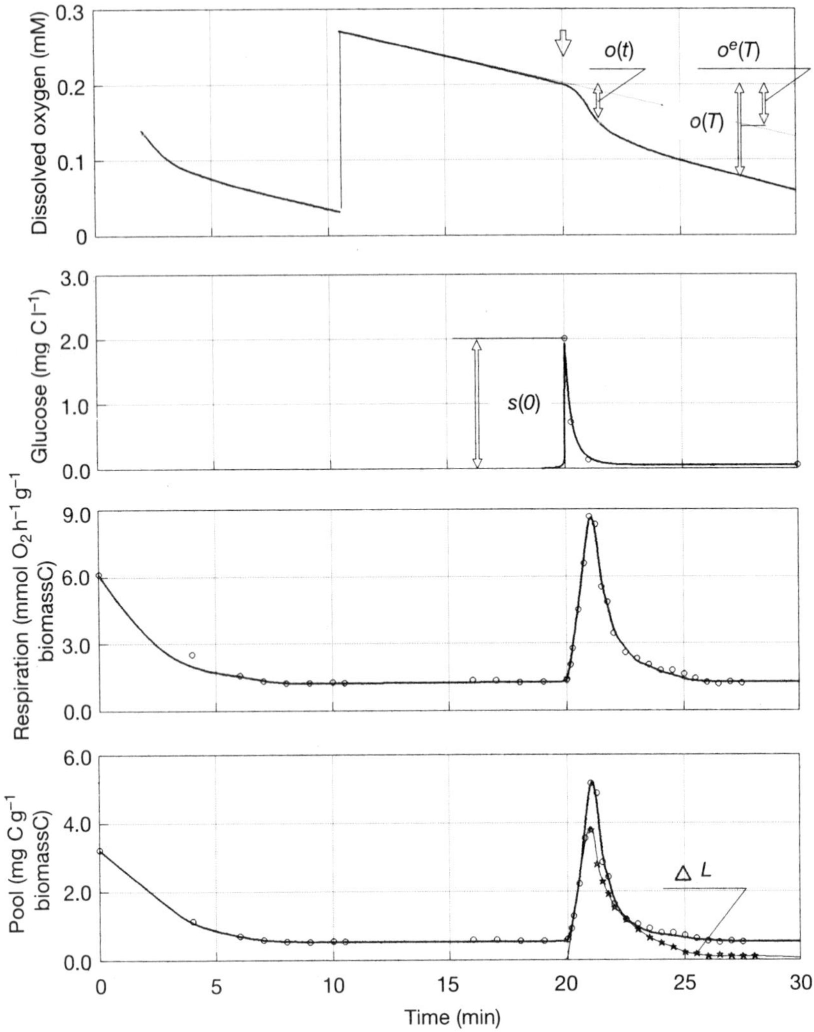

Figure 3.11 Dynamics of starvation–enrichment experiment used for determination of ΔL by dynamic-balance method. Cell suspension was withdrawn at $t = 0$ from *P. fluorescens* chemostat culture ($D = 0.08\ h^{-1}$, $s_0 = 2$ g glucose l^{-1}, $\bar{x} = 344$ mg l^{-1}); then at time indicated by arrow, 2 mg ^{14}C-glucose l^{-1} was added. The following variables were recorded and calculated: dissolved oxygen concentration $[O_2]$ by Clark electrode; residual ^{14}C glucose, s, in culture filtrate; specific oxygen uptake rate, q_{resp}; pool increment, ΔL, calculated from C balance according to Eqn. [3.21]; and total intracellular concentration of oxidizeable substrate, L, calculated according to Eqn. [3.23] (Dorofeyev and Panikov, 1994).

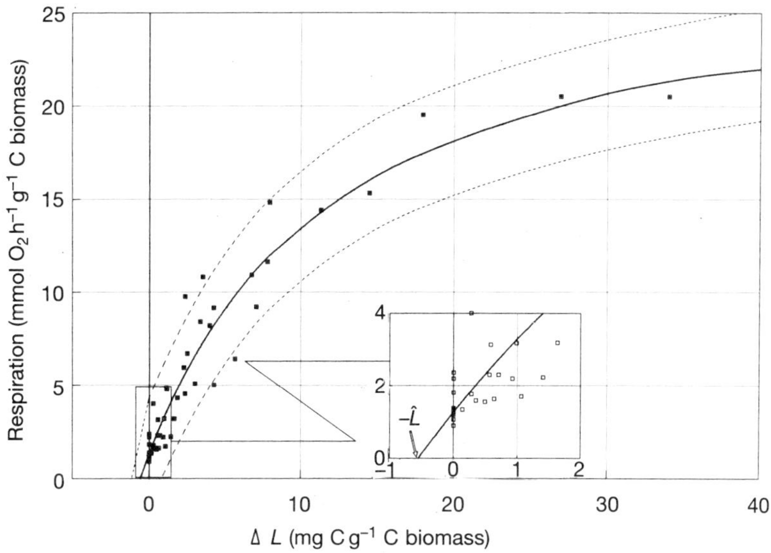

Figure 3.12 Relationship between specific respiration rate, q_{resp}, and increment ΔL of intracellular substrate pool. The data were obtained in short-term starvation–enrichment experiments with *P. fluorescens*; the curve was calculated from Eqn. [3.22]. Extrapolation of the curve to zero respiration rate gives estimation of pool sizes of starving cells before glucose addition (Dorofeyev and Panikov, 1994).

accounted for 3 ± 1 mg C (g biomass C)$^{-1}$ of the TCA-pool, close to the results obtained by Tempest *et al.* (1970) for the chemostat culture of other Gram-negative bacteria. After stopping the substrate feed, the TCA-pool rapidly decreased to 21 mg C (g biomass C)$^{-1}$, and then remained almost constant. A detectable decrease in pool sizes were observed only after long starvation. Hence, the TCA-pool consists of mobile and stable fractions, the former being no more than 10% of the total pool.

L determinations using the dynamic-balance method showed a pattern similar to the one observed for the mobile fraction of the TCA pool. Immediately after starvation, the *L* value fell from the chemostat steady-state value of 3.2 mg C (g biomass C)$^{-1}$ to the starvation background level of 0.4–0.5 mg C (g biomass C)$^{-1}$. From these data we may conclude that cold TCA is able to extract not only true respiratory cell substrates but also some relatively inert cell components. The chemical nature of the latter is obscure, and they may be spurious since TCA treatment could induce degradation of some polymeric cell components. Nevertheless, the close correlation between the kinetic methods and the size of the mobile TCA-pool does substantiate the suggested methodology.

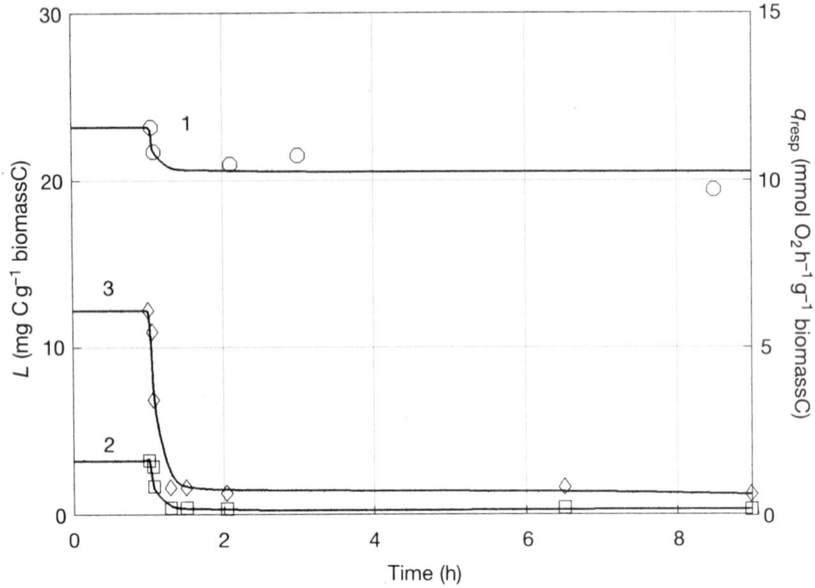

Figure 3.13 Comparison of pool size determination by traditional chemical procedure and by kinetic methods. Chemostat culture of *P. fluorescens* was run at $D = 0.088\ h^{-1}$ with ^{14}C-glucose in input medium. Just before and after shut-off (↓) pool size was determined as ^{14}C count in cold trichloroacetic acid extract (curve 1), and by a kinetic method (curve 2). Curve 3 is dynamics of respiration rate.

3.2.2 DIALYSIS CULTURE

Dynamics of microbial growth

A kinetic model of dialysis culture (Pirt, 1975) was presented in Table 2.2 as a set of two differential equations. If we accept the quasi-steady-state approximation ($ds/dt \approx 0$), and constancy of maintenance coefficient m, then integration of Eqn. [2.3] gives the following expression for x dynamics:

$$x = x_0\, e^{-Y^{max} mt} + \frac{F(s_0 - s_m)}{m}\left(1 - e^{-Y^{max} mt}\right), \qquad [3.24]$$

where x_0 is inoculum size, and $F(s_0 - s_m)$ is the flux of limiting substrate from the dialysis chamber to the microbial culture. According to Eqn. [3.24], biomass growth in dialysis culture is confined by the upper limit $x_\infty = F(s_0 - s_m)/m$, i.e. an increase of x should be accompanied by a progressive retardation of growth and finally by its complete termination when substrate input will totally balance the maintenance ration. It was of great interest to make observations on a real microbial culture as it enters this peculiar physiological state of maintenance (the absence of cell proliferation without loss of their activity).

However, we did experience some disappointment in that the maintenance state was not established in the dialysis culture of *D. formicarius*, the most carefully studied organism in our collection. The observed dynamics of x during the entire 74 days of the experiment showed an almost linear function of time (Figure 3.14). Expected upper limits, x_∞, calculated according to Eqn. [3.24] (m determined in chemostat culture) were surpassed at the very start of continuous growth. The only explanation for these dynamics is that the maintenance requirements of microbial cells may be significantly reduced. Under careful examination we can distinguish three waves of yeast growth. Every wave may be approximated by Eqn. [3.24], provided the m-value is abruptly decreased from one value to another. The reduction in maintenance requirements is correlated with the decrease in potential respiratory activity Q_{resp} measured in samples of culture amended with 1 g glucose l^{-1}.

Physiological transformations were accompanied by morphological changes. Yeast cells decreased in size and became more refractive. The relative amount of viable cells remained high (about 90%), while spores did not account for more than 5–9% of the total population. Thus, a decrease of m did not result in cell death or a transition to a resting state.

Linear growth was observed in semicontinuous dialysis cultures of other organisms (Figure 3.14), but without any waves. Linear approximation of biomass dynamics ($x = x_0 + YFt$) provides the general expression for μ and m (see Eqn. [1.34]):

$$\mu = \frac{FY}{x_0 + FYt}, \quad m = \mu\left(\frac{1}{Y} - \frac{1}{Y^{max}}\right). \qquad [3.25]$$

Linear growth implies that Y is constant and hence the coefficient m should be proportional to μ. The specific growth rate in dialysis culture decreases with cultivation time, approaching as low μ-values as one wants. For example, the specific growth rate of *D. formicarius* (even corrected to take into account non-viable and dormant cells) fell during the 74 days of dialysis cultivation to as low as 0.012 d^{-1} (equivalent to a generation time 58 days!) or 0.2% of μ_m. We did not observe any sign of the phenomenon termed 'minimum growth rate', which took place at low dilution rates of glycerol-limited chemostat culture of *Klebsiella aerogenes* (Tempest *et al.*, 1967). Probably this phenomenon of the dramatic loss of cell viability, as well as a drop in RNA and DNA content, may be caused by some apparatus effects inherent to chemostat but not to dialysis culture (e.g. discontinuity in supply of limiting substrate resulting in the substrate-accelerated death of starving bacteria).

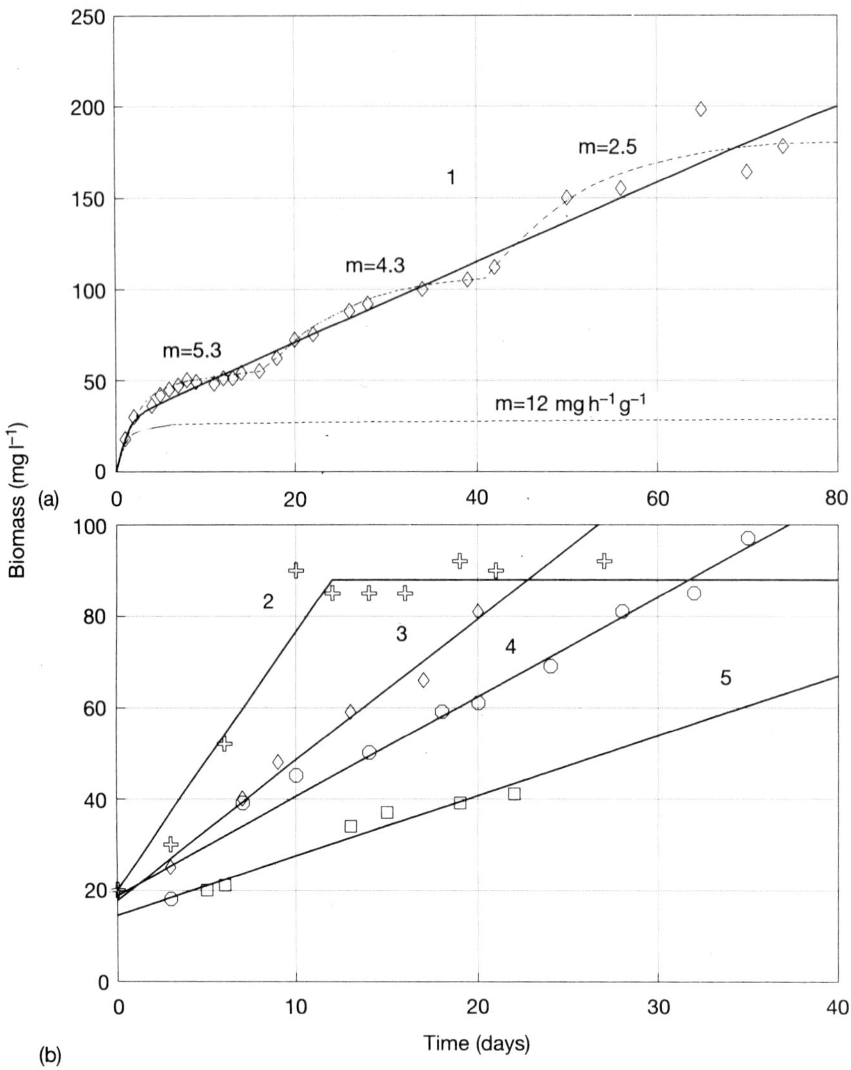

Figure 3.14 Dynamics of microbial growth in dialysis culture (Panikov *et al.*, 1982). (a) Growth of *D. formicarius* in continuous dialysis culture (curve 1). (b) Growth of *B. subtilis* (curve 1), *L. tetrasporus* (curve 2), *A. globiformis* (curve 3), and *P. fluorescens* (curve 4) in semi-continuous dialysis culture. The dotted curves were calculated according to Eqn. [3.24] with constant maintenance coefficient $m = 12$ mg glucose h^{-1} (g biomass)$^{-1}$ or by assuming that the m-value was abruptly decreased from one value to another.

Maintenance state in dialysis culture of *Bacillus subtilis*

The only exception from the general principles among the studied organisms was displayed by spore-forming bacteria (Figure 3.14, curve 2). In the first week of dialysis cultivation the growth of bacilli was linear. It then stopped abruptly at some critical biomass concentration x^*. Subsequently the culture density remained constant despite the glucose that was consumed and oxidized at the same rate. The non-replicating population was represented mainly by vegetative cells (the spore fraction did not exceed 10%), which retained viability and produced normal colonies when plated on meat–peptone agar. However, their morphology was significantly distorted with typical bacillar rods converted to shorter rods and almost spherical cells and numerous swollen cells with irregular shape. It is notable that these cells generally lost the ability to produce normal spores and after the cessation of the flow of substrate underwent rapid autolysis. The same type of dynamic behaviour was displayed not only by *B. subtilis*, but also other representatives of this genera, *B. mesentericus* and *B. cereus*.

Thus, in the case of *Bacillus* the maintenance state does take place in reality. However, it is most unusual. Equation [3.24] is evidently not appropriate to describe this phenomenon since it predicts smooth and monotonous approaching of the upper asymptote x_∞, whereas the actual x dynamics is abrupt with an inflection point at $x = x^*$. Mechanistic explanation and mathematical simulation of bacilli growth will be given later.

3.2.3 FED-BATCH CULTURE (FBC)

The dynamics of microbial biomass is shown in Figure 3.15. Biomass and limiting substrate were expressed in two ways: as the mass amount per entire culture vessel (X, S, mg), and as mass concentration per unit of culture volume (x, s, mg l^{-1}). Differential equations for all assigned entities were derived earlier (Table 2.2, Eqns [2.5] and [2.4]). The principal distinction of this model from that of the chemostat is that the dilution rate D is a time-dependent variable. In our experiments it decreased from 0.01 h^{-1} at $t = 0$ to as low as 0.0008–0.001 h^{-1} some 30–40 days later. The resulting rates of change of x and s were much lower than true rates of microbial growth and substrate consumption: $\dot{x} \ll \mu x$, and $\dot{s} \ll q_s x$. This implies that $\dot{x} \approx 0$, $\dot{s} \approx 0$ and $\mu \approx D$, i.e. fed-batch culture is characterized not by a true but at least a quasi-steady-state.

The principal difference from dialysis culture is that the volume of fed-batch culture is not constant; however, this feature is not essential if we are using variables X and S. The dynamics of x in dialysis culture and X in fed-batch culture are practically identical (compare Figures 3.14 and 3.15).

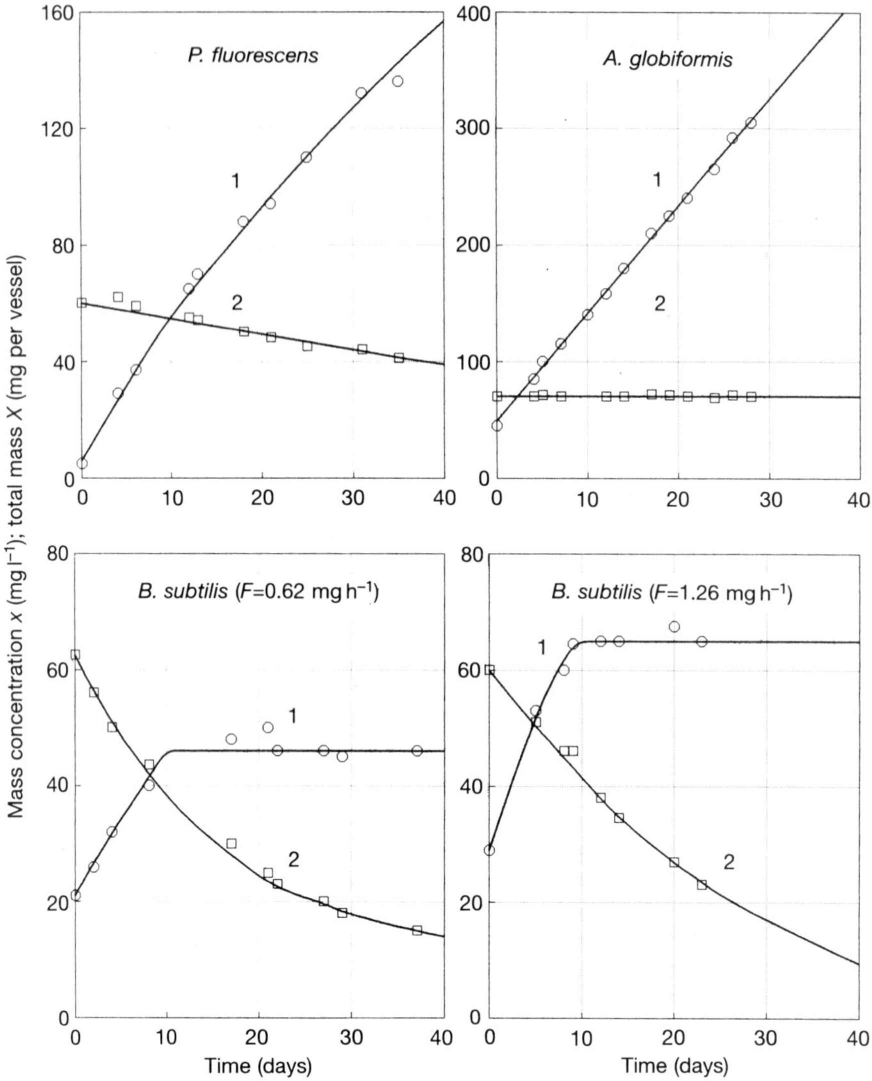

Figure 3.15 Microbial growth in fed-batch culture of *D. formicarius*, *P. fluorescens*, *A. globiformis* and *B. subtilis* (Dorofeyev and Panikov, 1988). Biomass is expressed as a total amount of cell mass in fermenter, X (curve 1), or as a biomass concentration per unit of culture volume, x, (curve 2). Substrate input rate, F, was 0.96 and 1.12 mg h^{-1} for 1st and 2nd organisms respectively. The curves were calculated according to model [3.17], dilution rate was given as $D = D_0/(1+D_0t)$ (D_0 is dilution rate at $t = 0$).

Other common features of the two cultivation systems are as follows: (a) there is no elimination (wash-out) of biomass and as a result, (b) the specific growth rate is reduced to extremely low values ($\ll 0.001\,h^{-1}$) during cultivation; (c) the growth of spore-forming bacteria is initially linear, and is then abruptly terminated.

At the same time the dynamic patterns of microbial growth in fed-batch and dialysis cultures were differentiated with respect to some subtle features. Firstly, it is possible to see curvature in the X dynamics in fed-batch culture (probably due to more accurate biomass measuring). This curvature originates from the absence of strong proportionality between m and μ (see Eqn. [3.25]), i.e. the reduction of maintenance requirements was slightly delayed as compared with the retardation of the specific growth rate. Another principal difference is that the microbial population in fed-batch culture remains diluted all of the time, so there are no negative cell-to-cell interactions or accumulation of non-dialysable metabolic products. This favourable property allows continuation of the discussion of the phenomenon of growth arrest in the culture of bacilli. We tested two different rates of substrate input, F_s, and each provided two significantly different threshold levels, x^*. It is interesting that the ratio F_s/x^* remained more or less constant at $1.62\,h^{-1}$ at various feed rates and in both cultivation systems. Thus, cessation of growth takes place in *Bacillus* culture independent of population density and product accumulation. Therefore, from the two possible informational signals (decrease of s and accumulation of metabolic products) the first should be the more important.

Cytomorphology of slowly growing microorganisms

Similar to dialysis culture, we observed the degeneration of cells into smaller units and the emergence of refractive forms. *Arthrobacter* and *Lipomyces* formed stable aggregates of 10–50 cells. The cells of *P. fluorescens* in the late stages of fed-batch culture lost their motility, but it was restored immediately after the addition of glucose. Viability of all organisms was about 90% or higher and the contribution of spores (endospores of *Bacillus* and ascospores of yeasts) to the total biomass of population was no more than 10–15%. The general principle of a decrease in cell size did not hold in the case of *L. tetrasporus* in that their cells became larger in the fed-batch culture. Vegetative cells of *B. subtilis* were small and aggregated into conglomerates surrounded by a common envelope. A small fraction of cells inside the aggregates had undergone autolysis while the remainder displayed a normal intracellular structure (Soina *et al.*, 1984).

Yield variations

The relationship between Y and μ is presented in Figure 3.7 together with the results obtained for chemostat culture. Y-values were calculated as the ratio $Y = \mu X/Fs_0$, and μ was estimated as the sum $\mu = D + \dot{x}/x$, where the derivative $\dot{x}$ was found from the slopes of the curves $\ln(x)$ *versus* time.

Pooled data obtained by all three cultivation techniques (chemostat, dialysis and fed-batch cultures) could be approximated by one and the same functional relationship which stems from a modification of the chemostat model [3.17]. Contrary to the classic chemostat model of Monod–Herbert, it implies that maintenance requirements are proportional to the amount of active cell components (expressed as product 'biomass' × 'potential metabolic activity' = $x \cdot Q_{resp}$) rather than to total biomass. This simple and natural assumption was sufficient to describe the yield variation at any D of chemostat culture, as well as at any time of dialysis and fed-batch culture under extremely severe growth limitation (generation times extend up to 2 months). The important consequence of this chemostat model modification was that at low $D=\mu$, biomass yield Y decreases not to zero, as predicted by Pirt's Eqn. [1.33], but to some positive value. The differences between the two models are shown on the inserts of Figure 3.7.

The influence of limiting substrate concentration on microbial growth rate

Residual glucose concentration in fed-batch culture was so low that it was impossible to determine it by available chemical procedures. Hence a kinetic approach was used. In short-term experiments the respiration response of microbial cells to a range of glucose amendments was measured, the Michaelis parameters K_s and Q_{resp} were estimated and then values of s were then calculated according to [3.13] using independent estimates of the culture respiration rate. The obtained results were in fairly good agreement with the modified chemostat model [3.17] (Figure 3.16). In the case of *B. subtilis* it was assumed that at some threshold substrate concentration, s^*, maintenance requirements were abruptly raised, this event corresponding to the cessation of bacilli growth and a drop of μ to zero. Other organisms had positive growth rates for all $s > 0$. Note that conventional kinetic equations which contain maintenance terms ([1.54] and [1.55]), give a basically different prediction of the $\mu(s)$ relationship in the range of very low s. They produce hyperbolic curves which cross the abscissa at $s_{min} = mY^{max}K_s/(\mu_m - mY^{max})$, implying a monotonous decrease of μ in the s range 0–s_{min} from zero to $-mY^{max}$. This prediction was not confirmed in any of our experiments. Even for *Bacillus* culture Eqns [1.54] and [1.55] did not hold since the observed $\mu(s)$ dependence was discontinuous at the point $s = s^*$, while the classical model requires a smooth monotonous functional relationship.

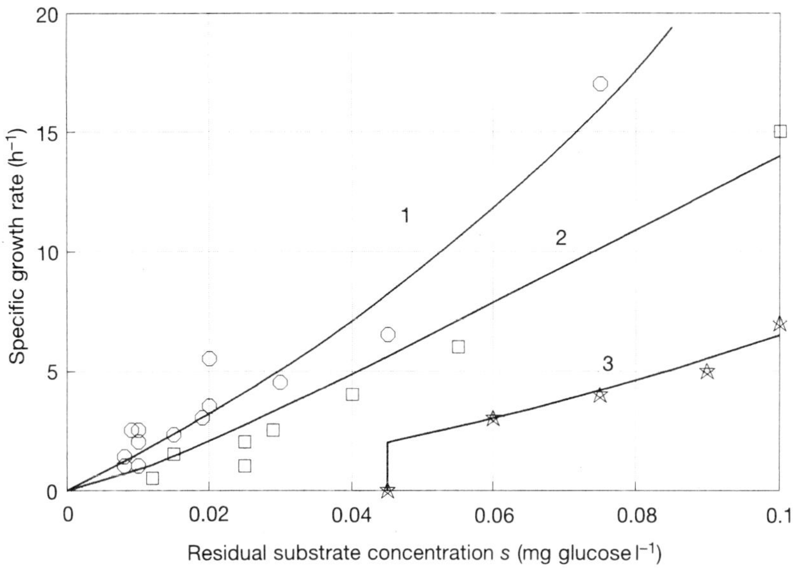

Figure 3.16 The effects of extremely low concentrations of limiting substrate on specific growth rate of *A. globiformis* (curve 1), *P. fluorescens* (curve 2) and *B. subtilis* (curve 3) (Panikov *et al.*, 1987). Bacteria were grown in fed-batch culture, μ-values were determined from biomass dynamics ($\mu = D + \dot{x}/x$), while residual substrate concentration, s, was calculated from respirometry data according to Eqn. [3.13]. The curves are fitted by Eqns [3.17]. In the case of *B. subtilis* the model was appended by the following condition: $m = m_0(Q_{resp} - Q^0_{resp}) + M$, where $M > 0$ if $s < s^*$, otherwise $M = 0$.

3.2.4 BATCH CULTURE

The lag phase, or phase of positive growth acceleration

When a culture medium is seeded with a small inoculum, and there is non-inhibitory excess of all nutrients and a minimum content of metabolic products, then we observe the so-called lag phase preceding the exponential growth of a batch culture. The dynamic pattern of microbial growth during this phase is determined mainly by the physiological state of the inoculum. To achieve the required degree of precision and reproducibility, we used as an inoculum microbial cells of perfectly defined physiological state which were grown under steady-state conditions (chemostat culture at various D) or quasi-steady-state conditions (fed-batch culture and long-term starving batch culture, the physiological state of which will be discussed below). Immediately after inoculation we followed the dynamics of biomass x by nephelometry as well as by the residual glucose, s, and respiratory activity, v_{resp}. There were two distinct groups of organisms differentiated with respect to the observed dynamic patterns.

1. Batch cultures of *Bacillus*, *Pseudomonas* and *Debaryomyces* displayed a clear-cut inverse relationship between the duration of lag phase and the growth rate of inoculum (Figures 3.17 and 3.18). The batch growth devaited significantly from a simple exponential pattern, when inoculum was taken from starving or greatly limited cultures. The higher the growth rate of the inoculum the shorter was the lag phase. During this phase the specific growth rate approached a maximum value, μ_m, the same for a particular species under all initial conditions. Dynamics of respiration rates were also described by monotonous smooth curves, exhibiting a parallel increase of biomass x and specific respiration activity, Q_{resp}.

2. Batch cultures of *Arthrobacter, Lipomyces* and *Rhodococcus* displayed biphasic growth (Figure 3.19). Immediately after inoculation there was very intensive consumption of glucose accompanied by a biomass increase. These processes subsequently slowed down until after 5–10 h the growth acceleration resumed. The second phase was basically identical with the main batch process of other organisms, i.e. growth acceleration was closely correlated with the preceding culture growth rates. However, the rate of transition was much lower, e.g. the adjustment of growth rate to its maximum level took several days in a culture of *A. globiformis versus* several hours in cultures of *P. fluorescens* and *B. subtilis*. The most intriguing phase is the first. Probably it was not true growth in the common sense, but no more than a refilling with exosubstrate (glucose) of some cell storage compartments. Hence, at this stage, the increase of cell biomass was mainly in the form of reserve compounds such as polysaccharides or cell wall components (Monakhova, 1984).

The last assumption was supported by the data on the dynamics of specific metabolic rates during the first phase of *Arthrobacter* batch culture (see insert in Figure 3.19): the respiration rate, Q_{resp}, was rather low (< 6 mmol $O_2\,h^{-1}$ (g biomass)$^{-1}$, while specific rates of glucose uptake and growth ($Q_s = 1.3\,g\,h^{-1}$ (g biomass^{-1}) and $\mu = 0.85\,h^{-1}$) were abnormally high. From the ratio Q_s/Q_{resp} we may easily calculate that at the start of the growth only a small fraction of glucose (less than 13%) was really oxidized, the remainder being incorporated into the cell. Subsequently, over an 8-hour period, the values of Q_s and μ rapidly dropped to low levels, an effect which corresponds to the normal physiological reaction of these bacteria, while Q_{resp} increased, exhibiting an intensification of balanced growth. The described phenomenon was called **false diauxie**, because it resembles the visible change from rapid to slow growth phases when a microbial culture is grown on a mixture of two substrates. The first phase of false diauxie (the refilling of the storage compartments) was well expressed in *Arthrobacter* cells when the inoculum was grown at some intermediate growth rate 0.01–0.05 h^{-1}. Quantitative interpretation and

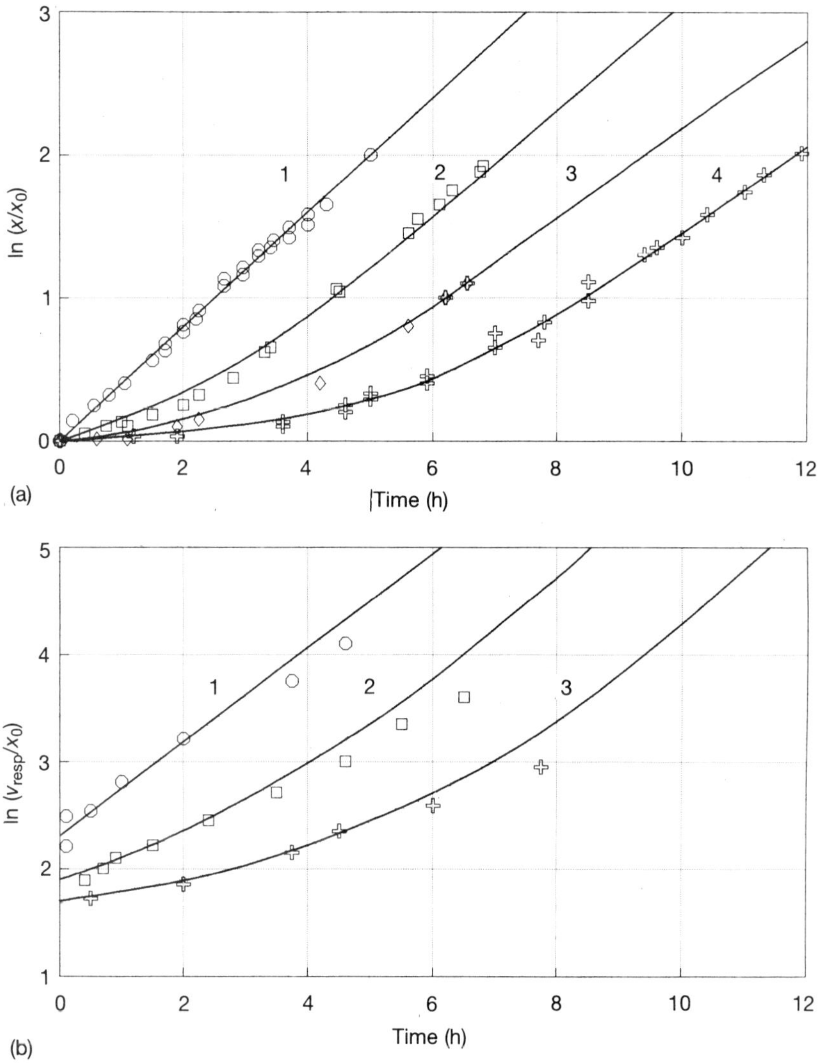

Figure 3.17 Batch growth of *P. fluorescens* as dependent on physiological state of inoculum (Dorofeyev and Panikov, 1991a). (a) Dynamics of biomass and (b) respiration rate. Cultivation conditions: s_0 500 mg glucose l^{-1}, inoculum size, x_0, 20 mg l^{-1}. To prevent growth limitation cultural liquid was diluted by fresh medium each time residual substrate concentration fell below 100 mg l^{-1}. Specific growth rates of inoculum (h^{-1}): 0.125 (curve 1), 0.0057 (curve 2), 0.0013 (curve 3), ~ 0 (curve 4, starving culture). The curves are calculated using a reduced synthetic chemostat model (SCM) (see below).

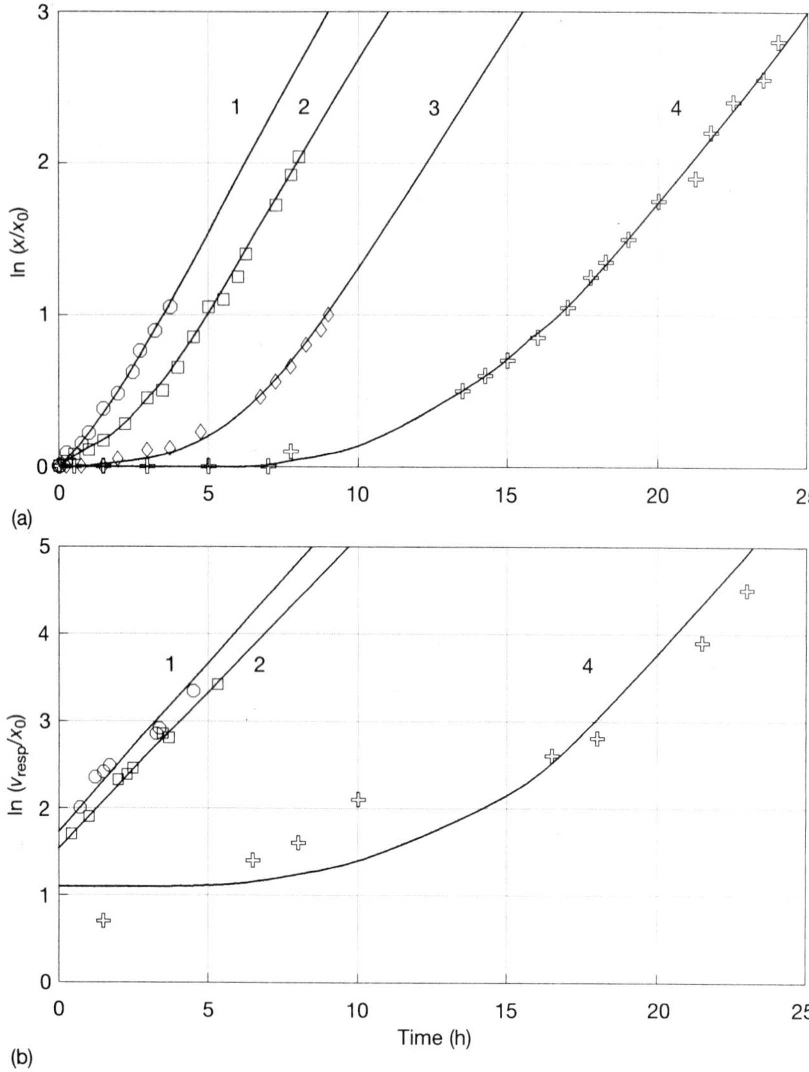

Figure 3.18 Batch growth of *B. subtilis* as dependent on physiological state of inoculum: (a) dynamics of biomass and (b) respiration rate. For cultivation conditions, see Figure 3.17. Specific growth rates of inoculum (h^{-1}): 0.25 (curve 1), 0.14 (curve 2), 0.03 (curve 3), ~0 (curve 4, starving culture). The curves are calculated using a reduced synthetic chemostat model (SCM) (see below).

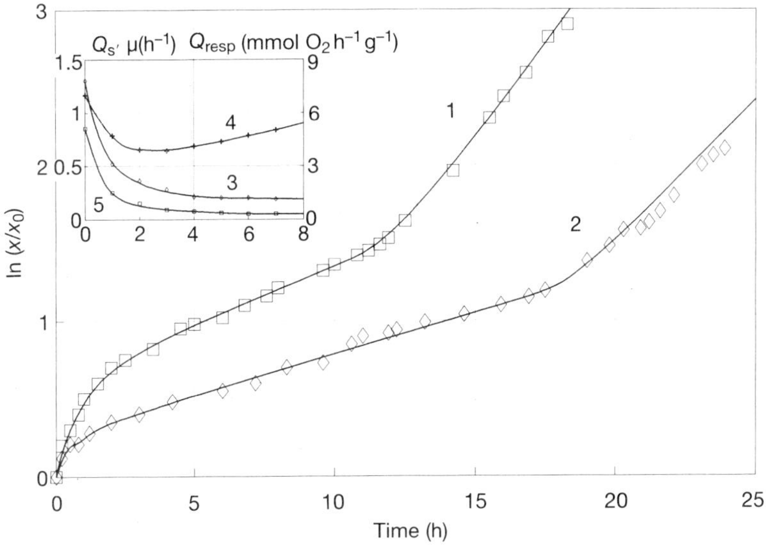

Figure 3.19 The batch growth of *A. globiformis* as dependent on physiological state of inoculum (Dorofeyev *et al.*, 1992). For cultivation conditions, see Figure 3.17. Specific growth rates of inoculum (h^{-1}): 0.08 (curve 1), 0 (curve 2, starving culture). The curves are calculated according to a modified synthetic chemostat model (SCM), Eqns [3.73–3.76]. The insert shows the first phase of *A. globiformis* biphasic growth: specific rates of respiration, Q_{resp} (curve 3), glucose uptake, Q_s (curve 4), and growth, μ (curve 5). The inocula growth rate was $0.05\ h^{-1}$.

simulation of lag phase is completely impossible within the known models of microbial kinetics. For these purposes we will use simple but effective structured models (see section 3.6).

The phases of exponential growth and negative acceleration of growth

These phases are commonly considered to be the simplest. Exponential growth with a constant specific rate $\mu = \mu_m$ and its deceleration caused by substrate exhaustion is simulated by the Monod model and for this particular case there is even an analytical solution in the form of Eqn. [1.27]. We attempted to use this equation for the simulation of batch growth with active inoculum, i.e. when lag phase is practically absent. An obvious deviation from our prediction was found for *Arthrobacter* and *Rhodococcus* cultures. In the case of other organisms Eqn. [1.27] was formally applicable, but there was one serious discrepancy. The parameter K_s identified from batch culture was one order of magnitude higher than the parameter obtained from chemostat data. Moreover, the saturation constant appeared to be dependent on s_0, in that the lower s_0 became, the closer was the apparent saturation constant K_s^{app} to the K_s value obtained in chemostat. Such a discrepancy with the Monod model could be explained by the combined effects of limitation and inhibition, i.e. the deceleration of batch culture growth is actually caused not only by substrate exhaustion but also by competitive product inhibition:

$$K_s^{app} = K_s\left(1 + \frac{p}{K_p}\right), \quad p = p_0 + Y_{p/x}(x - x_0), \qquad [3.26]$$

where K_p is the inhibition constant and p is the concentration of inhibitory exometabolites, which is assumed to be proportional to biomass increase. It is interesting to note that under some realistic assumptions ($x \gg x_0$, $Y_{p/x}/K_p \gg 1$) the combination of [1.27] and [3.26] produce the Contois equation [1.52], which is commonly suspected to be not justified theoretically (Pirt, 1975; Powell, 1967).

The death phase

After depletion of limiting substrate, which is the source of carbon and energy, the cells enter a starvation–survival state. The study of this phase is very important for the understanding of microbial ecology and for the development of new biotechnologies (Postgate and Gray, 1976; Nelson and Parkinson, 1978; Morita, 1982, 1990). Over the last decade these studies have been significantly advanced due to the progress in molecular biology and genetics. As a result new hypotheses have been put forward, which describe molecular mechanisms responsible for survival (Reeve *et al.*, 1984; Kaprelyants *et al.*, 1993). However, there is a serious gap in the research at the level of whole cells or cell populations, which may only be filled within the framework of a kinetic approach.

Unfortunately, the kinetics of microbial starvation–survival is still in its juvenile stages. One of the reasons is the absence of an adequate quantitative technique. So here we will start by considering methodological problems and will suggest a system of dynamic variables which permits the integrated interpretation of major starvation–survival events.

Dynamics of batch culture before and after complete depletion of glucose was followed by measuring the following variables:

- the total concentration of solids, x, by dichromate and dry weight determinations;
- the concentration of viable cells, x_{via}, by plating with the parallel measurement of CFU sizes (Eqn. [3.1]) and from the ratio OD_{600}/x (Eqn. [3.4]);
- the necromass (concentration of dead cells), $x_{necro} = \theta x_{via}/(1 - \theta)$, where θ is the fraction of non-growing cells determined by an agar film technique;

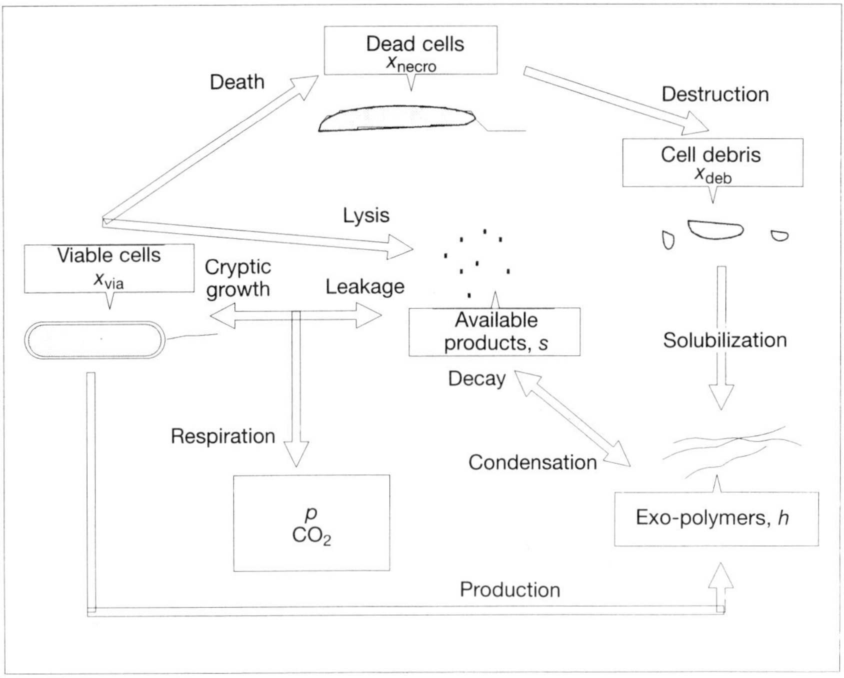

Figure 3.20 Flowchart of elementary processes in starving microbial culture.

- the concentration of cell debris (cell fragments and their secondary aggregates, which are not ascribed as cells), $x_{deb} = x - x_{via} - x_{necro}$;
- the concentration of soluble organic matter (SOM), which consists of readily available components, s, and refractive exometabolites (such as glycocalyces), h;
- the products of microbial respiration, p (CO_2+HCO_3^-);
- the potential respiration rate, Q_{resp}, which is determined in the presence of added glucose;
- the RNA content in the cells. RNA and Q_{resp} are parameters characterizing physiological state of the population.

The interrelationship between all the assigned variables is shown in Figure 3.20. It is essential that in a starving microbial culture destructive reactions (death, lysis, physical destruction) are accompanied by biosynthetic processes (cryptic cell growth, coupled respiration, synthesis of glycocalyces and other exometabolites), the latter effects being responsible for population survival. Biosynthetic reactions in starving cells are concealed by net decay. The most essential and rewarding objective of kinetic analysis is the estimation of true (as opposed to apparent) rates of all essential 'elementary' processes. As an example of such an analysis, we will consider a starving batch culture of *P. fluorescens* (Figure 3.21).

Two methods of x_{via} measurement gave practically the same results. The fraction of viable biomass x_{via}/x fell from 0.97 at $t = 0$ to 0.17 at $t = 38$ days. The fraction of non-growing cells estimated by an agar-film technique, θ, increases over the same period from 0.05 to 0.4. Thus, we may calculate necromass and detritus concentrations at $t = 38$ days: $x_{necro} = 0.4 \times 16/(1 - 0.4) = 10.7\ \text{mg l}^{-1}$ and $x_{det} = 94 - (10.7 + 16) = 67\ \text{mg l}^{-1}$, where 16 and 94 are the values of x_{via} and x respectively. Thus, in the long-term starving batch culture the bulk of the solids was represented by detritus, while the amounts of viable and dead cells were roughly the same. The concentration of SOM and amount of $CO_2 + HCO_3^-$ increased during starvation with progressively diminishing rates. The real respiration activity (which is in fact endogenous respiration, not to be confused with potential respiration, Q_{resp}!) fell from 18 μmol CO_2 $\text{h}^{-1}\,\text{l}^{-1}$ at the onset of starvation to 0.43 μmol CO_2 $\text{h}^{-1}\,\text{l}^{-1}$ after 1 month. The accuracy of the measurements was confirmed by the convergence of C balance.

Both indexes of physiological state, Q_{resp} and cell RNA content, fell drastically during the first 5 days and then stabilized at about 20% of their initial values. There are two ways to interpret this observation. The first is as follows: some part of the starving population undergoes lysis and death, while surviving cells adjust their metabolic machinery to starving conditions during so-called cryptic growth. This adaptation is accompanied by synthesis of some new cell constituents and a reduction of those which were required for intensive growth (i.e. RNA).

The second interpretation (Postgate and Gray, 1976) states that starving cells endogenously utilize their own components (RNA, proteins, polysaccharides) to meet maintenance requirements. This concept dominates the literature. However, we discern here some internal contradictions which stem from the straightforward identification of a single microbial cell as a starving multicellular organism. If that was really the case then the decline in x would not be accompanied by a decrease of cell number, starvation would be exhibited as 'emaciation' but not extinction of individuals. However, the experimental data contradicts this prediction (see curves 1 and 2, Figure 3.21, showing a parallel decrease of biomass measured dichromate oxidation and plating). More relevant is the analogy between a **population** of starving bacteria and a single primitive metazoan macroorganism deprived of food. In both cases some cells are sacrificed to be used as nutrients for other surviving cells.

Survival dynamics of *A. globiformis* (Figure 3.22) was characterized by the same pattern as *P. fluorescens*, the only quantitative difference being that survival was longer by several orders of magnitude. Better survival was correlated with a more rapid change in the parameters of physiological

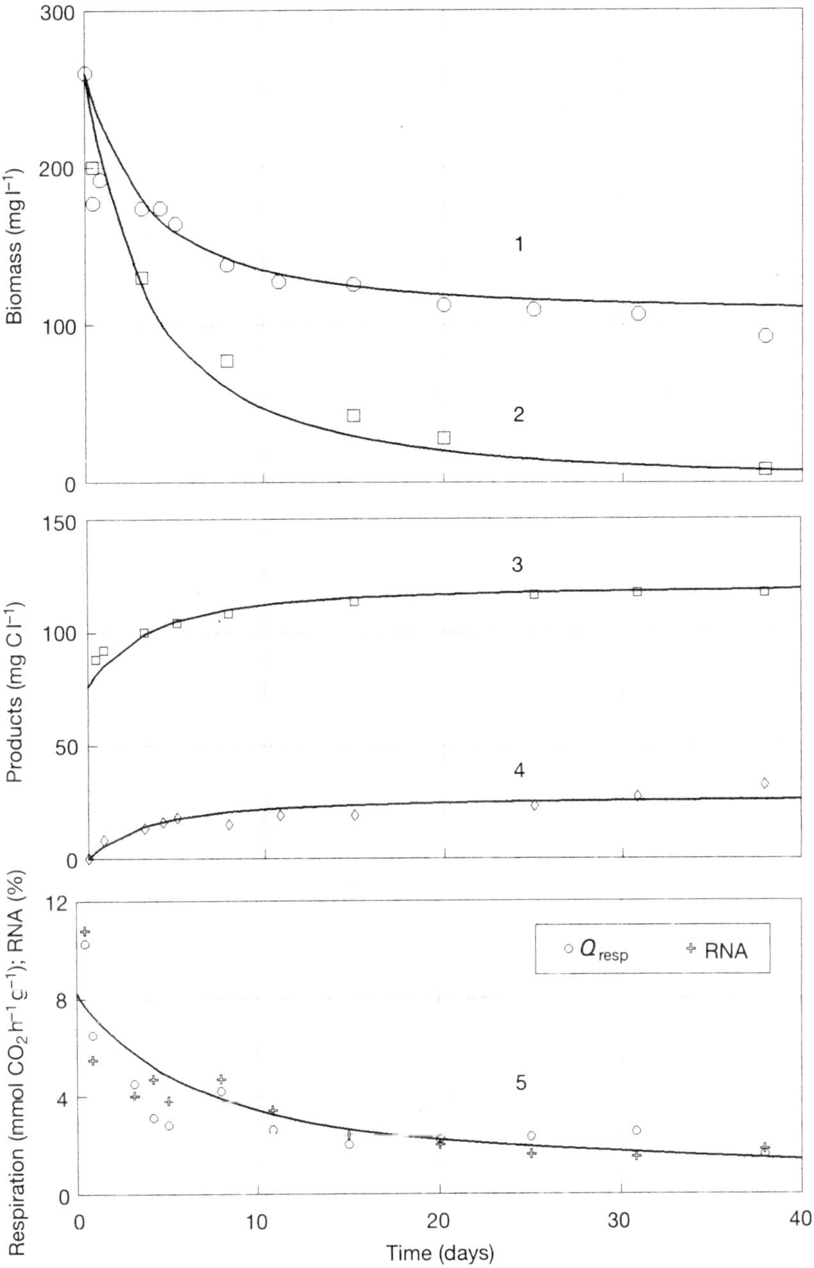

Figure 3.21 Survival dynamics of *P. fluorescens* 1472 in a batch culture after complete depletion of glucose (Panikov *et al.*, 1988b). Measured variables: total biomass, x (curve 1), biomass of viable cells, x_{via}, by plating and following calculation using Eqn. [3.1] (curve 2), $CO_2+HCO_3^-$ (curve 3), total soluble organic matter, $s+h$ (curve 4), Q_{resp} and RNA content (curve 5). The curves are fitted by synthetic chemostat model (SCM) (Table 3.10).

state, i.e. Q_{resp} and RNA content fell in *Arthrobacter* culture more dramatically as compared with *P. fluorescens*.

In order to elucidate the dependence of survival dynamics on the physiological state of the initial population, we used continuous culture (chemostat, dialysis and fed-batch). Starvation was initiated by stopping the substrate feed. It is evident from Figure 3.23 that the slower the preceding growth the longer is the survival. The same principle was revealed for the rest of studied organisms (Dorofeyev and Panikov, 1991b).

Thus, we obtained detailed information on the death phase of batch culture, including data on the relative contribution of viable cells, dead cells, detritus and exometabolites to C balance, on respiration and the change of culture physiological state, as well as on the effects of prehistory. All this persuades us that the decline phase of batch culture could be studied quantitatively with the same precision as the exponential phase. The mathematical simulation of survival in mechanistic terms is possible only by structured models (see section 3.6). However, an approximation of starvation dynamics may be achieved by very simple semiempirical equations. Let us assume that the specific death rate $a = dx/dt(1/x)$ decreases with starvation time: $a = a_0/(1 + t)$. Then the dynamics of the residual viable biomass is given as

$$\dot{x} = -ax = -\frac{a_0}{1+t}x, \quad a_0 = \text{constant}. \qquad [3.27]$$

Integration produces

$$\ln x = \ln x_0 - a_0 \ln(1+t), \quad \text{or} \quad x = x_0(1+t)^{a_0}.$$

Equation [3.27] ignores mechanistic considerations but does fit the wealth of experimental data on survival dynamics. As an example we demonstrate the results obtained for several soil bacteria (Figure 3.24). Among them were non-spore-forming species with either a very good survival ability (*Arthrobacter, Rhodococcus*) or a high extinction rate (*Pseudomonas*, *Micrococcus*, *Rhizobium*), as well as *Bacillus* producing spores immediately after glucose depletion.

Batch growth of oligotrophic bacteria

Data on the growth and survival of different microorganisms are very extensive. Here we give just a few examples demonstrating the diversity of observed patterns (Figure 3.25). Oligotrophic bacteria *Caulobacter*, *Labrys* and *Prosthecobacterium* revealed very low growth rates during the exponential phase (0.006–0.04 h^{-1}) and notable false diauxie (refilling of storage compartments), which was discussed above in relation to

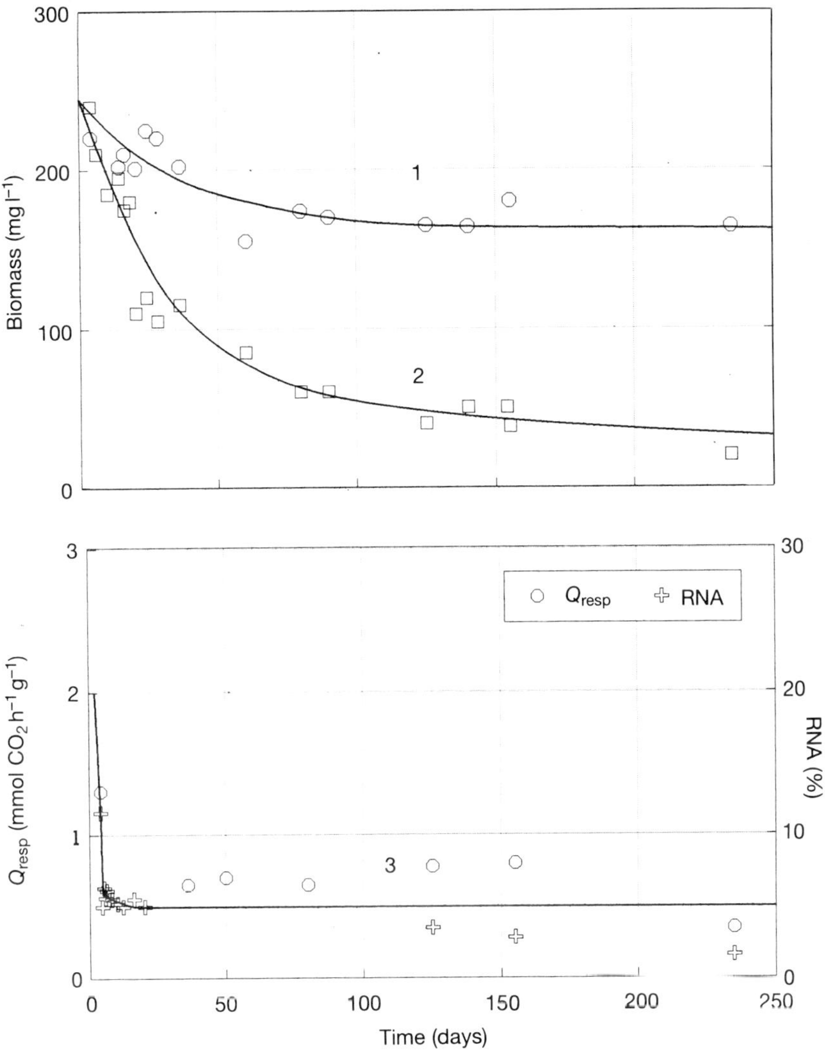

Figure 3.22 Survival dynamics of *A. globiformis 658* in a batch culture after complete depletion of glucose (Panikov, 1991a). Measured variables: x (curve 1), x_{via} (curve 2), Q_{resp} and RNA content (curve 3). The curves are synthetic chemostat model (SCM) calculations.

Arthrobacter. After substrate depletion they retained biomass at a high and almost constant level (a complete absence of net-decay). This stabilization is probably of a steady-state nature, i.e. slow death is balanced by growth due to scavenging of volatile organic substances from the

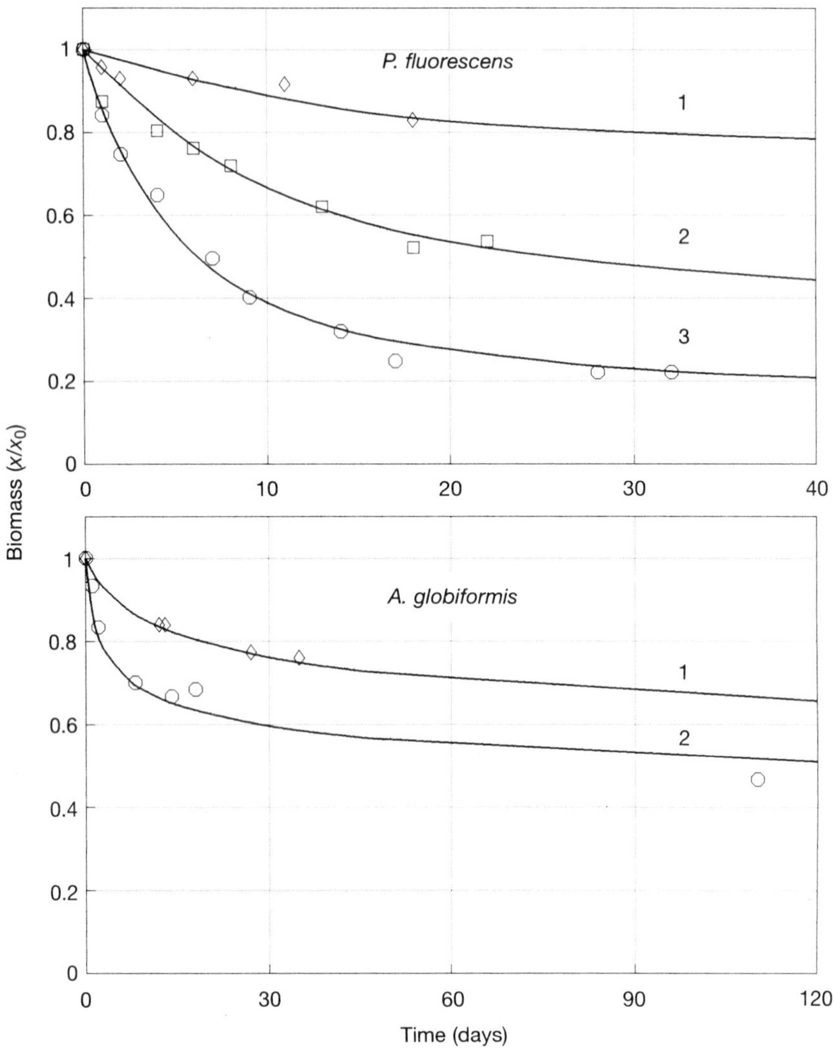

Figure 3.23 Survival dynamics as dependent on growth rate before starvation (Dorofeyev and Panikov, 1991b). Continuous cultures were run at different dilution rates in glucose-limited chemostat or fed-batch culture. At $t = 0$ cultural liquids were transferred aseptically to shaking flasks and incubated on rotary shakers. Total biomass, x, was measured by dichromate oxidation. Preceding growth rates (h^{-1}) are 0.0013 (curve 1), 0.01 (curve 2) and 0.4 (curve 3) for *Pseudomonas* and 0.0025 (curve 1) and 0.25 (curve 2) for *Arthrobacter*. The curves are synthetic chemostat model (SCM) calculations.

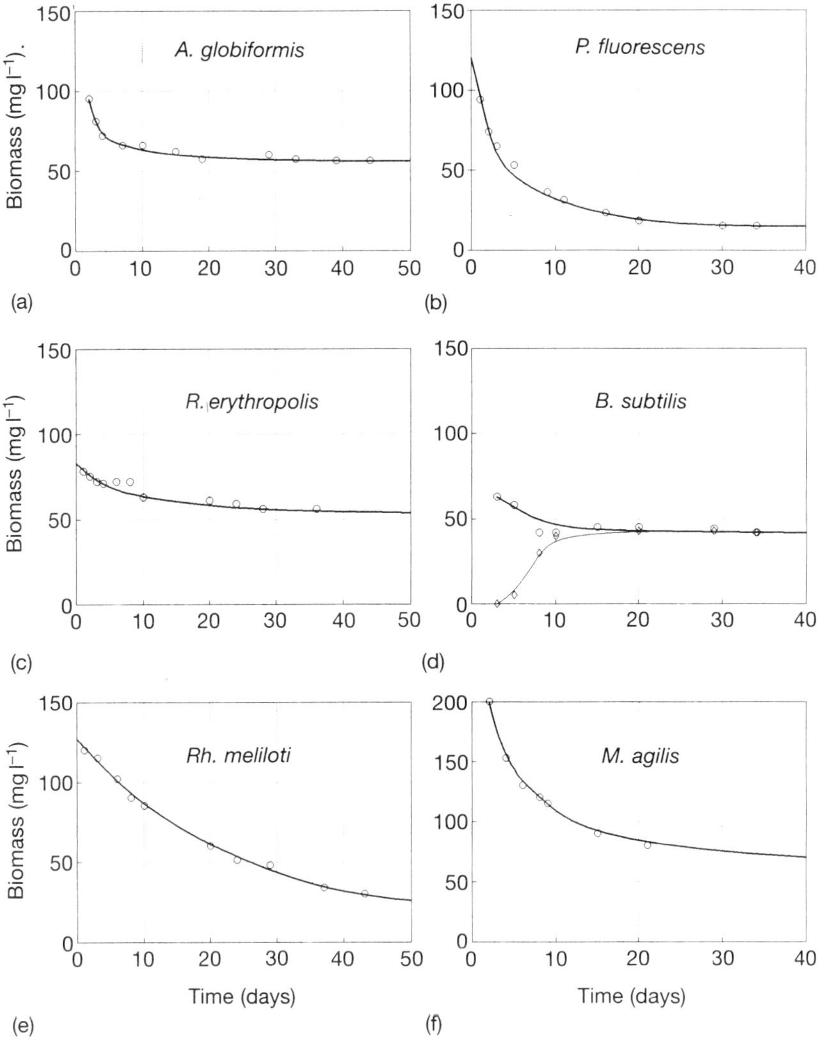

Figure 3.24 Survival dynamics of some soil bacteria in declining phase of batch culture: (a) *A. globiformis* C-4; (b) *P. fluorescens* 312; (c) *R. erythropolis;* (d) *B. subtilis*; (e) *R. meliloti*; (f) *M. agilis*. The curves are calculated using Eqn. [3.27]. (Panikov, 1991a.)

laboratory air. This is obvious in view of the fact of an almost linear production of some extracellular products (curves 3, Figure 3.25) resulted in continuous increase of the sum 'biomass + exometabolites' during starvation. Preliminary determinations by NMR indicated that extracellular products are polysaccharides.

Needless to say, the mathematical simulation of the entire dynamics of batch culture is impossible with simple kinetic models. Instead we will use

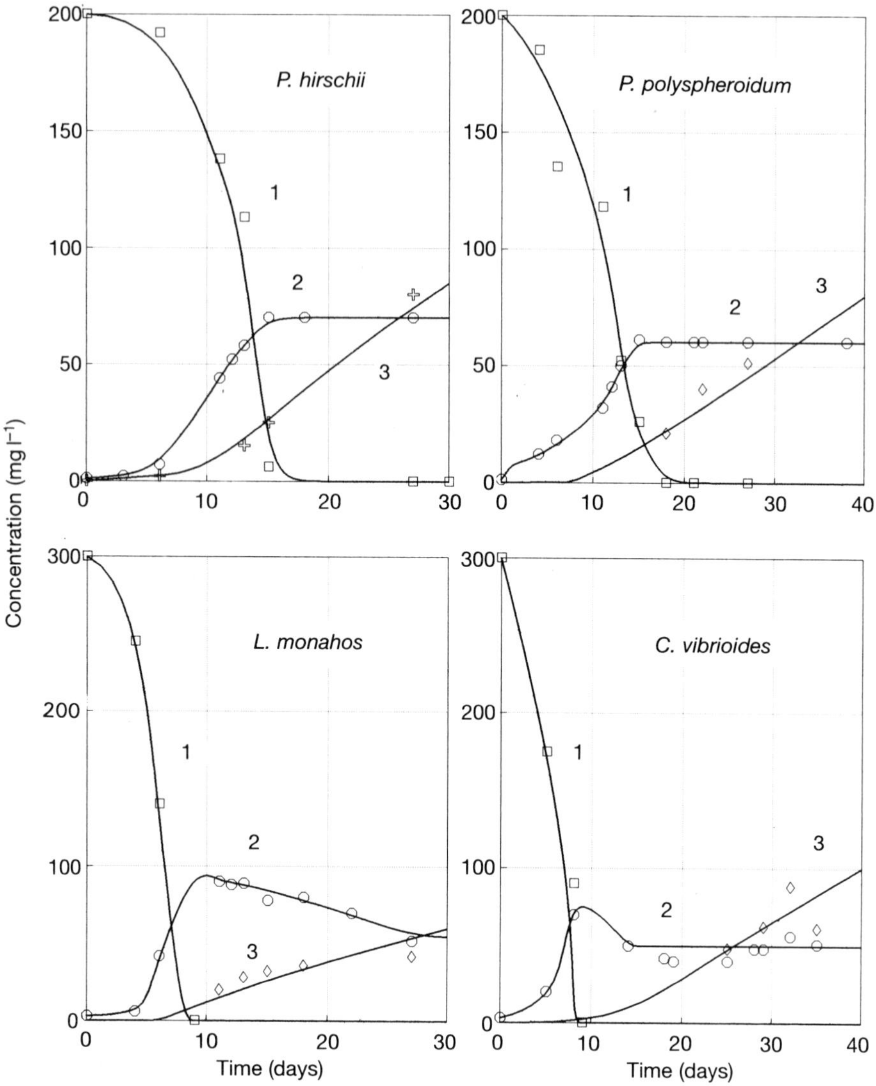

Figure 3.25 Dynamics of biomass (curve 1), residual substrate (curve 2) and exometabolites (curve 3) in the course of batch cultivation of several oligotrophic bacteria.(Panikov, 1991a)

structured models (see section 3.6), which will describe, by a single equation, batch growth from its very beginning (lag phase) until the end (complete extinction of starving bacteria or their conversion to dormant cells), particular phases being produced automatically in the course of simulation.

3.3 MICROBIAL GROWTH LIMITED BY CONSERVED SUBSTRATES

We have already discussed the term 'conserved substrates' (CS) in section 1.4. These are the opposite of catabolic substrates (sources of energy). Uptake of CSs results in their incorporation into *de novo* synthesized cell constituents. (Some insignificant fraction of CS may escape the conservation, being incorporated into exometabolites. However, it does not break the general rule since exometabolites are reutilizable and should be considered functional as normal exosubstrates.) Under starvation conditions respective cell compounds can be degraded and the limiting substance reused for synthesis of new biomass. Contrary to CS, the catabolic substrates are consumed **irreversibly**, since they are dissipated into waste products, which could not be used again as an energy source (e.g. glucose is irreversibly oxidized to CO_2 and H_2O). Let us see what consequences this dissimilarity between conserved and catabolic substrates will have on microbial growth kinetics. CSs in the present study were exemplified by the sources of nitrogen and phosphorus.

Chemostat cultures limited by conserved substrate

D-dependent variations of steady-state biomass $\bar{x}$ and substrate concentrations $\bar{s}$ are shown in Figure 3.26. The values of $\bar{x}$ decreased as $\mu = D$ increased, reflecting a linear or almost linear drop of cell yields per unit of nitrogen and phosphorus, Y_N and Y_P respectively (Figure 3.27). As mentioned previously, yield variations were also observed in glucose-limited chemostat cultures, but the trend was different. Y-values decreased nonlinearly and at low $\mu = D$ due to maintenance requirements. In the case of CS-limitation, D-dependent changes of Y are explained by the fact that growth acceleration requires elevated intracellular content of many biogenic elements including P and N. Let us clarify this situation.

The mass balance equation for a biogenic element, E, which is a cell component is as follows:

$$\underset{\Delta E}{\text{total consumption}} = \underset{\Delta E_{\text{cell}}}{\text{incorporation into cells}} + \underset{\Delta E_{\text{exo}}}{\text{incorporation into exometabolites}}$$

Division by Δx produces

$$\frac{1}{Y} = \sigma + \frac{\Delta E_{\text{exo}}}{\Delta x}, \quad \text{or} \quad \sigma = \frac{1-\theta}{Y} \approx \frac{1}{Y}, \qquad [3.28]$$

where σ is the intracellular E content, and θ is the fraction of exometabolites from total E consumption (it is taken into account that $\theta \ll 1$).

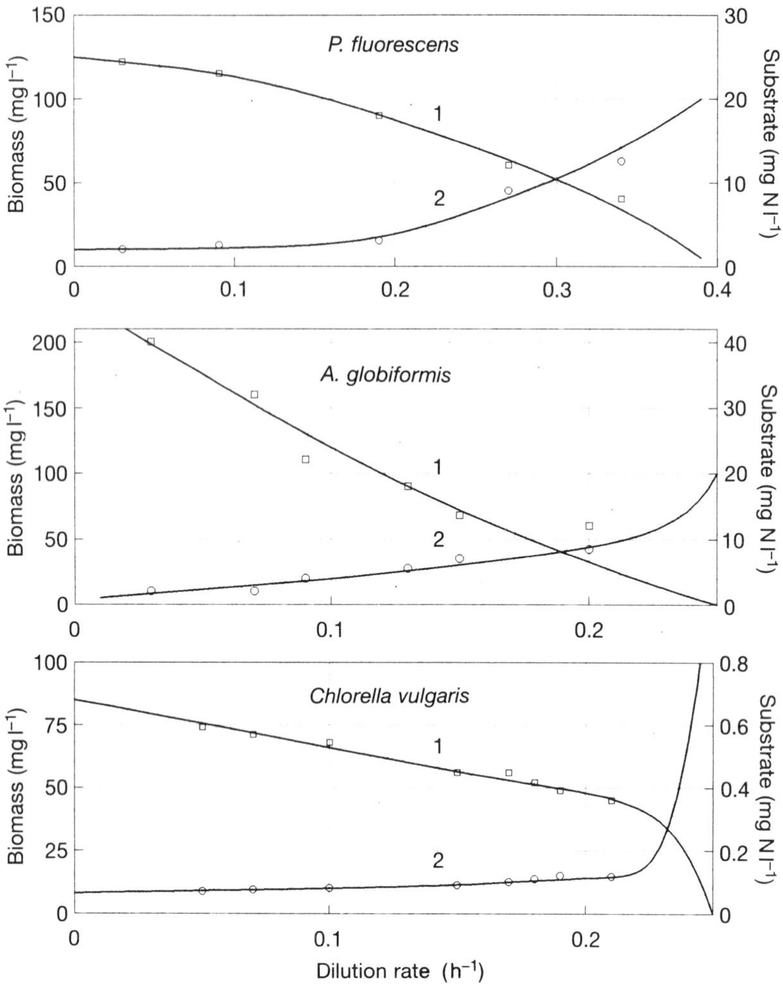

Figure 3.26 Steady-state growth of *P.fluorescens*, *A. globiformis* and *Chlorella vulgaris* limited by nitrogen source: biomass (curve 1) and limiting substrate concentration (curve 2). (Panikov, 1991a.)

Thus, an increase of σ does lead to a decrease of Y, the D-dependent Y variation being caused mainly by changes in cell composition.

The effect of conserved substrate concentration on the specific growth rate of bacteria was adequately approximated by a simple Monod equation or double hyperbolic equation reflecting the involvement of high- and low-affinity transport systems (data are not demonstrated). In the case of algae, the Monod equation was inadequate. It overestimated the observed growth rates at low s and underestimated at high s values (Figure 3.28).

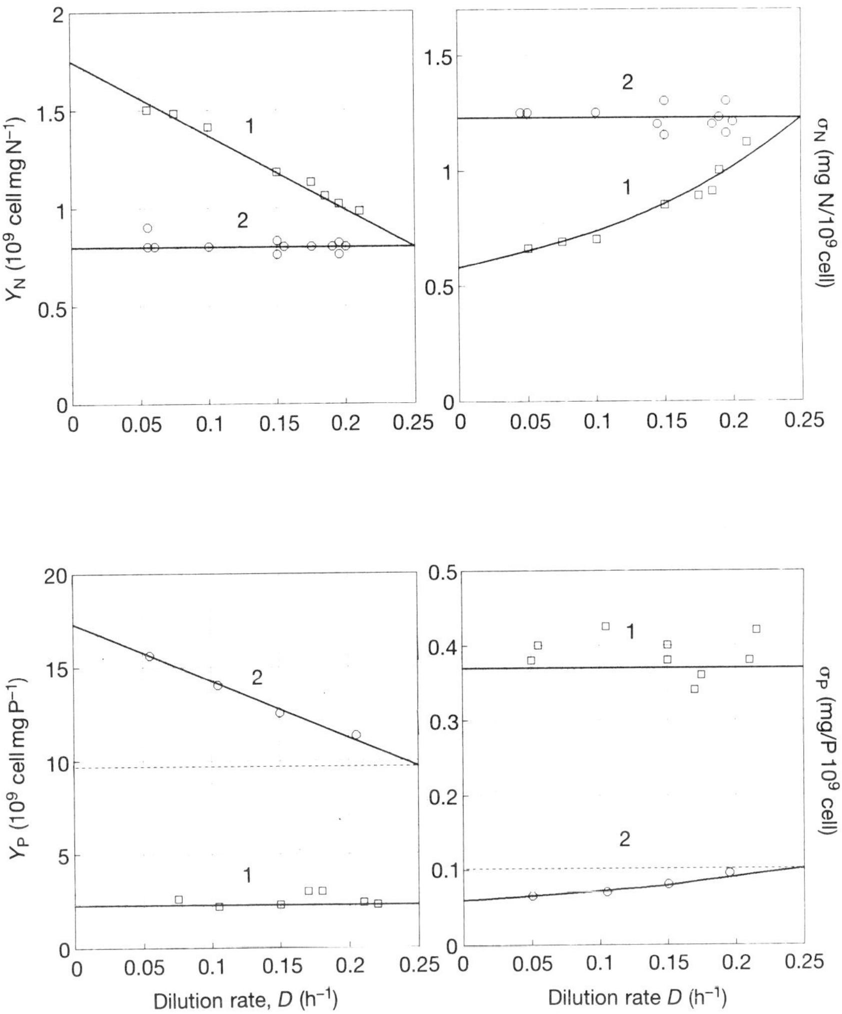

Figure 3.27 Relationship between stoichiometric parameters Y and σ and specific growth rates of *Chlorella vulgaris* in chemostat culture limited by nitrogen (curve 1) and phosphorus (curve 2) (Panikov and Pirt, 1978). The curves are calculated using Eqn. [3.34] using the parameters shown in the table:

Parameter	*N-uptake*	*P-uptake*	*Parameter*	*N-uptake*	*P-uptake*
μ_m, h^{-1}	0.25	0.25	s^*, mg l^{-1}	0.065	0.008
Q_s, mg h^{-1} 10^{-9} cell	0.31	0.026	s^{**}, mg l^{-1}	–	0.2
σ_s, mg per 10^9 cell	1.235	0.103	K'_s, mg l^{-1}	0.094	0.116
σ_0, mg per 10^9 cell	0.584	0.058	$\lambda = K_s/K_\sigma$	1.0	0.116
$L = \sigma_0/\sigma_m$	0.47	0.56	n	5.5	1.32
K_s, mg l^{-1}	0.031	0.018			

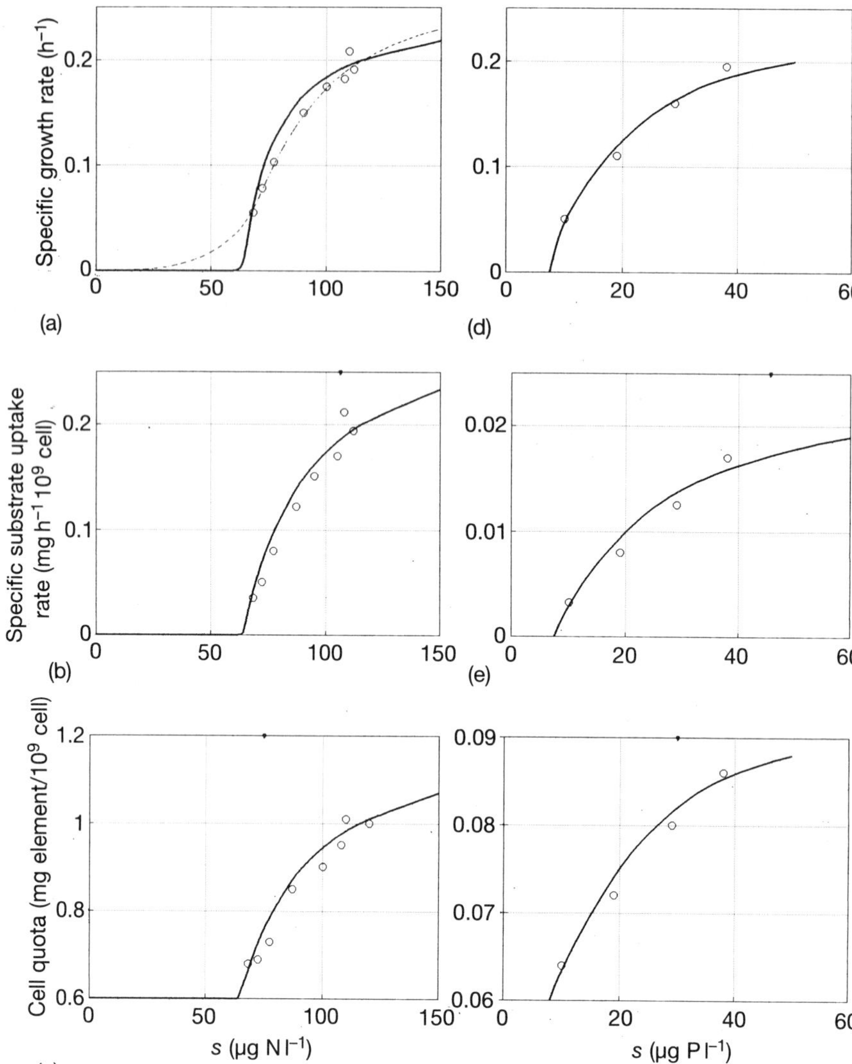

Figure 3.28 Effects of concentration of limiting substrate, urea (a)–(c) and phosphate (d)–(f) on the steady-state growth of *Chlorella* in chemostat culture. The curves were calculated using Eqn. [3.29] and in one case (dotted line) using Eqn. [3.30]. For parameters, see Figure 3.27.

Fair agreement was attained using the following assumptions.

- Some fraction of limiting substrate, s^*, is non-available for microbial uptake (perhaps as a result of chelation or other types of complexation with exometabolites), then

$$q_s = Q_s \frac{s - s^*}{K_s + s - s^*}. \qquad [3.29]$$

- The 'bottle-neck' is occupied by enzymes with cooperative kinetics, then

$$q_s = Q_s \frac{s^n}{K_s' + s^n}, \qquad [3.30]$$

where n is the apparent degree of cooperativity ($n > 1$ in the case of positive cooperative effects, formally it corresponds to n interacting subunits of the bottle-neck enzyme). Both expressions [3.29] and [3.30] produce good agreement with experimental data (compare the solid and dotted curves in Figure 3.28). Since it is impossible to differentiate between the two postulated mechanisms, we shall use the simpler expression [3.29] for our simulation.

Let us modify the chemostat model to account for all the observed experimental findings. First of all we need to formalize the D-dependent variation of σ. It has been stated already several times that equations of this kind should contain as an independent variable the steady-state substrate concentration (but not D!). Let us put it as the following empirical expression similar to [3.29]:

$$\sigma = \sigma_0 + \frac{(\sigma_m - \sigma_0)(s - s^*)}{K_\sigma + s - s^*} = \frac{\sigma_m(LK_\sigma + s - s^*)}{K_\sigma + s - s^*}. \qquad [3.31]$$

where σ_0 and σ_m are the lower and upper limits of σ variation ($\sigma \to \sigma_0$ when $\bar{s} \to 0$, and $\sigma \to \sigma_m$ when $\bar{s} \to \infty$), and $L = \sigma_0/\sigma_m$.

An expression for $\mu(s)$ is obtained by the combination of [3.28], [3.29] and [3.31]:

$$\mu = \frac{q_s}{\sigma} = \mu_m \frac{(s - s^*)(K_\upsilon + s - s^*)}{(K_s + s - s^*)(LK_\sigma + s - s^*)}, \qquad \mu_m = \frac{Q_s}{\sigma_m}. \qquad [3.32]$$

Eliminating $s - s^*$ from [3.31] and [3.32], we obtain the relationship between μ and σ:

$$\mu = \mu_m \frac{\sigma_m(\sigma - \sigma_0)}{\sigma[\lambda(\sigma_m - \sigma_0) + \sigma - \sigma_0]}, \qquad \lambda = \frac{K_s}{K_\sigma}. \qquad [3.33]$$

If $K_s = \mathrm{K}_\sigma$ ($\lambda = 1$), then last expression is simplified:

$$\mu = \frac{\mu_m}{1 - L}\left(1 - \frac{\sigma_0}{\sigma}\right) = \frac{\mu_m}{Y_m - Y_0}(Y_m - Y),$$

$$Y_{\mathrm{m}} = \frac{1}{\sigma_0}, \quad Y_0 = \frac{1}{\sigma_{\mathrm{m}}}. \qquad [3.34]$$

Equation set [3.29–3.34], appended by mass balance equation [1.28], provides a modified chemostat model which provides an adequate description of steady-state continuous growth limited by conserved substrate. For the majority of studied organisms the condition $K_s = K_\sigma$ was satisfied ($\lambda = 1$), so that in agreement with the prediction of [3.34], the yield of biomass, Y, decreased linearly with an increase of $\mu = D$ (Figures 3.27 and 3.29). However, some organisms are characterized by λ-values significantly different from unity. Thus, the growth of *Aerobacter aerogenes* limited by P, K and Mg was approximated by Eqn. [3.33], with λ varied within the narrow range 3.4–3.9 (Figure 3.29). *Chlorella* growth was adequately described by the modified chemostat model with respect to the two conserved substrates, N and P, under conditions of both growth-limitation and excessive supply (Figure 3.27). The only *ad hoc* adjustment was made in respect of the calculation of σ_{P} to account for the accumulation of polyphosphates in the culture limited by CS other than P: $\sigma_{\mathrm{m}} = \sigma_{\mathrm{m1}}$ under P-limitation and $\sigma_{\mathrm{m}} = \sigma_{\mathrm{m2}} > \sigma_{\mathrm{m1}}$ under N-limitation, the difference $\sigma_{\mathrm{m2}} - \sigma_{\mathrm{m1}}$ being attributed to the maximum content of polyphosphate in P-sufficient cells.

Multiple growth limitation in chemostat culture

It is common practice to supply the microbial culture with medium which contains only one deficient nutrient, so that continuous growth is limited by one particular component. *A priori* we may assume that at some ratio between concentrations of input nutrients there will be multiple limitation, i.e. several substrates simultaneously affecting the growth rate, μ. In the following simple experiment we examined whether or not these conditions are actually achievable. Chlorella was grown at constant D and an input of N (as urea) at a concentration $s_{01} = 4.6\,\mathrm{mg\ N\,l^{-1}}$, while the input in P concentration, s_{02}, was increased stepwise. Every increase in the value of s_{02} resulted in the smooth attainment of the new steady-state values of biomass, $\bar{x}$, as well as residual substrate concentrations $\bar{s}_1$ and $\bar{s}_2$, urea and phosphate respectively (Figure 3.30).

The whole plot of $\bar{x}$ *versus* s_{02} contains three domains. At low s_{02} (0.05–0.2 mg $\mathrm{P\,l^{-1}}$) biomass increases linearly, indicating strict P-limitation. At high s_{02} (> 1 mg $\mathrm{P\,l^{-1}}$) steady-state biomass is constant, therefore it is in the domain of strict N-limitation. At intermediate values of s_{02} both residual concentrations s_1 and s_2 were close to the lowest limiting levels, but the increment $\mathrm{d}\bar{x}/\mathrm{d}s_{02}$ was smaller as compared with the

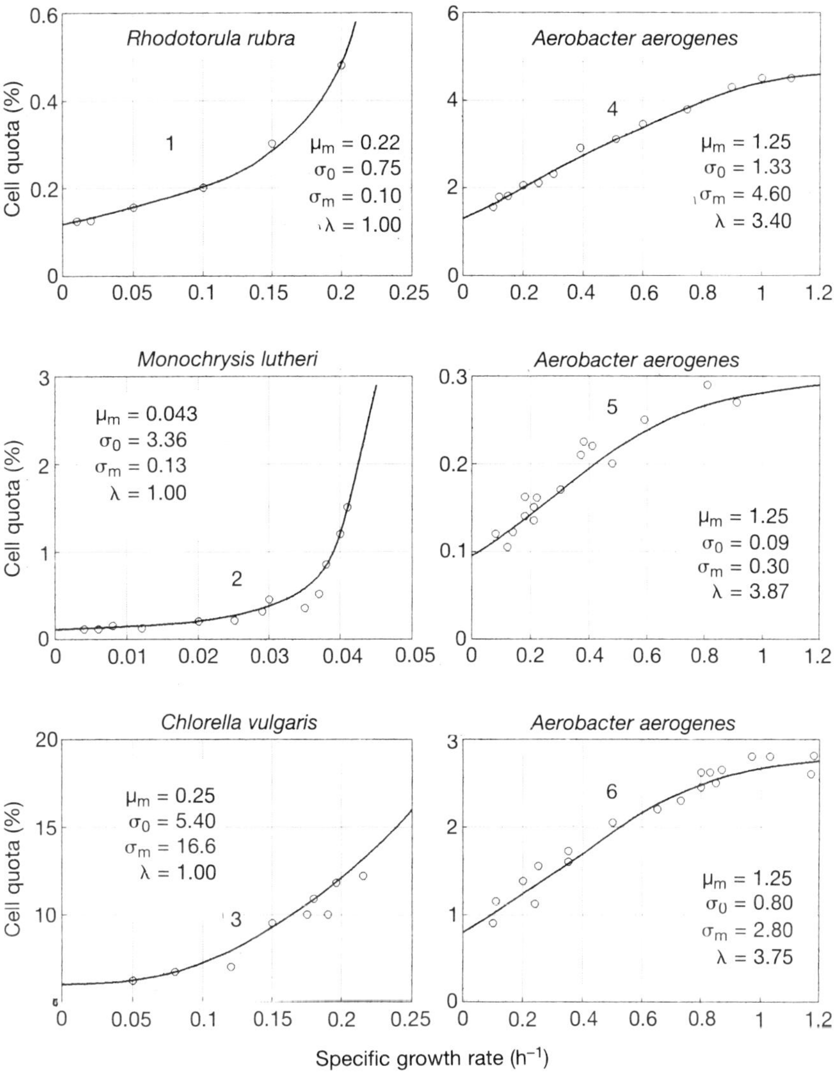

Figure 3.29 Relationship between specific growth rates and the cellular content of deficient element in chemostat cultures limited by various conserved substrates: P (curves 1, 2 and 4), N (curve 3), Mg (curve 5) and K (curve 6). Redrawn from the original publications of Robertson and Button (1979) (1); Burmaster (1979) (2); Panikov and Pirt (1978) (3); Dicks and Tempest (1966) (4); and Tempest *et al.* (1965, 1966) (5,6). Calculations were made using Eqn. [3.33].

domain of strict P-limitation, indicating a difference in σ_P. It is in this domain that double limitation occurs. The lower the dilution rate, D, the wider is this domain (compare the top and bottom parts of Figure 3.30). Thus, the main distinctive feature of double limitation is that the microbial

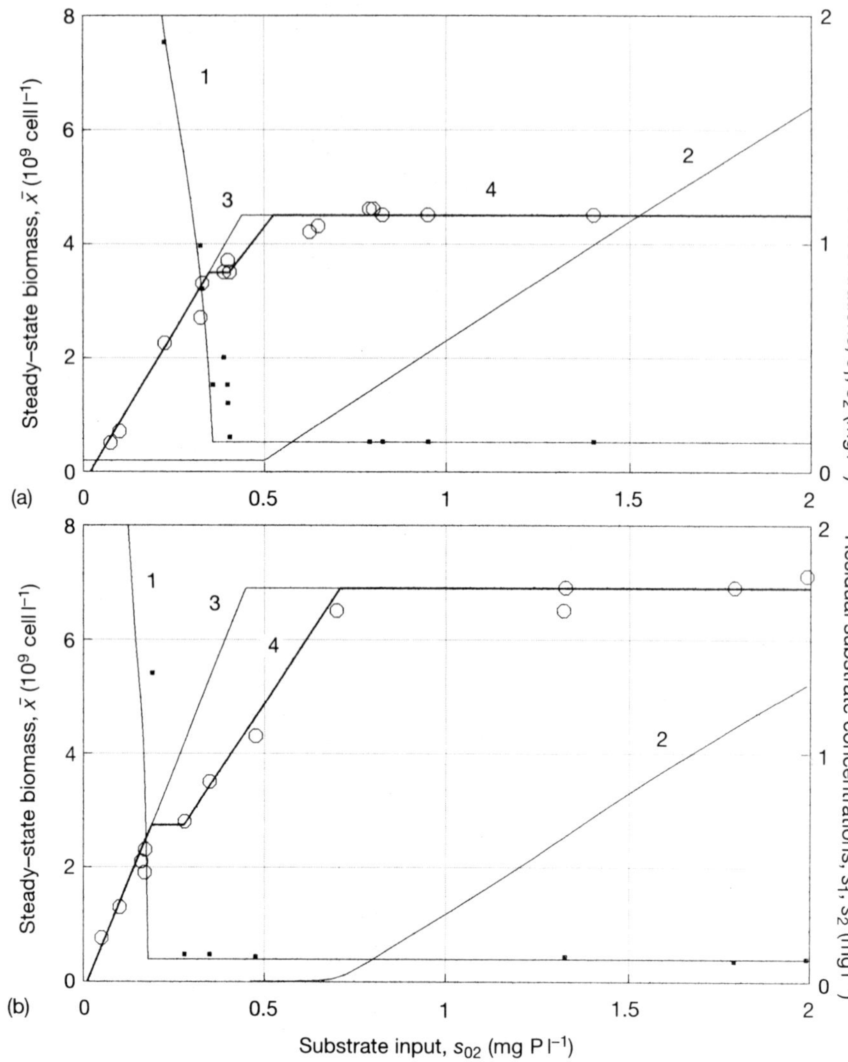

Figure 3.30 Double-limited growth of *Chlorella vulgaris* in chemostat culture (Panikov, 1979). Algae were grown at constant $D = 0.195\ h^{-1}$ (a) or $D = 0.053\ h^{-1}$ (b) and constant input of urea concentration $s_{01} = 4.6$ mg N l^{-1}, while the input in P concentration, s_{02}, was increased stepwise. Curve 1 is s_1, curve 2 is s_2, curves 3 and 4 are cell concentration calculated according to models [3.37] and synthetic chemostat model (SCM) (Table 3.9) respectively.

culture responds to concentrations of both substrates, experimentally revealed by the dependence on s of both steady-state biomass and the content of deficient element in the cells (variable σ).

Now let us try to translate our finding into a mathematical form. The chemostat model derived for conserved substrates (Eqns [3.29–3.34]) should be rewritten for two substrates as follows:

$$\dot{s}_1 = D(s_{01} - s_1) - q_1 x\,, \quad \dot{s}_2 = D(s_{02} - s_2) - q_2 x\,.$$

$$\sigma_1 = \sigma_{1m} \frac{L_1 K_1 + s_1 - s_1^*}{K_1 + s_1 - s_1^*}\,, \quad \sigma_1 = \sigma_{2m} \frac{L_2 K_2 + s_2 - s_2^*}{K_2 + s_2 - s_2^*}\,,$$

$$\dot{x} = [\mu(s_1, s_2) - D]x\,. \qquad [3.36]$$

There are two known ways to present a specific growth rate as a function of two limiting substrate concentrations $\mu(s_1, s_2)$ (Bader *et al.*, 1975):

$$\mu = \mu_m \min\left(\frac{s_1 - s_1^*}{K_1 L_1 + s_1 - s_1^*}, \frac{s_2 - s_2^*}{K_2 L_2 + s_2 - s_2^*}\right), \qquad [3.37]$$

$$\mu = \mu_m \frac{s_1 - s_1^*}{K_1 + s_1 - s_1^*} \times \frac{s_2 - s_2^*}{K_2 + s_2 - s_2^*}\,. \qquad [3.38]$$

The first expression, [3.37], is in fact a formalization of Libig's 'law of minimum', which implies an instantaneous switch from one controlling factor to another. The second expression, [3.38], is called a multiplicative or interactive model, and implies the concurrent effects of both s_1 and s_2 on the microbial growth rate. There is also an additional conservation condition:

$$\bar{x} = \frac{s_{01} - s_1}{\sigma_1} = \frac{s_{02} - s_2}{\sigma_2}\,. \qquad [3.39]$$

However, none of the suggested models was successful in the simulation of the observed microbial growth within the domain of double limitation. Both [3.37] and [3.39] overestimated the real values, although the multiplicative version gave marginally better results. The complete and adequate description of the phenomenon of multiple limitation we will give elsewhere (see section 6.5).

Thus, we may conclude at the present stage that multiple limitation is quite feasible. It may be attained in a real chemostat culture if the mixing ratio of individual substrates in the feed medium is close to the respective cell quota values

$$s_{01} : s_{02} : \ldots : s_{0n} \approx \sigma_1 : \sigma_2 : \ldots \sigma_n\,.$$

If the range of σ variation is wide, then the probability of encountering multiple limitation is rather high. We suspect that frequently double and triple limitation occurs undetected in dense chemostat cultures when the investigator has to apply a high input concentration of medium component which is supposed to be a single limiting substrate.

Chemostat model and the concept of 'cell quota'

The equations derived earlier ([3.29–3.38]) are no more than modifications of the classic chemostat model of Monod. The only independent variable in the steady-state formulae is the concentration of limiting substrate(s). From this point of view, we may identify $\mu(s)$, $q_s(s)$ and $x(s)$ as **primary** functional relationships while $\mu(\sigma)$ may be described as a **secondary** one, which is derived from the primary functions. In other words, Eqns [3.33] and [3.34] display **interconnection** between μ and σ, but by no means **the effects** of σ on μ! Now we return to our review (section 1.4) to examine carefully the 'cell quota concept' which still dominates kinetic literature.

In our notation, 'cell quota' is no more than the variable σ, while the whole concept states that the relationship $\mu(\sigma)$ is a primary one, and that there is no clear relationship between μ and s (Droop, 1974; Nyholm, 1976, 1977). Not to be biased we will quote one of the best papers explaining the concept (Nyholm, 1976, p.1044): 'During recent years kinetic studies of nutrient-limited algal growth have shown that the specific growth rate depends on *the amounts of limiting nutrients within the cells* and no direct relationship exists between the growth rate and the concentration of extracelluar substrate'. Hence, (ibid, p.1054) '. . . there is no need for a kinetic model of substrate uptake'.

The deficiency in this theoretical reasoning is obvious (it is amazing why it was not opposed for many years!), as it stems from the wrong interpretation of structured models. Indeed, the introduction of the state variable, **the amounts of limiting nutrients within the cells**, implies the necessity to construct the simplest structured model describing the conversion of exo-substrate S to biomass components X′ via this particular entity of 'intra-cellular substrate concentration, L: $S \rightleftarrows L \rightleftarrows X'$. However, at least two mistakes were made: (a) the intermediate pool L was replaced by the intercellular content of the deficient element, σ; and (b) the first reaction of the reversible $S \rightleftarrows L$ transport was completely ignored, in spite of the fact that it is an absolutely essential source of L. If this transport stage is very rapid and could not be rate-limiting, then the bottle-neck is really located within the second stage $L \rightleftarrows X'$. However, at any transport capacity of microbial cells, the pool size is two to four orders of magnitude lower than the σ value, and in addition depends on the transport rate and hence the extracelluar substrate concentration. These arguments are valid for steady-state growth. Under transient conditions there is a possibility for a temporary lack of a straightforward relationship between CS concentration in the

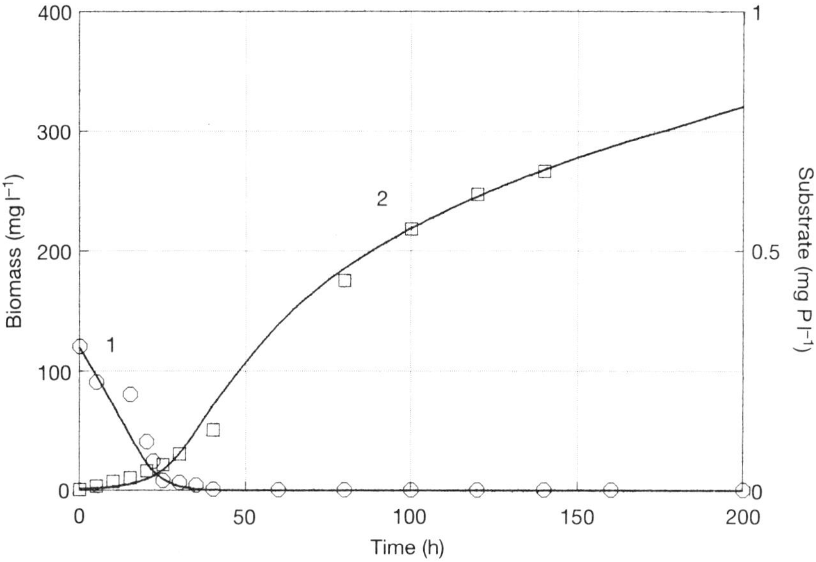

Figure 3.31 Batch growth of *Selenastrum capricornutum* limited by phosphate. Residual PO_4^{3-} concentration (curve 1) and cell biomass (curve 2). Experimental data are redrawn from Nyholm (1977). Curves are calculated according to synthetic chemostat model (SCM) (Table 3.9) with the following parameters: $Q_s = 0.05$, $Q_L = 0.03$, $K_L = 0.9$, $K_s = K_r = 0.5$, $a = 0.2$, $\sigma_0 = 0.00005$, $\sigma_m = 0.01$.

medium and the instantaneous value of the microbial growth rate. This conclusion was drawn from the data obtained for batch culture.

Batch culture

Experimental study of batch growth limited by CS is best realized with phototrophic organisms. In the case of heterotrophic microbes, there are specific obstacles stemming from the use of concentrated glucose concentration, substrate inhibition, accumulation of catabolic intermediates in the medium, a drop in pH, etc. In addition, the reutilization of CS makes it quite probable that double limitation will occur or even a switch from CS to energy limitation.

Phototrophic growth has been studied very thoroughly, so we have an opportunity to analyse published data (Figure 3.31). The characteristic feature of the dynamic curves is the lack of stringent coupling between CS uptake and the increase of cell biomass. P consumption was completed in an algae culture in 40 hours. Then cell growth did not cease but accelerated, with a subsequent gradual decline during 60–160 hours. It is absolutely clear that after 40 hours the growth was realized due to a decrease in σ_P, i.e. at the expense of intracellular phosphorus. Note that this includes not only specialized reserve compounds such as polyphosphates, but also normal functional cell constituents. It follows from the data on cell composition (a drop in σ_P is accounted for by a decrease in RNA, nucleotide and phospholipid content), as well as from the fact that this type of growth pattern is observed with deficient elements which are not stored as reserved compounds (N, Mg, K, Fe).

Thus, there are two stages of batch culture growth: (a) active accumulation of the deficient element with the concomitant increase of σ while growth is rather slow; and (b) growth at the expense of the intracellular deficient element due to its redistribution between mother and daughter cells (accompanied by a decrease in σ). We can see a definite similarity between the batch growth of phototrophs and the false diauxie of *Arthrobacter*. In both cases the growth was highly unbalanced with time delay of biomass increase relative to substrate uptake.

Even more dazzling deviations from traditional growth laws were observed in the experiments with batch cultures of heterotrophic organisms limited by conserved substrates. To avoid complications caused by a high initial concentration of glucose (substrate and product inhibition), we used the following experimental set-up. Cells from an unlimited batch culture were washed by centrifugation and transferred to the medium lacking nitrogen and containing an elevated glucose concentration ($5\,g\,l^{-1}$) as well as fortified phosphate buffer ($KH_2PO_4 + K_2HPO_4$, $6\,g\,l^{-1}$). Hence, batch growth was realized on those N-compounds which come from laboratory air (mainly ammonia, since N_2 fixation was impossible either because of absence of nitrogenase or its inhibition by intensive aeration) and reagent impurities. The dynamics of batch culture was recorded by determinations of total cell biomass, σ, cell viability by agar-film technique, residual glucose and pH of the liquid. Usually glucose consumption was $<50\%$, and pH remained neutral, but even so, slow net growth of total biomass was accompanied by dramatic loss of cell viability (Figure 3.32). The apparent absence of growth of the starving bacteria after their plating on complete recovery medium is called 'substrate-accelerated death' (Postgate and Gray, 1976). This phenomenon was observed during the long-term starvation of bacteria in a medium devoid of energy source (Luscombe and Gray, 1974). Under starvation caused by the lack of CS (we tested N, P and Mg), substrate-accelerated death is already detectable during the first hour of incubation.

Quantitative interpretation and mathematical simulation of batch culture limited by CS is impossible even by the modified chemostat model (Eqns [3.29–3.34]). The reason for this is obvious. The derived model does not contain any dynamic equations aimed at the description of transient growth. Again we advise our reader to refer to section 3.6, where such a model will explain all observations including the mysterious phenomenon of substrate-accelerated death.

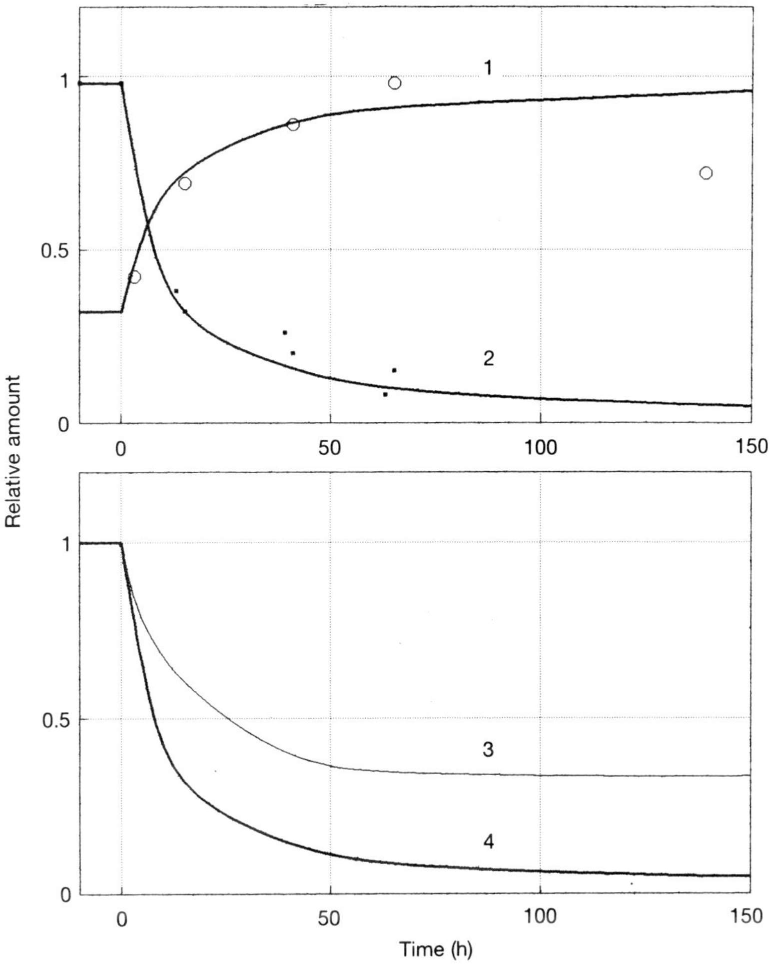

Figure 3.32 Illustration of the phenomenon of substrate-accelerated death. *P. fluorescens* cells from an exponential phase of batch culture were washed by centrifugation and transferred to the N-free medium 1 (section 3.1.2) with $5\,g\,l^{-1}$ glucose and $6\,g\,l^{-1}$ KH_2PO_4 + K_2HPO_4; biomass was measured by dichromate oxidation (curve 1); fraction of viable cells was measured by an agar-film technique (curve 2). The values of σ_N (curve 3), and r-variable (curve 4) were calculated using synthetic chemostat model (SCM) (Table 3.9) with the following parameters: $Q_s = Q_L = 0.1\,g\,N\,h^{-1}\,g^{-1}$, $K_s = K_r = 0.05\,mg\,N\,l^{-1}$, $\sigma_0 = 0.05$, $\sigma_m = 0.12\,g\,N\,g^{-1}$ biomass, $r_0 = 0.975$.(Evdokimova *et al.*, 1994)

3.4 DIAUXIE

Diauxie is defined as growth on a mixture of substrates of different availability. Until now, we have considered microbial growth on minimal media, every component of which contains a single source of energy or biogenic element. Now we turn to the more complicated problem of growth on a mixture of several competing substrates. Since we are focusing on the general problem, and a detailed discussion is beyond the scope of this book, this section will be restricted to two examples.

3.4.1 GROWTH OF *PENICILLIUM FUNICULOSUM* ON A MIXTURE OF STARCH AND GLUCOSE

Polymeric substrates (represented here by starch) are not available for immediate uptake. Their utilization is performed according to following scheme:

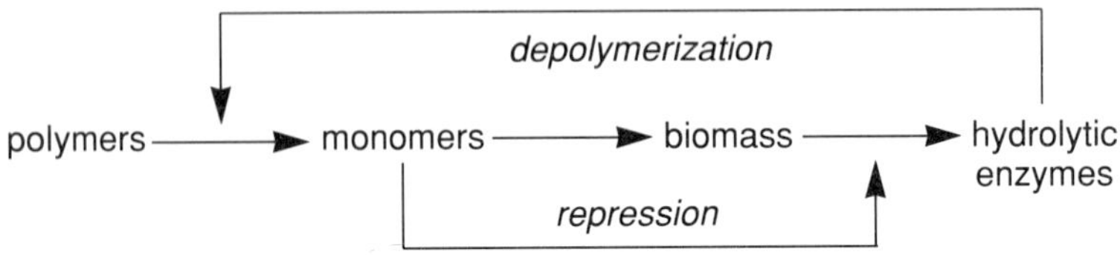

[3.40]

Depolymerization results in the formation of low-molecular weight compounds which are used by microorganisms for their growth and maintenance as well as for the biosynthesis of hydrolytic enzymes. Monomers exert control on enzyme formation and activity. Our objective is to understand this controlling mechanism in quantitative terms (i.e. the effects of monomer concentrations, limitations by conserved substrates, cost of enzyme synthesis, etc.).

Batch culture

P. funiculosum was grown on a nutrient medium containing starch ($1000\ \mathrm{mg\,l^{-1}}$) and glucose ($500\ \mathrm{mg\,l^{-1}}$). The dynamics of batch growth were followed by determinations of total microbial biomass (x) by dichromate oxidation, residual glucose (s') with hexokinase, and starch (s) by iodometry. The concentration of starch degradation products (l) were estimated from the mass balance: $l = s_0 + s'_0 - s - s' - (x - x_0)/Y$ (only during growth phase!). The observed dynamic pattern was a typical diauxie (Figure 3.33). The first phase of rapid growth ($\mu = 1.6\ \mathrm{day^{-1}}$) was accompanied by an uptake of glucose alone, starch degradation being completely suppressed. The second phase of slow growth on starch ($\mu = 0.3\ \mathrm{day^{-1}}$) was initiated after a decline in glucose concentration down to $10–50\ \mathrm{mg\,l^{-1}}$. Increase of biomass was delayed as compared with starch depolymerization, thus giving rise to the transient accumulation of low-molecular weight products. Thus, it was 'normal' diauxie with the following characteristic features: the consecutive uptake of two substrates, the abrupt change from high to low growth rate, and a reversible repression exerted

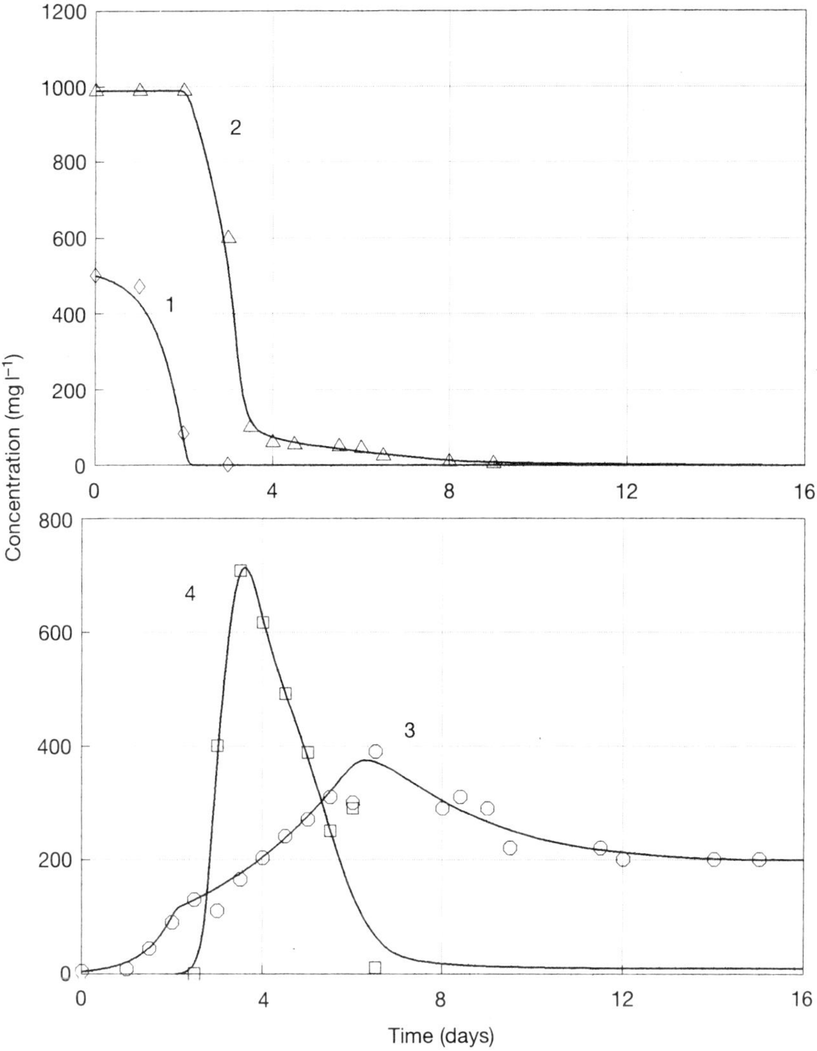

Figure 3.33 Batch growth of *Penicillium funiculosum* on the mixture of starch and glucose: residual glucose (curve 1), residual starch (curve 2), biomass concentration (curve 3), and total concentration of extracellular products (curve 4). (Panikov and Evdokimova, 1989.)

by the preferential substrate (transformation of starch was initiated at once after depletion of glucose).

Under conditions of mineral limitation, glucose repression may be protracted and even irreversible. Fungi were grown on monosubstrates (only glucose or only starch) on a mineral base solution diluted 10, 100 and 1000 times. Higher growth retardation was observed at a dilution of 100 (a decrease in μ and Y, increase of time-lag) caused mainly by the deficiency of N. Batch growth on starch was more susceptible to limitation as compared with growth on glucose, because some fraction of the limiting substrate should be incorporated to extracelluar enzymes. Batch growth on two substrates (glucose + starch) was characterized by a prolonged second lag phase of 3–10 days. In some cases starch remained intact even longer. Derepression was attained by small N-amendments or by high concentrations of NaCl (Figure 3.34), other salts, e.g. KH_2PO_4, $MgSO_4$, being inefficient.

We shall limit our quantitative analysis of batch culture data to growth stoichiometry. Mass balance of N and C is outlined as follows:

Growth on glucose

$$\underset{\Delta N}{\text{total N uptake}} = \underset{\Delta x \sigma_N}{\text{N incorporation into cells}}$$

$$\underset{\Delta C}{\text{total C uptake}} = \underset{\Delta x \sigma_C}{\text{for growth}} + \underset{\Delta x Y_{p/x}}{\text{for respiration}}$$

Growth on starch

$$\underset{\Delta N}{\text{total N uptake}} = \underset{\Delta x \sigma_N}{\text{N incorporation into cells}} + \underset{\Delta e \sigma_E}{\text{N incorporation to enzymes}}$$

$$\underset{\Delta C}{\text{total C uptake}} = \underset{\Delta x \sigma_C}{\text{for growth}} + \underset{\Delta x Y_{p/x}}{\text{for respiration}} + \underset{\Delta e / Y_{e/s}}{\text{for enzymes synthesis}}$$

where σ_N and σ_C are the contents of N and C in cells respectively, σ_E is the N content in amylases, and $Y_{p/x}$ and $Y_{e/s}$ are the respective yields.

A comparison of cell yields on starch and glucose (0.15 and 0.25 g biomass g^{-1} substrate respectively) allowed us to conclude that biosynthesis of extracelluar enzymes is rather costly in terms of energy expense. Enzyme yield was calculated to be 3.11 g of protein C per g of N consumed and 0.4 g of protein C per g of carbon of utilized starch.

Under conditions of N-limitation microorganisms consume N from the medium (down to threshold level N*) and reutilize intracellular N, decreasing the value of σ down to the minimum level, σ_0. We may distinguish three possible situations in batch culture grown on the mixture glucose + starch at the instant of complete glucose depletion:

1. $N > N^*$, $\sigma > \sigma_0$, the culture is limited by C source, so that exhaustion of

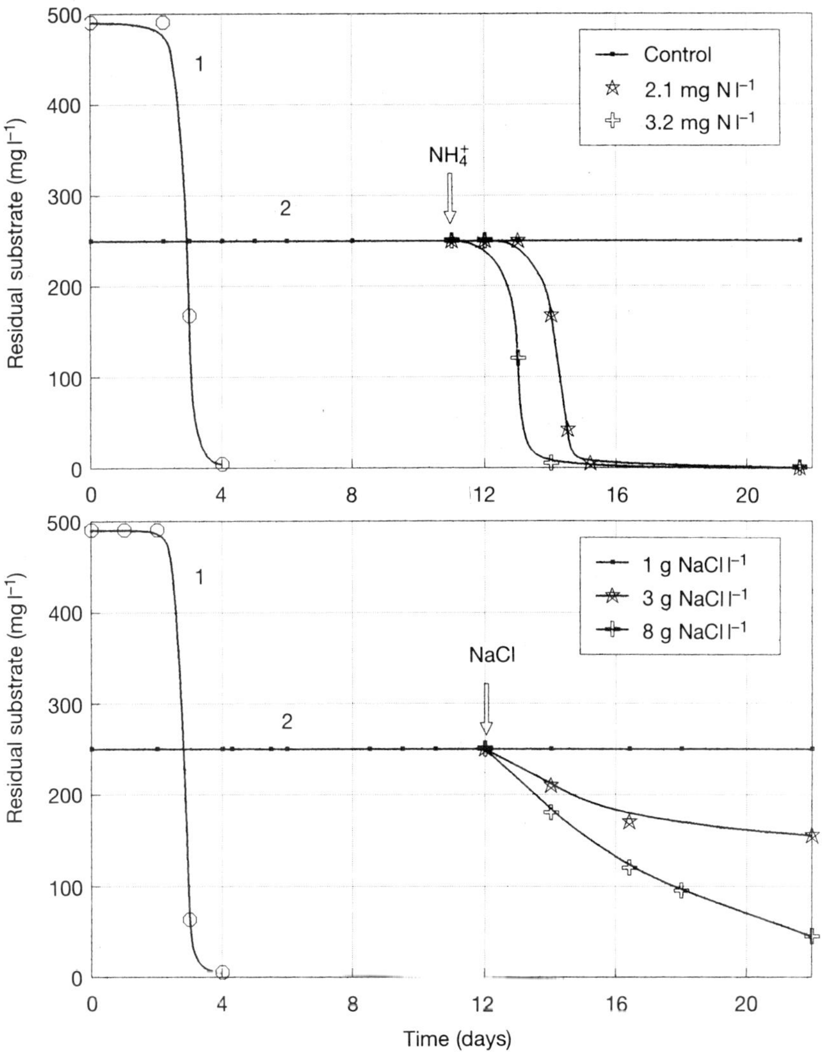

Figure 3.34 Dynamics of starch utilization in N-limited batch culture of *Penicillium funiculosum*: residual glucose (curve 1), starch (curve 2). Initial concentration of $(NH_4)_2SO_4$ in the medium was 2.1 mg l^{-1}. Arrows indicate the time of addition of mineral salts. (Panikov and Evdokimova, 1989.)

glucose should be followed by an immediate derepression of amylase biosynthesis.

2. $N \sim N^*$, $\sigma > \sigma_0$, there is a double C and N limitation, the main source of N is intracellular reserves, derepression is delayed producing prolonged lag phase before the start of starch utilization.

3. $N \sim N^*$, $\sigma \sim x_0$, the culture is in a state of N starvation, starch utilization is impossible, microbial cells (mycelium) lose their viability and are lysed, released N-compounds may be used by survivors, so that starch utilization may proceed slowly but without net growth of the culture.

In cases 2 and 3 there is a high probability of observing the effect of substrate-accelerated death. Thus, glucose repression may become truly irreversible, because of the complete absence of metabolically active cells.

Chemostat culture

Diauxie could not be observed in continuous culture. At low D, the steady-state concentration of the preferential substrate (glucose) is as low as $1\ \mathrm{mg\,l^{-1}}$, so that amylase activity is expected to be fully expressed, and both substrates, glucose and starch, to be utilized simultaneously. At high D, residual glucose concentration is increased to $10^2\ \mathrm{mg\,l^{-1}}$, preventing amylase production, so we may expect the cessation of starch utilization. This is just a hypothetical experiment; however, numerous results obtained with another substrates combination (e.g. glucose–lactose) support these speculations (Powell *et al.*, 1967). So our objective could no longer be diauxie *per se*, but instead we will take the opportunity offered by chemostat culture to clarify the underlying phenomenon, i.e. the controlling mechanisms responsible for the regulation of extracelluar enzyme biosynthesis.

P. funiculosum was grown continuously on medium with starch as the only source of C and energy, under C- and N-limitations. Steady-state concentrations of biomass ($\bar{x}$) and starch ($\bar{s}$) were determined by dichromate oxidation and iodometry respectively, and degradation products ($\bar{s}'$) were found from the difference in these two entities. In addition, residual mineral N (NO_3^- and NH_4^+) was estimated colorimetrically. Amylase activity was assayed as a rate of starch decrease during incubation of samples of the culture liquid without separation of mycelium. The presence of microbial biomass was necessary because a significant fraction of enzymes was in the bonded form.

Results are presented Figure 3.35 for a C-limited chemostat culture. Amylase activity was highest at moderately low dilution rates $0.01–0.05\ \mathrm{h^{-1}}$, the further increase or decrease of D resulting in a decline in enzymatic activity. Simulation of obtained data was performed with kinetic models given in Table 3.5. The set of equations was no more than an account of the mass balance according to scheme [3.40]. The maintenance coefficient, m, and potential metabolic activities, Q_s and Q_N, were accepted as constants. A qualitative stability analysis of this model reveals some interesting features. Contrary to the simple chemostat model there exists the possibility of bistability. At every D we have three stationary

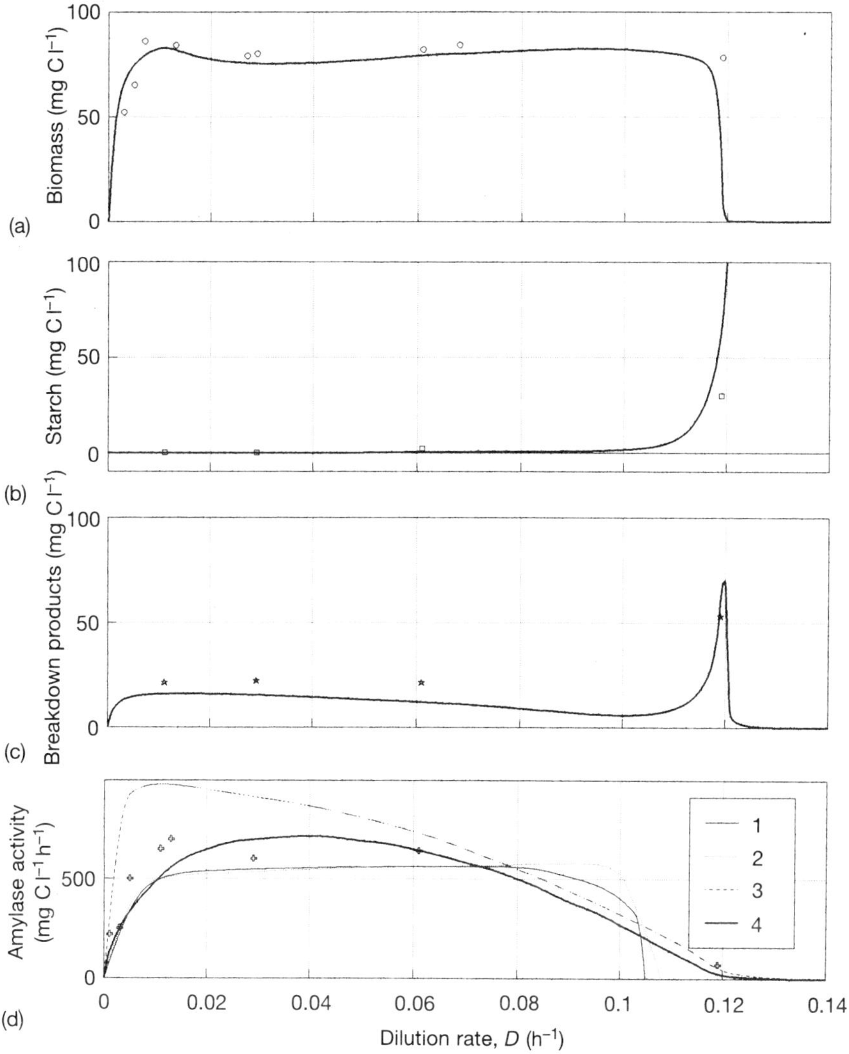

Figure 3.35 The steady-state growth of *Penicillium funiculosum* in chemostat culture limited by starch. (d) Amylase activity in the cultural liquid as dependent on D, simulation was made using the model (Table 3.5) with different assumptions about the control of enzyme formation: Eqn. [3.41] (curve 1), Eqn. [3.42] (curve 2), Eqn. [3.42] (curve 3), and Eqn. [3.44] (curve 4). The remainder of the curves were simulated by Eqn. [3.44]. (Panikov *et al.*, 1989c.)

states. One is unstable (with low $\bar{x}$ and $\bar{e}$) and two are stable (the first one is trivial at zero, while the second corresponds to high positive values of $\bar{x}$ and $\bar{e}$). The biological interpretation of this stability analysis is as follows. Microbial growth on polymers is influenced not only by substrate concen-

Table 3.5 Chemostat model of microbial growth on polymeric substrate exemplified by starch

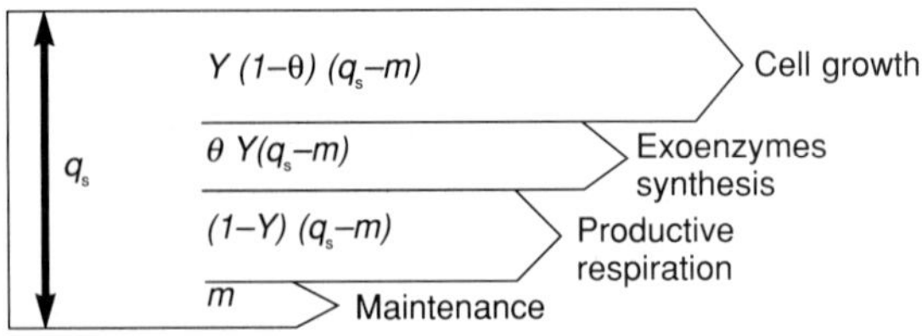

Variable	Equation
Starch (mg C l^{-1})	$\dot{s} = D(s_0 - s) - q_e e\,,$
	$q_s = Q_e \dfrac{s}{K_m + s}$
Monomers (mg C l^{-1})	$\dot{s}' = q_e e - q_s x - Ds'\,,$
	$q_s = Q_s \dfrac{s'}{K_s + s'}$
Amylases (mg C l^{-1})	$\dot{e} = \theta Y(q_s - m)x - De$
Biomass (mg C l^{-1})	$\dot{x} = (1 - \theta)\,Y(q_s - m)x - Dx$

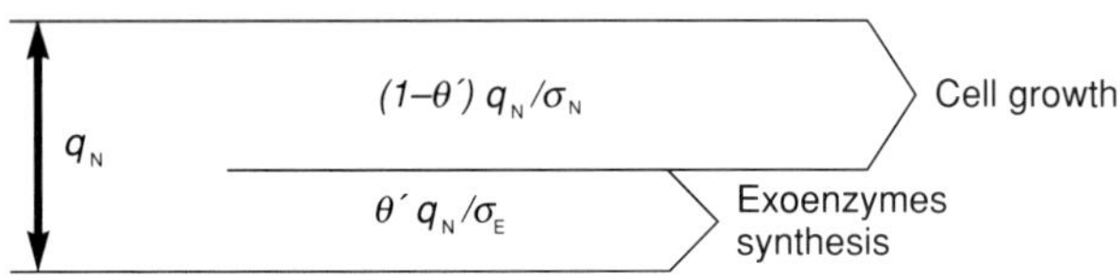

Variable	Equation
N source (mg N l^{-1})	$\dot{N} = D(N_0 - N) - q_N x\,,$
	$q_N = Q_N \dfrac{N}{K_N + N}$
Amylases (mg N l^{-1})	$\dot{e} = \dfrac{\theta' q_N x}{\sigma_E} - De$
Biomass (mg C l^{-1})	$\dot{x} = \dfrac{(1 - \theta')q_N x}{\sigma} - Dx$
N cell quota (g N g^{-1} biomass)	$\dot{\sigma} = \mu\left[\dfrac{(\sigma_m - \mu_0)N}{K_N + N} - (\sigma - \sigma_0)\right]$

tration, but also by the amount of hydrolytic enzymes. A steady-state with low $\bar{x}$ and $\bar{e}$ corresponds to those situations when growth is controlled only by the amount of enzymes. Since μ is not dependent on s, then the stabilizing negative feedback characteristic of the chemostat ($d\mu/ds < 0$) no longer operates and this state is unstable. Under small perturbations, the culture is washed out or converges to a stable state with high $\bar{x}$ and $\bar{e}$.

To describe the metabolic control of amylase synthesis we introduced variable $\theta \leqslant 1$ which acts as a valve separator. It separates the total flux of monomeric substrates into two subfluxes, one used for enzyme biosynthesis (proportional to θ), and another for the synthesis of cell biomass (proportional to $1 - \theta$). A similar variable θ' was used to describe the control of N-fluxes. The numerical values of θ and θ' were chosen to correspond to a particular mode of metabolic control:

- product inhibition of amylase activity:

$$q_e = Q_e \frac{s}{K_m(1 + s'/R_1) + s}, \quad \theta - \text{constant}, \quad R_1 - \text{constant}; \quad [3.41]$$

- control of enzyme production by its level outside the cells:

$$\theta = \frac{\theta_0}{1 + R_2 e}, \quad \theta - \text{constant}, \quad R_2 - \text{constant};$$

- repression of enzyme biosynthesis by monomeric products:

$$\theta = \frac{\theta_0}{1 + R_3 s'}, \quad \theta - \text{constant}, \quad R_1 - \text{constant}. \quad [3.43]$$

All parameters of the model except θ and R_1–R_3 were identified from independent experimental data not affected by the choice of the particular hypotheses [3.41–3.43]. Y and m were found from relationship between q_{resp} and $\mu = D$, K_s and Q_s were determined in short-term enrichment experiments with glucose. K_m and Q_e were identified from the rates of starch degradation in the absence of microbial growth. N-limited chemostat culture with glucose as a source of C and energy was used to identify the parameters σ_0, σ_m, K_N, K'_N and Q_N. An estimation of parameters θ and R_1–R_3 was made as a last step to minimize the sum of the square residuals in respect to all measured variables shown in Figure 3.35. The identification procedure revealed that the best agreement with experimental data was given by hypothesis [3.43]. The only discrepancy was found at low D, where the model severely overestimated the observed amylase activity. A refinement reduces this deficiency by accounting for spontaneous enzyme inactivation:

$$\dot{e} = \theta Y(q_s - m)x - (D + R_4)e, \quad R_4 - \text{constant},$$

$$\dot{x} = (1 - \theta)\, Y(q_s - m)x + R_5 R_4 e - Dx, \quad R_5 - \text{constant}. \qquad [3.44]$$

These equations imply that the decay in amylase activity is a first-order reaction, and inactivated proteins are reutilized by microorganisms. Equation [3.44] in combination with [3.43] and the basic model (Table 3.5) produced the best simulation of the experimental curves. The rest of the suggested hypotheses were discarded as being inadequate because they strongly underestimated the critical dilution rate and generated the wrong relationship between amylase activity and D (Figure 3.35).

In the case of N-limitation (data not shown) derepression of amylase activity took place only at very low D where actually double N- and C-limitation occurs.

3.4.2 THE GROWTH OF *PSEUDOMONAS* SP. ON A MIXTURE OF GLUCOSE AND PHENOL

These bacteria were isolated from oil-polluted soil and are able to utilize with almost equal efficiency both glucose and phenol. Batch culture of *Pseudomonas* sp. was grown on a glucose–phenol mixture with the usual concentration of mineral components. Glucose was assimilated first, then growth ceased and during 2–3 days bacterial numbers gradually declined. After lysis of about one-third of the cells there was the second growth phase on phenol (Figure 3.36). Repeated addition of the glucose–phenol mixture resulted in a very interesting effect of change in substrate preference: phenol was taken up first followed by glucose.

Thus, diauxie on the glucose–phenol mixture was characterized by the following features:

- a long midway lag phase preceding the consumption of the second substrate, during which a significant fraction of cells grown on the first substrate perished from starvation or toxicity exerted by the second one;
- substrate preference is not absolute and depends on the prehistory of the culture inoculum (i.e. the substrate that had been used to obtain it).

It is clear that the transition from one substrate to another is accomplished via the modification of the metabolic machinery of all (or at least the majority of) cells of the populations rather than by an autoselection of mutants. Otherwise, we would not observe such a steep increase in biomass during the second growth phase. Bacteria grown in chemostat culture consumed both substrates simultaneously (data not shown).

Thus, the two considered pairs of substrates (glucose–starch and glucose–phenol) revealed two types of diauxie. Quantitative interpretation of the first type is straightforward and is allowed by relatively simple

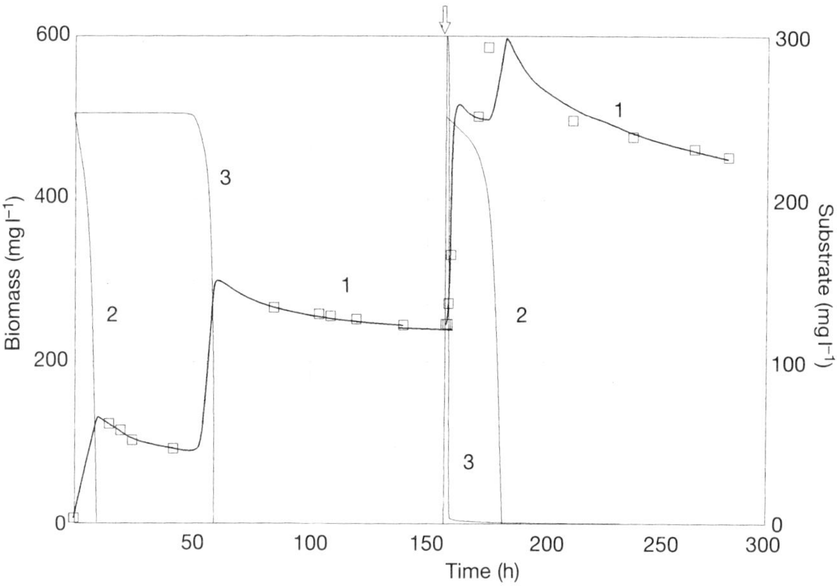

Figure 3.36 Batch growth of *Pseudomonas* sp. on a mixture of glucose and phenol: cell biomass (curve 1), residual glucose (curve 2), and residual phenol (curve 3). The medium containing 250 mg l^{-1} glucose and 250 mg l^{-1} phenol was inoculated by cells initially grown on glucose (note that glucose was the preferential substrate). At the time indicated by the arrow the glucose–phenol mixture was added for the second time (note that in this case phenol was preferential substrate). The curves were simulated by the synthetic chemostat model (SCM) (Table 3.12) with the following parameters: $Q_1 = 0.72$, $Q_2 = 0.5$, $Q_L = 50$, $Q_m = 0.03\ h^{-1}$; $k_1 = 0.87$, $k_2 = 1.67$, $K_L = 1.0$, $K_m = 0.01$ g/g; $K_1 = K_2 = 0.5$, $K_r = K_\rho = 0.2$, $s_1^* = s_2^* = 0.001$ mg l^{-1}; $r^* = \rho^* = 0.34$; at $s_1 > s_1^*$ $a_1 = 0$ and $\beta_1 = 1.0$; at $s_2 > s_2^*$, $a_2 = 0$ and $\beta_2 = 1.0$; at $s_1 < s_1^*$, $a_1 = 0.05$ and $\beta_1 = 2.0$; at $s_2 < s_2^*$, $a_2 = 0.02$ and $\beta_2 = 5.0$. (Panikov, 1991a.)

kinetic models. The second type of diauxie is much more complicated and requires structured models for its explanation. They should take into account the profound modifications in the physiological state of cells which occur during the intermediate lag phase.

3.5 MORPHOLOGY AND SIZES OF CELLS

It was noted as early as the start of the century that microbial growth in terms of **biomass increase** is not identical to cell multiplication in terms of **cell number** increase. Rod-shaped bacteria, mainly of the *Enterobacteriaceae* family, provided a solid database for the conclusion that average cell volume increases parallel to growth rates in batch culture or chemostat (Herbert, 1961b; Stepanova *et al.*, 1980). Our plan now is to survey the other morphological types of bacteria, including spherical, ellipsoid, and Y- and V-shaped, branched cells.

Cell shape

The trend in changes of bacterial morphology induced by a progressive increase of D in chemostat culture is outlined in Figure 3.37. The rod-shaped bacteria *Pseudomonas* and *Bacillus* increased their length, preserving practically constant diameter. *Arthrobacter* displayed a morphogenetic transition from cocci to simple rods as well as V- and Y-shaped cells. *Rhodococcus* cells behaved in the same way, but at low D instead of simple cocci they formed irregular ellipsoid cells. *Micrococcus* exhibited a transition from regular cocci to triangle cells or 'grape–grain' type of cells. Nodule bacteria displayed the highest degree of pleomorphism, forming spherical, globular, branched and rod-shaped cells with a general tendency for cell size to increase with growth acceleration. Under N-limitation, cell diameters were bigger than under C-limitation, the difference being well expressed at low D and almost undetectable at high D. In the continuous culture of coryneform bacteria we observed an intensive branching of cells (up to 5–15 tips) during transition from one D to another. However, these unusual cells disappeared after approaching steady-state.

Cell size distribution

To quantify the observation we measured the sizes of 300–400 cells at each D on photoprints obtained under known magnifications with scanning electron microscopy. An example of the generated distribution of bacteria with respect to cell lengths l is shown in Figure 3.38 in the form of a cumulative frequency, $F(l)$.

Advanced statistical analysis of cell size distributions has been done in many excellent studies (Eakman *et al.*, 1966; Romanovsky *et al.*, 1984). Here we will suggest a simplified treatment which is well matched to the low precision of ordinary manual measurements. Let us assume that there are two sources of gross l variation: (a) random variation due to unknown reasons which is governed by Gaussian law; and (b) regular variation due to the fact that every individual cell passes through a cell cycle. Let us assume further that every individual cell during its cell cycle increases roughly with constant rate from l_0 (length of daughter cell at birth) to $2l_0$ (mother cell at the instant of division). Then the second type of size variation in a completely asynchronous culture should produce a uniform distribution of l between l_0 and $2l_0$. The combined effect of random and regular variations should not alter the mean, since both normal and uniform distributions are symmetrical. Therefore a cell with the mean

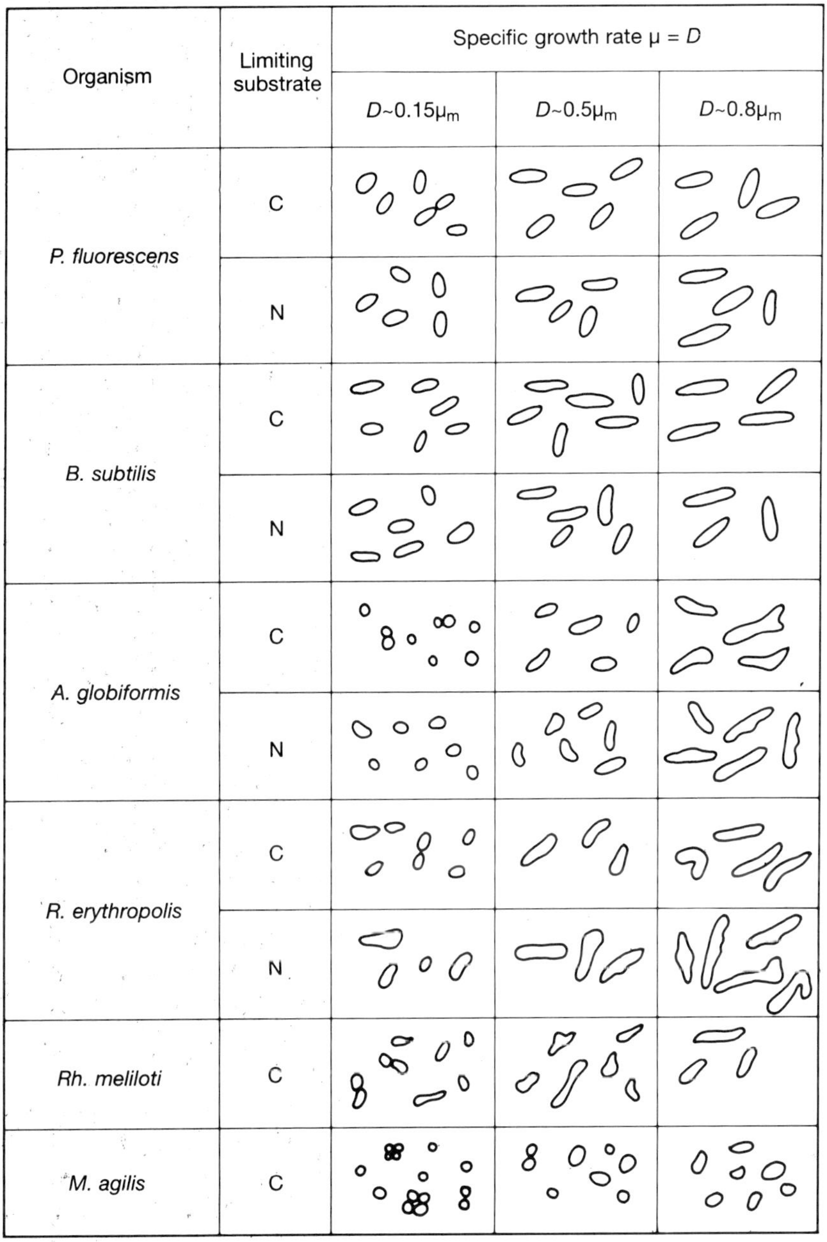

Figure 3.37 Variations of the cell shape as dependent on dilution rate in chemostat cultures limited by glucose or N.

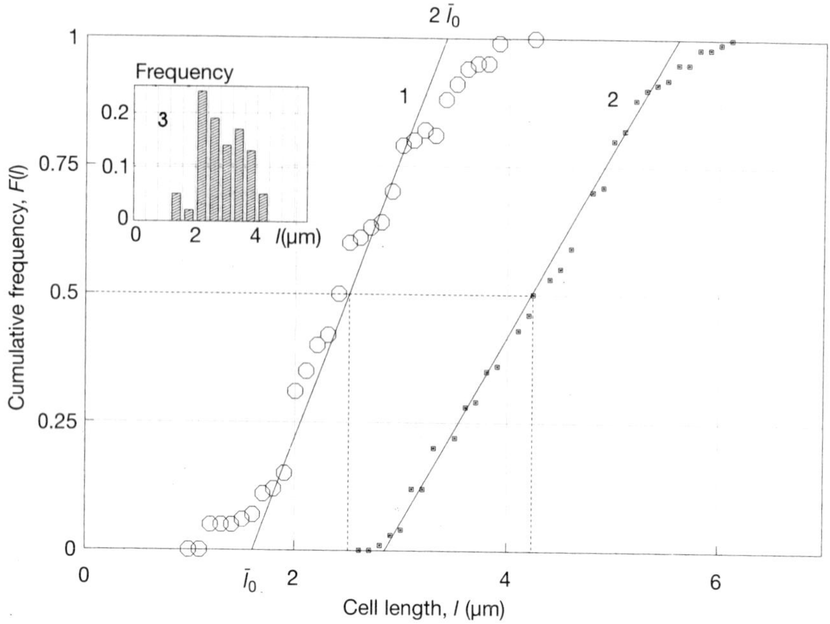

Figure 3.38 Examples of cell lengths distribution of *B. subtilis* grown at $D = 0.029$ (curves 1 and 3) and $D = 0.33\,h^{-1}$ (curve 2) under N-limitation. Curves 1 and 2 are cumulative frequencies used to assess the average length of cell at birth, l_0. (Bondarenko *et al.*, 1985.)

length $\bar{l}$ should occupy the position in the middle of the cell cycle, and the average value of the cell length at birth may be found very simply by

$$\bar{l}_0 = \frac{2}{3}\,l \qquad [3.45]$$

The real distribution of rod-shaped bacteria (Figure 3.38) was indeed symmetrical. For example, the mean cell length of bacilli at $D = 0.029\,h^{-1}$ was $2.52 \pm 0.68\,\mu m$, practically the same as the median, $2.4\,\mu m$. The pattern of the frequency distribution (see insert 3 in Figure 3.38) resembles the combination of a normal and a uniform one. Thus, the described simplistic model is adequate in respect of rod-shaped bacteria. As far as other morphological types are concerned we would suggest the use of a **single rod equivalent** (SRE). The length of such an equivalent for branched bacteria having n tips and a total length L is calculated as follows:

$$l = \frac{L}{n-1}. \qquad [3.46]$$

The determined l value is equal to that imaginative length of branched bacteria, if, instead of branch initiations they would divide. V- and Y-shaped cells of *Arthrobacter* and *Rhodococcus* are the particular case of branched bacteria when $n = 3$ (corresponding to two SRE). In a cocci population the length of a SRE should be equal to the diameter (truly spherical isometric cells) only in the particular cases of new-born daughter cells or those cocci which produce four daughter cells. Otherwise, cocci should have an ellipsoid shape between consecutive binary divisions, the length of the SRE being assigned as an ellipse major axis.

The dependence of cell size variations on *D*

With the given assumptions, we analysed the distributions of the studied bacteria with respect to average length of SRE, calculated the average length of a new-born cell, l_0, and plotted them *versus* the normalized specific growth rate $(\mu = D)/\mu_m$ (Figure 3.39). Apart from length, the cell diameters were measured at every D and average values d were calculated. For all the tested organisms l_0 increased linearly with growth rate, thus purely empirically we have

$$\bar{l}_0 = \bar{l}_0^{\min} + \zeta\mu\ , \qquad \zeta = \frac{l_0^{\max} - l_0^{\min}}{\mu_m} = \text{constant}\,. \qquad [3.47]$$

Parameter ζ reflects the degree of size variability, the highest ζ values being observed for pleomorphic bacteria, *Arthrobacter*, *Rhodococcus* and *Rhizobium* (8–23 μm h^{-1}). Since cell length increases by l_0 μm every generation time, $g = (\ln 2)/\mu$, then the rate of cell elongation is calculated as

$$K_R = \frac{\bar{l}}{g} = \frac{\bar{l}_0 D}{\ln 2}\,. \qquad [3.48]$$

Maximum values of K_R occur under unlimited growth with μ_m, varying from 0.39 μm h^{-1} in *Micrococcus* up to 2.4 μm h^{-1} in *Bacillus* culture.

Morphological variability of oligotrophic microorganisms

The most intriguing morphological transformations occur in the populations of *Prosthecomicrobium* and *Caulobacter* which, according to the paradoxal Parkinson's law, were cultivated in chemostat with the greatest difficulties. Contrary to other bacteria the length of SRE declined with the increase of growth rate. Under oligotrophic conditions these bacteria form prosthekae which significantly enlarge their absorbing surface and transport capacity. On the rich media prostekae disappear and the cells acquire a more common shape. This type of anomaly is beyond the scope of the present study and should be analysed specifically. It may be worthwhile to

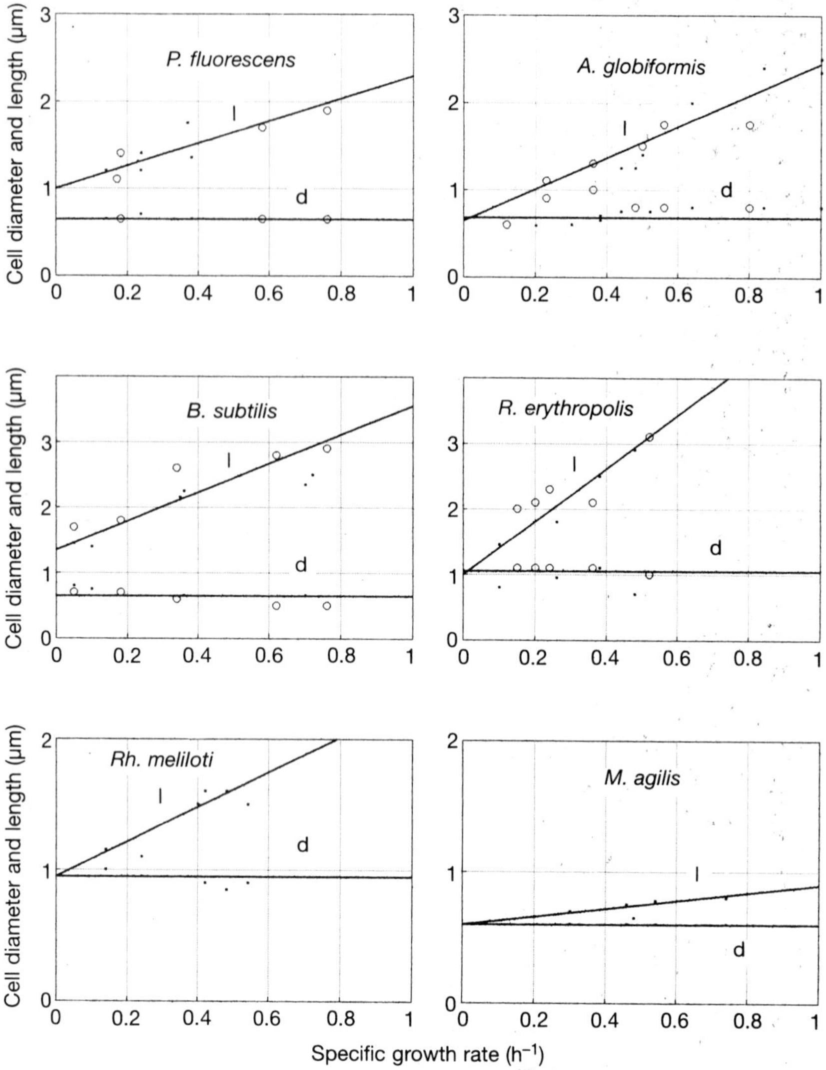

Figure 3.39 The relationship between cell size and specific growth rate in chemostat cultures. Ordinate is the length of a 'single rod equivalent' (SRE) for average cell at birth (curve 1), and average cell diameter (curve 2). The filled circles represent glucose limitation; open circles represent N-limitation. (Bondarenko *et al.*, 1985.)

use the aforementioned anomaly as a formal identification characteristic of oligotrophy. The same trend to increase cell size under severe substrate limitation (in dialysis, prolonged fed-batch and spend batch cultures) was observed in the population of the soil yeast *Lipomyces tetrasporus*.

3.6 THE DEVELOPMENT OF A GENERAL KINETIC MODEL

Sections 3.2–3.5 dealt with a wide spectrum of growth phenomena. Some experimental results have been formally described by kinetic (mathematical) models of different degrees of complexity. However, we are not fully content with these models for the following reasons:

1. They are empirical or semiempirical, being introduced without substantial theoretical justification.
2. They remain valid for only a restricted range of experimental conditions, mainly for steady-state growth, and in the case of transient processes or unbalanced growth they fail miserably.
3. The suggested models preserve the rudimentary deficencies inherited from chemical and enzymological kinetics in the sense that they treat a microbial culture as no more than a 'multiplying enzyme'. They ignore the optimal nature of microbial regulatory processes, do not simulate adaptive 'teleonomic' behaviour of microorganisms and fail to predict the qualitative changes of microbial biocatalysts resulting from microbial adaptation to a changing environment.

The last point seems to be the most critical. To fill the gap we shall concentrate first of all on optimization models, which most naturally are associated with such terms as adaptation, regulation and control.

3.6.1 OPTIMIZATION MODELS OF CELL GROWTH

The premise that nature in all its manifestations attempts optimal decision making is one of the oldest principles of theoretical science (Fermat, Mopertue, Pontryagin). Thus, the laws of geometrical optics (refraction and reflection) were derived from Fermat's principle, which states that in travelling from one point to another a ray of light will take the path of minimum or maximum time as opposed to adjacent arbitrary paths. The introduction and use of an optimization principle is justified only by the fact that it almost always produces elegant solutions to complicated problems and gives constructive, practical results. This has been already confirmed for some particular aspects of microbial stoichiometry and kinetics. Optimization principles have been successfully used to calculate gross biomass yield, Y, from the data on individual biochemical reactions constituting the cell's metabolic network (Skurida *et al.*, 1984). We will now attempt to use an optimization principle for the generalization of the dynamic data obtained in sections 3.2–3.5.

Model description

Let us portray cell growth as being the result of exosubstrate transformation into cell components via a **metabolic network** which is in fact a highly complex combination of branched and alternative metabolic pathways and cycles of interdependent syntheses. Each particular route from exosubstrates to an entire cell has its own 'cost' in terms of limiting substrate expenditure per biomass production rate (expressed in the case of chemostat culture as $Ds_0/\mu/x$), but there is only one route with the lowest cost: $Ds_0/(\mu x) \rightarrow$ min. For a given D and s_0 this route corresponds to the maximum productivity of a chemostat culture $\mu x \rightarrow$ max. We postulate that it is this route which is selected by microbial cells, due to the operation of natural mechanisms of metabolic control.

Let us start with a relatively simple example of steady-state microbial growth in chemostat culture limited by a conserved substrate, e.g. by the source of N. Alternative metabolic pathways are given by the flow diagram (Figure 3.40). There are six state variables altogether: limiting exosubstrate (NH_4^+ in the medium), intracellular pool of monomeric N-compounds, L (NH_4^+ in cytosol, amides, amino acids, purine and pyrimidine derivatives); transport proteins, P_1; enzymes catalyzing the synthesis of non-protein constituents, P_2; ribosomes R (the entire multienzymatic complex involved in the synthesis of proteins); and inert macromolecular cell components, U (structural elements of biomass, including the cell envelope, reserved compounds etc.). A pool of monomers, L, is formed from the transported exosubstrate, then monomeric precursors enter a branched pathway to be used for the biosynthesis of four polymeric components, P_1, P_2, R and U. The first bifurcation point subdivides the L flux into two subfluxes. One is directed towards protein synthesis (proportional to the ribosomal content, R) and the second leads to non-protein components (proportional to P_2). The next separation of L subfluxes is under the control of two cybernetic variables, α and β, which occupy the second and the third bifurcation points.

Regulators α and β may vary within the open interval [0, 1],

$$0 < \alpha < 1\,, \quad 0 < \beta < 1\,,$$

so that they operate like a mechanical valve or a water tap changing the L flux between zero to full strength. Each pair of particular values α and β determines one, and only one, set of all six assigned steady-state variables since the end-products, P_1, P_2 and R, exert positive feedback on the starting and first bifurcation points. Thus α and β may be used as tools of optimization control. They may be interpreted as a simplistic model (or even a caricature!) of the *gen*-operator which sets the frequencies of messenger RNA's replication and initiates the synthesis of particular enzymes.

The flow diagram is translated into a set of differential equations compiled in Table 3.6 (for details of the derivation see the Appendix). It should be stressed that the obtained structured model no longer contains simple expressions for macrokinetic and macrostoichiometric parameters. Biomass yield is derived from Eqn. [3.28] and Table 3.6 as

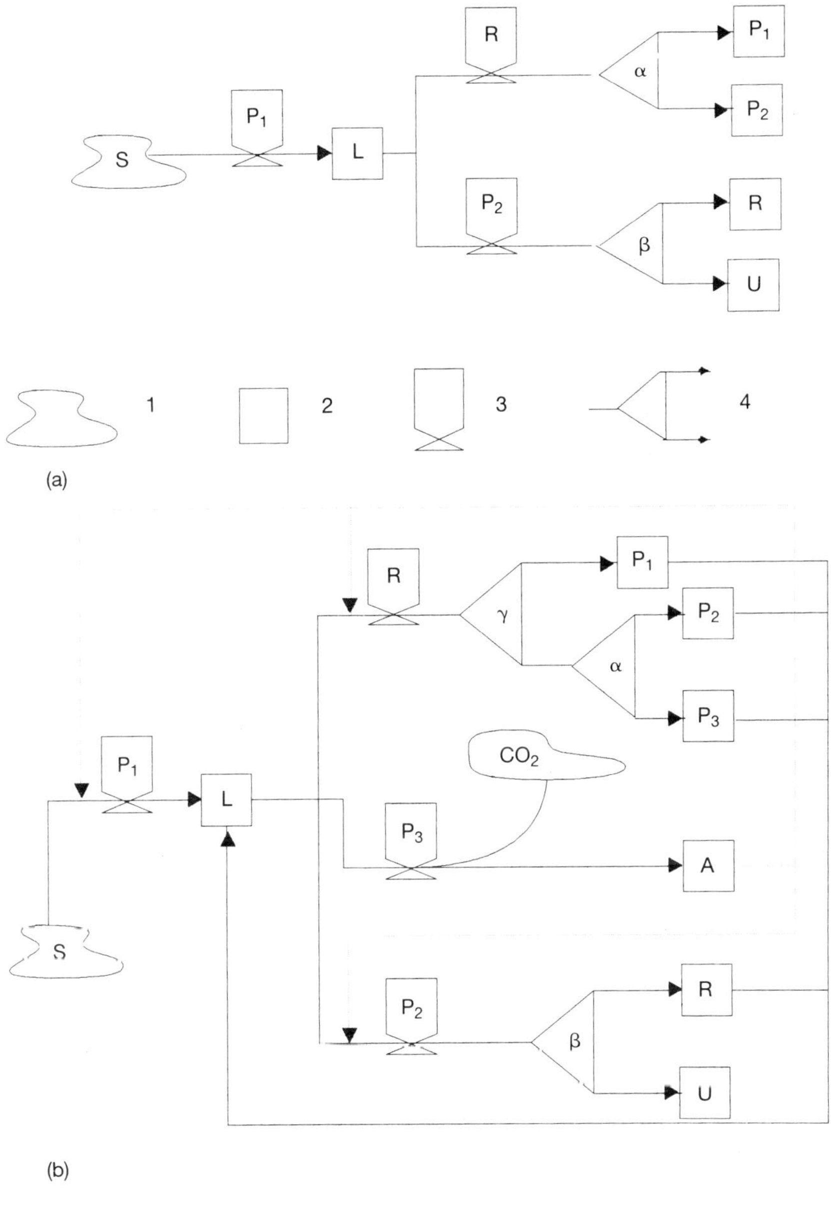

Figure 3.40 Flow diagram of microbial growth used to formulate optimization of structured models. (a) Growth limited by N source. (b) Growth limited by the source of C and energy. 1, Extracellular elements; 2, intracellular components; 3, fluxes; 4, regulators. For definitions of P_1, P_2, P_3, R, U, A, S and L, refer to Tables 3.6 and 3.7.

Table 3.6 Optimization model of microbial growth in chemostat culture limited by N source

Variable	Equation	
Limiting substrate (mg N l^{-1})	$\dot{s} = D(s_0 - s) - q_s P_1 x$, $q_s = k_s \dfrac{s}{K_s + s}$	[A]
Monomeric pool (g N g^{-1} biomass)	$\dot{L} = q_s P_1 - q_L(R + P_2) - \mu L$, $q_L = k_L \dfrac{L}{k_L + L}$	[B]
Transport proteins (g N g^{-1} biomass)	$\dot{P}_1 = \dfrac{\alpha q_L R}{\sigma_P} - \mu P_1$	[C]
Enzymes of non-proteinaceous synthesis (g N g^{-1} biomass)	$\dot{P}_2 = \dfrac{(1-\alpha) q_L R}{\sigma_P} - \mu P_2$	[D]
Ribosomes (g N g^{-1} biomass)	$\dot{R} = \dfrac{\beta q_L P_2}{\sigma_R} - \mu P_3$	[E]
Structural cell components (g N g^{-1} biomass)	$\dot{U} = \dfrac{(1-\beta) q_L P_2}{\sigma_U} - \mu R$	[F]
Biomass (mg l^{-1})	$\dot{x} = \left[q_s P_1 - q_L R \left(1 - \dfrac{1}{\sigma_P} \right) + q_L P_2 \left(1 - \dfrac{1-\beta}{\sigma_U} - \dfrac{\beta}{\sigma_R} \right) \right] x - Dx$	[G]

$$Y_s = (\sigma_N)^{-1} = (L + \sigma_x x' + \sigma_P(P_1 + P_2) + \sigma_R R)^{-1} . \qquad [3.49]$$

Even more complicated is the expression for μ:

$$\mu = q_s P_1 - q_L \left[R \left(1 - \frac{1}{\sigma_P} \right) + P_2 \left(1 - \frac{1-\beta}{\sigma_x} - \frac{\beta}{\sigma_R} \right) \right] . \qquad [3.50]$$

The first major term of this expression is the specific rate of substrate uptake, $q_s P_1$, while the others are related to intracellular fluxes. In spite of its terrifying image this expression may be readily used for the calculation of the $\mu(s)$ relationship by the simultaneous solving of the entire equation set A–G (Table 3.6). For some particular combination of parameters, the resulting plot $\mu(s)$ would resemble those which are generated by the simple formulae of non-structured models (Eqns [1.25], [1.50–1.52], [3.29], etc.); however, it is utterly unreasonable to reduce the correct expression [3.50] to such oversimplified formulae as a final stage of model derivation (see instances of this general practice in Romanovsky *et al.* (1984), since this reduction is equivalent to the model's desecration).

In the case of growth limitation by the source of carbon and energy, the flow diagram is augmented by the respiratory enzymes (P_3), an ATP pool (A), and the third cybernetic variable, γ (Figure 3.40; Table 3.7). The variable A does not participate in the mass balance since ATP belongs to L pool, but it is constrained by energy conservation equation which accounts for all sources and sinks for ATP (see Eqn. [H] in Table 3.7). Molar ATP yield was assumed to be 155 mmol (g substrate)$^{-1}$ corresponding to P/O = 2. Other parameters were identified from the data summarized by Stouthamer (1977; Table 3.8).

In addition, we took into account the fact that some cell polymers (P_1–P_3 and R) are turned over with specific rate 0.06 h^{-1} (Pirt, 1975), and that some fraction of ATP is required for maintenance (m_{ATP}). All rates were supposed to obey simple Michaelis–Menten kinetics. For the sake of simplicity, the effects of the second substrate concentration (mainly ATP) in 2-substrate reactions was accounted by the threshold conditions, e.g.: $q_L \geqslant 0$, if $A > 0$, otherwise $q_L = 0$.

Computer simulation

Under steady-state conditions the differential equations of the structured models (Tables 3.6 and 3.7) were reduced to sets of algebraic equations. They are susceptible to exact analytical solution, if we provide numeric values for both biological constants such as maximum rates, saturation constants, etc. (see the complete list in Tables 3.6 and 3.7) and cybernetic parameters α, β and γ. To find the required optimal values of α, β and γ we should minimize the 'cost' of growth, $Ds_0/(\mu x) \rightarrow \min$, for the entire range of permissible D variation, $0 \leqslant D \leqslant D_{washout}$. This problem is equivalent to the search of for a maximum value of $\bar{x}$ for a given s_0 and D:

$$\bar{x}(\alpha,\beta,\gamma) \rightarrow \max\,, \quad 0 \leqslant \alpha \leqslant 1,\, 0 \leqslant \beta \leqslant 1,\, 0 \leqslant \gamma \leqslant 1\,.$$

The numerical solution was performed with a simple computer routine based on random search (Rafales-Lamarka and Nikolayev, 1971).

Simulation results are shown in Figures 3.41–3.43. Let us first comment on N-limited growth. The optimization principle allowed the successful simulation of the observed pattern of D-dependent variations of steady-state biomass, $\bar{x}$, and N content of cells, σ (Figure 3.41). Moreover, it

Table 3.7 Optimization model of microbial growth in chemostat limited by the source of carbon and energy

Variable	Equation	
Limiting substrate ($mg\ l^{-1}$)	$\dot{s} = D(s_0 - s) - q_s P_1 x$, $q_s = k_s \dfrac{s}{K_s + s}$	[A]
Monomeric intermediate ($g\ g^{-1}$ biomass)	$\dot{L} = q_s P_1 - q_L(R + P_2 + P_3) + a(P_1 + P_2 + P_3 + R) - \mu L$	[B]
Transport proteins ($g\ g^{-1}$ biomass)	$\dot{P}_1 = \gamma q_L R - (\mu + a)P_1$, $q_L = k_L \dfrac{L}{K_L + L}$	[C]
Enzymes of non-proteinaceous synthesis ($g\ g^{-1}$ biomass)	$\dot{P}_2 = \alpha(1 - \gamma)q_L R - (\mu + a)P_2$	[D]
Respiratory enzymes ($g\ g^{-1}$ biomass)	$\dot{P}_3 = (1 - \alpha)(1 - \gamma)q_L R - (\mu + a)P_3$	[E]
Ribosomes ($g\ g^{-1}$ biomass)	$\dot{R} = \beta q_L P_2 - (\mu + a)R$	[F]
Inert polymeric components ($g\ g^{-1}$ biomass)	$\dot{U} = (1 - \beta)q_L P_2 - \mu U$	[G]
ATP ($mmol\ g^{-1}$ biomass)	$\dot{A} = Y_{ATP} q_L P_3 - q_L[R a_P + \beta P_2 a_R + (1 - \beta)P_2 a_U] - q_s P_1 a_T - m_{ATP} - \mu A$	[H]
Biomass ($mg\ l^{-1}$)	$\dot{x} = (\mu - D)x, \quad \mu = q_s P_1 - q_L P_3$	[I]

explains clearly this variation by adaptive modifications of cell composition. At low $D = \mu$, minimization of the 'growth cost' was attained by enlarging the inert cell components U with low N content ($\sigma_U = 40$ mg N g^{-1}). Functionally active components R and P_i with high N content ($\sigma_P = \sigma_R = 160$ mg N g^{-1}) were, under these conditions, at their minimum. However, at high $D = \mu$ their contribution progressively increased,

Table 3.8 ATP requirements for cell biosynthesis and function. (After Stouthamer, 1977)

Cell component or function	Fractional mass	ATP requirements, mmole g^{-1}		
		Of biomass	Of cell component	Average for group
Polysaccharides	0.166	2.05	12.35	
Lipids	0.094	0.15	1.0	$a_U = 11.13$
DNA	0.032	1.05	32.81	
Proteins	0.524	20.5	39.12	$a_P = 39.12$
RNA	0.157	5.86	37.32	$a_R = 37.32$
Transport of organic substrates	–	12	–	
Transport of mineral substrates	–	12	–	$a_T = 60$
'Membrane energization'		36		

since growth acceleration was impossible without the build up of the intracellular content of ribosomes and key enzymes.

Simulated *D*-dependent variations of cell composition are in reasonable qualitative agreement with experimental data on cell RNA content (Herbert, 1961a; Chmel, 1970), activity of the enzymes of ammonium assimilation (Gräzer–Lampart *et al.*, 1986), and reserved C-polymers such as starch (Pirt and Pirt, 1977). Always, an increase of *D* induces the enlargement of the relative activity of those cell components which are imitated by our variables R and P (RNA, enzymes of primary metabolic pathways), and the diminishing of inert components equivalent to U (reserved polymeric compounds).

To simulate the observable changes of cell composition at a quantitative level, we have to augment our structured model (Table 3.6) because of one obvious point: the synthesis of some cell components should be performed constitutively under any growth conditions. It was accepted that the constitutive compartment amounts to 30% of the total cell mass and includes DNA (4%), carbohydrates (5%), lipids (6%) and proteins including cell wall peptidoglycan (15%). The computer routine was modified to give top priority to the biosynthesis of these constitutive components. Figure 3.42 demonstrates reasonable agreement of such modifications with the available experimental data.

However, the most striking result was the simulation of the specific growth rate dependence on *s*. Although the individual concentrational relationships were Michaelis–Menten hyperbolic forms, the resultant curve

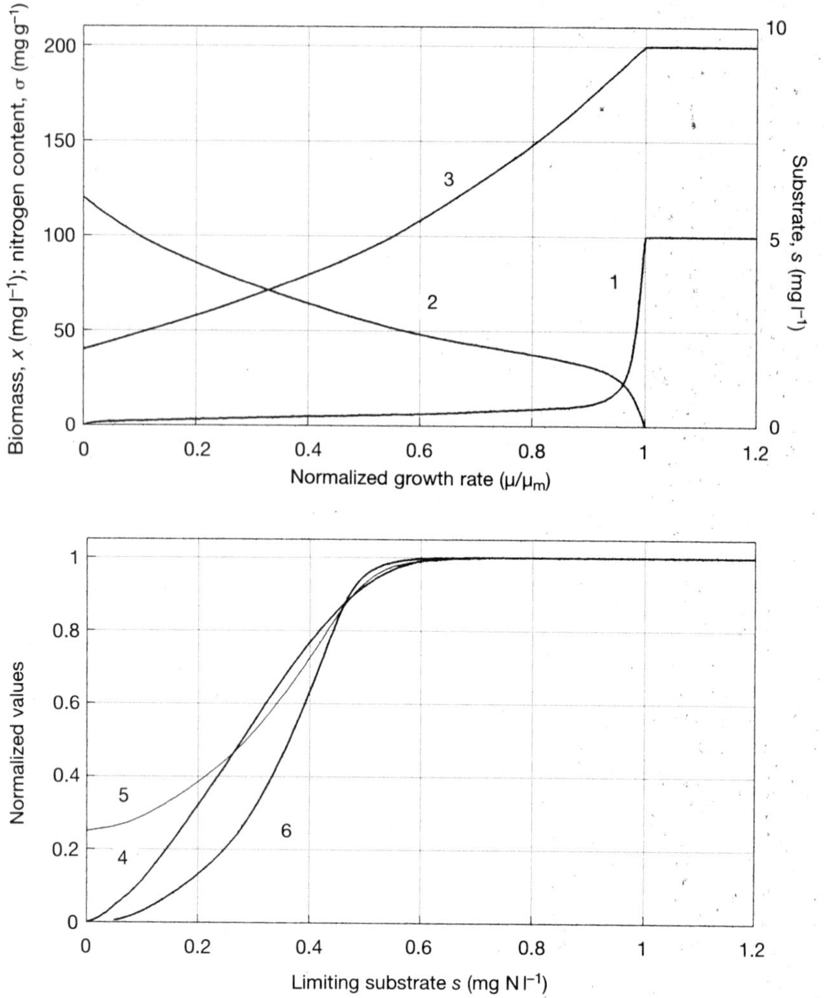

Figure 3.41 Simulation of steady-state N-limited growth: limiting substrate (curve 1), biomass (curve 2), N-content in cell (curve 3), μ/μ_m (curve 4), σ_N/σ_N^m (curve 5), P_1/P_1^{max}, R/R^{max} (curve 6). For notation, see Table 3.6.

$\mu(s)$ was sigmoid. The sigmoidal relationship was even more pronounced in the case of the stoichiometric variables σ_N, P_1 and R (Figure 3.41), indicating their responsibility for this phenomenon. We may infer that significant deviation from a simple hyperbolic relationship takes place when the content of biocatalysts in the cells increases with an increase of s which is steeper than that expected from hyperbolic or even linear law. It is this type of $\bar{s}$-dependency which provides the maximization of the growth rate. This suggested explanation is somewhat uncertain but is more

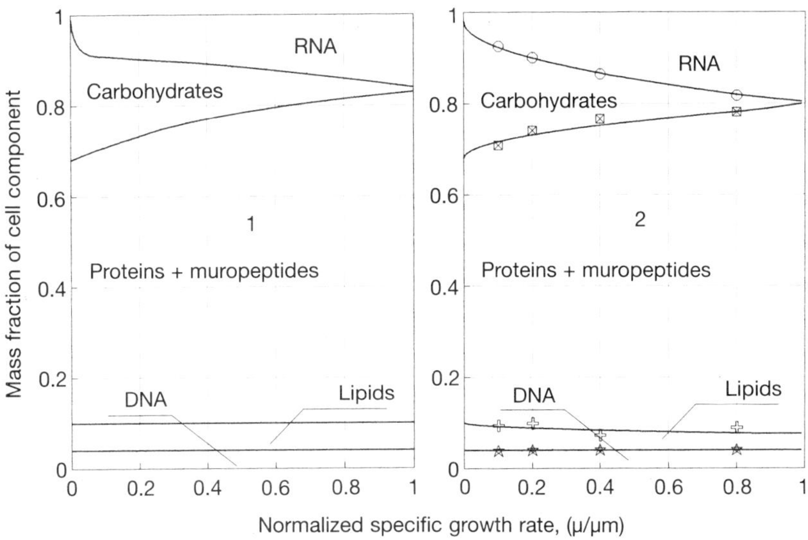

Figure 3.42 Simulation of *D*-dependent changes in cell composition of *A. aerogenes* grown in NH_4^+-limited chemostat culture: simulation by optimization model (curve 1) and by synthetic chemostat model (SCM) (curve 2). Experimental points are redrawn from Tempest *et al.* (1965).

trustworthy and less perplexing than the two hypotheses formulated above as Eqns [3.29] and [3.30] (respectively nutrient leakage and cooperativity effects of a bottle-neck enzyme).

Growth limited by the source of C and energy was also simulated adequately. The explanation of *D*-dependent variations of steady state biomass (Figure 3.43) is trivial, since a decrease of $\bar{x}$ at low D associated with maintenance was imitated by much simpler models. More important is the fact that the macroscopic parameter Y was reproduced with reasonable accuracy from the microstoichiometric parameters a_U, a_R, a_T, P/O, etc. *D*-dependent variations in macrocomposition were basically the same as in the case of N-limited culture (the complete data are presented in Panikov, 1991a). That is, the contribution of P- and R-components (transport proteins, respiratory and anabolic enzymes, ribosomes) progressively increased with the increase of $D = \mu$. The identity of the possible outcomes at different (in biochemical terms) optimization goals – economizing on nitrogen substrates under N-limiting growth and on glucose under C and energy limitation – is explained by the fact that the biosynthesis of nucleic acids and proteins both require an enhanced supply of N and energy as compared with inert cell constituents.

New unexpected conclusions from the simulation of C-limited growth were as follows:

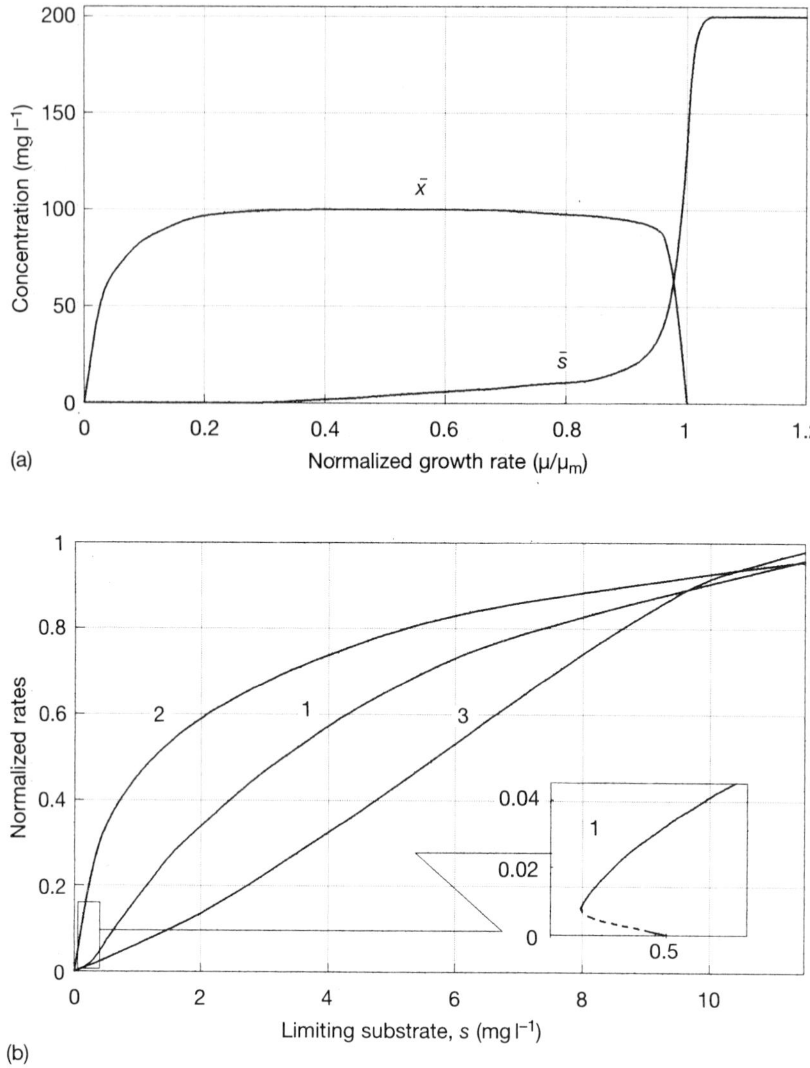

Figure 3.43 Simulation of steady-state C- and energy-limited growth. (a) Limiting substrate (s) and biomass (x) *versus* normalized specific growth rate. (b) The simulated effect of limiting substrate concentration on μ/μ_m (curve 1), P_2/P_2^{max} (curve 2), and R/R^{max} (curve 3). For notation, see Table 3.7.

1. At all D, maximization $\bar{x}(\alpha, \beta, \gamma) \rightarrow$ max was equivalent to the **minimization of the ATP pool size** ($\bar{A}(\alpha, \beta, \gamma) \rightarrow$ min). Thus high ATP content appeared to be detrimental for cell metabolism.
2. Contrary to the case of N-limitation, the dependencies of μ, R and P_1-P_3 on s were **not sigmoidal** (Figure 3.43, bottom panel), but of a

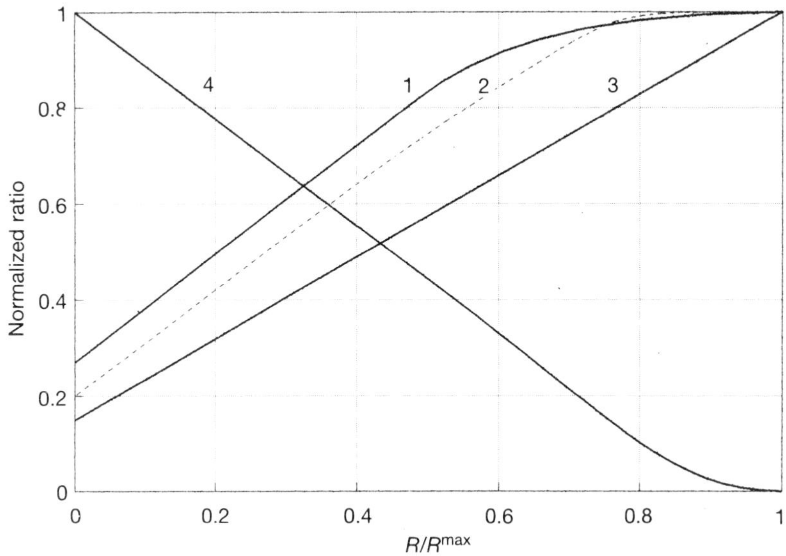

Figure 3.44 The demonstration of almost linear relationship between intracellular content of macromolecular components. Extract from simulation with optimization model (Table 3.7).

hyperbolic type. There is not a single and straightforward explanation of this difference; however, it is clear that the absence of a sigmoidal dependence under C-limitation is associated a with narrower range of cell composition variation.

3. At very low D the model predicts **growth instability**, which is manifested as a decrease of μ with an s increase (see dotted line on the insert to Figure 3.43). If $d\mu/d\bar{s} < 0$ then the state is unstable, i.e. under small perturbations the culture should be washed out or converge to a stable state with higher μ. The critical value $d\mu/ds = \infty$ corresponds to $D \sim 0.01\ \mu_m$. There is a temptation to define (following Tempest *et al.*, 1967) this point as a 'critical minimum dilution rate'. What may be the biological interpretation of this simulation? Probably, at very low s the chosen optimization goal (maximization of growth rate) fails to harmonize the metabolic network, adaptation to an oligotrophic environment requiring other types of optimal strategy.
4. The model simulates the **multistability of steady-states** producing two or more equitable local maxima of the function $\bar{x}(\alpha, \beta, \gamma) \rightarrow \max$ in the α,β,γ-space (note that the set of the two cybernetic variables α and β provides only one unique solution of the optimization problem). Therefore, complex structured models with many controlling parameters are expected to generate several ways to attain the strategic goal.

The common feature of N- and C-limited growth was that the particular indexes of cell composition **intercorrelated** with each other. Normalized values of all entities (P_1/P_1^{max}, P_2/P_2^{max}, R/R^{max}, X'/X'^{max}) were related to each other in an almost linear fashion (Figure 3.44). This remarkable property of the optimized model will be used in the next, more manageable model (see section 3.6.2).

Prospects for the application of optimization models

Until now we have described only steady-state growth. To widen the scope of this approach we have to solve the rather difficult problem of the numerical integration of the differential equations with the simultaneous optimal control of intracellular fluxes, e.g. by maximization of the function $\mu x(\alpha, \beta, \gamma) \rightarrow$ max at every integration step. This problem is aggravated by the awkward nature of those differential equations which describe the most interesting growth effects, i.e. they are non-linear and stiff, requiring numerical integration to proceed in small steps (see supplement for details).

Even greater difficulties arise if we try to simulate slightly more complicated situations than uptake of a single substrate. For instance, an account of growth dependence on two substrates results in the following complications:

- an increase in the number of equations from 5–7 to 15–20;
- an increased number of cybernetic variables $\alpha, \beta, \gamma, \ldots$ by two to three times with a concomitant amplification of alternative solutions and uncertainty;
- the complication of the kinetic expressions for metabolic rates which should include the intracellular concentrations of two to four substrates;
- the absence of an analytical solution to calculate $\bar{x}$ or any other complex entity used as an optimization function.

Thus, the realization of the optimization model is possible but the price would be very high and nobody can guarantee that it would be worthwhile. In fact, the complicated model (especially based on the optimization principle) seems no better for gaining understanding than the archetype itself (the real microbial culture). From this statement we derive the pessimistic conclusion that in the near future optimization models will not be effective tools in microbial kinetics. Instead, we turn to another approach based on deductions that follow from the optimization principle.

3.6.2 SYNTHETIC CHEMOSTAT MODEL (SCM)

The name coined for this structured model (Panikov, 1991a,b) stems from two considerations. Firstly, SCM embraces all known chemostat models, including classic versions (Monod, 1950; Herbert *et al.*, 1956; Ierusalimsky, 1967), new developments based on the cell quota concept (Nyholm, 1976), structured models (Guthke and Knorre, 1980; Esener *et al.*, 1982; Romanovsky *et al.*, 1984; Domach *et al.*, 1984), cybernetic models (Kompala *et al.*, 1986; Turner *et al.*, 1989) and, finally, the optimization models which have been just discussed. Secondly, SCM is used for the **synthesis** of empiric information on microbial behaviour under various conditions and for any cultural system. Formally, SCM contains the term D, dilution rate, which is set in chemostat culture in the range $0 < D < \mu$. For other cultures we define the D-value as follows:

simple batch culture $D = 0$

dialysis culture $Ds_0 = \text{constant}, \quad Ds = Dx = 0$

fed-batch culture $D = \dfrac{F}{V(0) + Ft},$

where t is cultivation time, $V(0)$ is the starting culture volume and F is the substrate feed rate.

According to the basic SCM microbial growth is described by the flow diagram

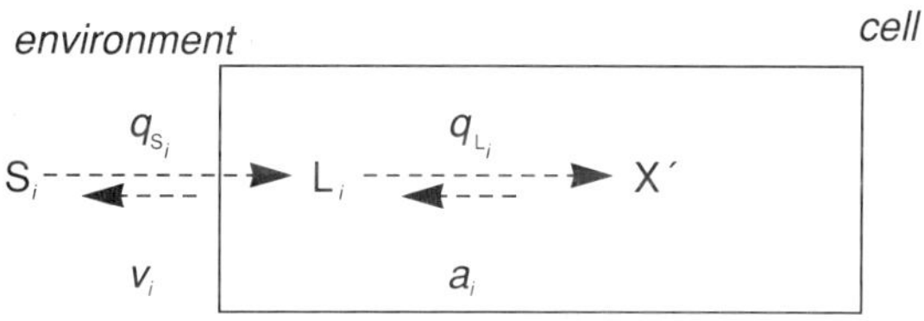

[3.51]

where S_i are substrates ($i = 1, \ldots, n$), L_i are low-molecular weight intracellular intermediates, and X′ is the macromolecular cell components. Arrows stand for specific rates of: S transport (q_{S_i}), L leakage (v_i), L conversion into $X'(q_{L_i})$, and X′ turnover (a_i). The X′-compartment accounts for more than 95% of the total biomass (the monomeric pool does not normally exceed 40 mg g^{-1}, see section 3.2.1), and consists of various cell components. The essence of the SCM is the solution of the problem, how to cope with this severe heterogeneity.

Account of variation of cell composition

According to the SCM all macromolecular cell constituents are divided into two groups. One is represented by **primary** components absolutely necessary for **growth** (P-components), while the second are essential for cell **survival** under any kind of growth restriction (U-components). The characteristic examples of P-components are ribosomes or their constituents – rRNA and ribosomal proteins, other RNA fractions, key enzymes of the primary metabolic pathways. Their intracellular content increases parallel to the increase of growth rate. For example, total RNA content varies from 1–2% of cell dry weight in starving populations and comprises up to 20% in intensively growing culture.

U-components are responsible for persistence or survival of microorganisms under any kind of growth suppression and limitation. Their contribution to the total cell biomass decreases with the increase of growth rate. Examples are the enzymes of secondary metabolism, protective pigments, reserved substances, transport systems of high affinity, etc.

Let us express the concentrations of P- and U-constituents as fractions of total cell mass (g component/g biomass) and denote concentration terms as P and U respectively, then

$$\sum_{k=1}^{n} P_k + \sum_{j=1}^{m} U_j = X' = 1 - \sum_{i=1}^{v} L_i = 1\,, \quad n, m, v\text{: integer}\,, \qquad [3.52]$$

where L_i and X' are also expressed as fractions of total biomass.

Any kind of metabolic activity q_k and q_j may be expressed as a function of cell composition and environmental conditions:

$$q_k = k_{\text{cat}}\, P_k \frac{L_i}{K_{m_i} + L_i}\,, \qquad q_j = k'_{\text{cat}}\, U_j \frac{L_j}{K_{m_i} + L_j} \qquad [3.53]$$

where k_{cat} is a rate (catalytic) constant characteristic for each cell component of an enzymatic nature, its value being dependent on temperature, local pH, tonicity and other environmental factors. The pool size, L, also depends immediately on growth conditions responding quickly to substrate concentration(s) in the medium (as we have shown in section 3.2.1, the response time is typically about 1 min). Contrary to pool size, the quantities P and U remain constant until *de novo* synthesis occurs (a response time of about hours and days depending on the long-term growth conditions).

From this consideration we may draw two conclusions. Firstly, any specific potential metabolic activity $Q_k = k_{\text{cat}} P_k$ or $Q_j = k'_{\text{cat}} U_j$, measured under constant environmental conditions and non-inhibiting excess of substrates ($L_i \gg K_{L_i}$), may be formally treated as a characteristic of cell composition, because it is proportional to the intracellular content of the respective enzymes P_k or U_j. Secondly, the physiological state of the microbial culture (as a set of relatively stable functional properties related to past history) is determined mainly by its macromolecular cell composition. Thus, we may define **the vector of the cell's physiological state, $\boldsymbol{\phi}$**, in the following matrix notation:

$$\boldsymbol{\phi} = \boldsymbol{P} + \boldsymbol{U} = \begin{bmatrix} P_1 \\ P_2 \\ \cdots \\ P_n \end{bmatrix} + \begin{bmatrix} U_1 \\ U_2 \\ \cdots \\ U_m \end{bmatrix} \qquad [3.54]$$

The physiological adaptations resulting from repression–derepression of particular biosynthetic reactions are displayed as P and U variations. Let us confine this variation by upper ($P^{\max}$, $U^{\max}$) and low ($P^{\min}$, $U^{\min}$) limits and assume the following linear approximation of $\boldsymbol{\phi}$:

$$\boldsymbol{\phi} = \begin{bmatrix} P_1^{\min} \\ P_2^{\min} \\ \cdots \\ P_n^{\min} \end{bmatrix} + \begin{bmatrix} U_1^{\min} \\ U_2^{\min} \\ \cdots \\ U_m^{\min} \end{bmatrix} + r \begin{bmatrix} P_1^{\max} - P_1^{\min} \\ P_2^{\max} - P_2^{\min} \\ \cdots \\ P_n^{\max} - P_n^{\min} \end{bmatrix} + (1-r) \begin{bmatrix} U_1^{\max} - U_1^{\min} \\ U_2^{\max} - U_2^{\min} \\ \cdots \\ U_m^{\max} - U_m^{\min} \end{bmatrix} \qquad [3.55]$$

The first two columns of [3.55] symbolize constitutive cell components formed irrespectively of environmental conditions. The third and fourth columns indicate the changeable part of the cell, its biosynthesis being under the metabolic control of repression and induction–derepression. The variable r is no longer a vector, being a simple scalar number. The meaning of the linear transformation [3.55] is illustrated by the following scheme:

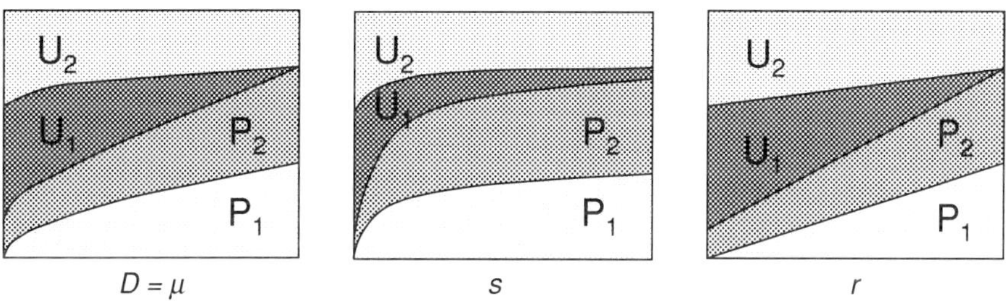

The left-hand drawing portrays D-dependent variations of several cell components, P and U, typically observed in chemostat culture. On the centre figure the abscissa is the steady-state substrate concentration established at respective D values. Finally, on the right-hand drawing the P and U quantities are plotted *versus* the function which converts the curves into straight lines. Is it feasible to find this uniform function in reality?

As far as some particular single relationships, $P_i(\bar{s})$ or $U_j(\bar{s})$, are concerned, there is no doubt. It follows from [3.55] that

$$r = \frac{P_k - P_k^{\min}}{P_k^{\max} - P_k^{\min}} = F(\bar{s}). \qquad [3.56]$$

We have used already as a suitable approximation of simple algebraic expressions giving hyperbolic and sigmoid curves (Eqns [3.14] and [3.29], Figure 3.10 and Figures 3.28–3.30). Similar patterns were simulated by optimization models (Figures 3.41 and 3.43). So that the most general expression among empirical ones may be

$$F(\bar{s}) = \frac{s^n}{K_r + s^n}, \quad K_r, n - \text{constant}. \qquad [3.57]$$

The parameter n increases the flexibility of the approximation simulating sigmoid curve at $n > 1$ (found in almost all instances of growth limitation by conserved substrates) and simple hyperbola at $n = 1$.

It is much more difficult to answer the question, will the one functional relationship hold for all assigned P- and U-components? Alternatively, is there proportionality between all their changeable parts, $P_k - P_k^{min}$ and $U_j - U_j^{min}$ ($k = 1, \ldots, n, j = 1, \ldots, m$)? We are inclined to answer 'yes' in view of the following three sources of argument. Firstly, from the outcome of growth simulations by optimization models, some cell components did correlate positively and some negatively with growth rates, and interrelationships between P_1–P_3, R and X′ tended to be linear (Figure 3.44). However, this argument is not very strong since the optimization paradigm itself could not be proven to be absolutely legitimate.

The second line of reasoning comes from the more solid experimental field. There is a great body of empirical facts that state that various indices of the physiological state of microbial cultures, grown continuously at different D, display close cross-correlation (see as an example the massive amounts of data enclosed in the volumes on *Continuous Cultivation of Microorganisms*, 1964–1980). Most frequently this has been documented in relation to the positive relationship of RNA (ribosomes) content with specific respiratory activity (especially of its CN-sensitive fraction), transport capacity and protein biosynthesis rate. Negative correlation was always found between RNA content, on the one hand, and antibiotic production rate, the content of reserved compounds (starch, glycogen, poly-β-oxybutyrate, polyphosphate), some pigments (carotinoid, melanoid), activity of transport systems of high affinity, as well as resistance to starvation, heat, radiation, respiratory inhibitors (cyanide, azide, etc.) and other deleterious factors, on the other. One particular microbial species retains a phenomenally constant stoichiometric ratio K : Mg : PO_4^{-3} : RNA, although it varies within a wide range for different species (Pirt, 1975).

Thirdly, we have found that it is always possible to use a combination of [3.55–3.57] for the approximation of the real data by an accurate choice of n and m, i.e. by the application of the correct assignment procedure. This procedure is a compromise between our ambitions and the quality of experimental data (how many details are available and are intended to be simulated), but it should as well obey the following boundary conservation conditions derived from [3.52] and [3.55]:

$$\sum_{k=1}^{n} P_k + \sum_{j=1}^{m} U_j = \sum_{k=1}^{n} P_k^{\max} + \sum_{j=1}^{m} U_j^{\min} \sum_{k=1}^{n} P_k + \sum_{j=1}^{m} U_j$$

$$= \sum_{k=1}^{n} P_k^{\min} + \sum_{j=1}^{m} U_j^{\max} = 1 - L \approx 1\,. \qquad [3.58]$$

A reliable test for the applicability of the SCM approximation is the constancy of ratios obtained simultaneously for all assigned P (or U) components:

$$\frac{P_1 - P_1^{\min}}{P_1^{\max} - P_1^{\min}} = \frac{P_2 - P_2^{\min}}{P_2^{\max} - P_2^{\min}} = \ldots = \frac{P_n - P_n^{\min}}{P_n^{\max} - P_n^{\min}} = r\,. \qquad [3.59]$$

As an example we may demonstrate again the *D*-dependent variations of cell composition observed in chemostat culture of *Aerobacter aerogenes* (Figure 3.42). We may state that SCM simulates the experimental data even better than the optimization model.

However, we would not like to overestimate the actual weight of the relationship [3.55]. It remains no more than an approximation of reality. Such approximations are justified only if they provide a better understanding of the studied phenomenon and produce new knowledge. Encouraging examples are found in physics, in the exploitation of the notations of 'ideal gases', 'absolutely black body', 'Newtonian liquid' and other idealized but extremely useful substitutes of reality.

In our case the described formalism greatly simplifies the use of structured models because the variations of any cell components now can be expressed via two constants (upper and lower limits) and one single 'master' variable, r, christened as the **variable of physiological state** (or simply r-variable):

$$P_k = P_k^{\min} + r(P_k^{\max} - P_k^{\min}),$$

$$U_j = U_j^{\min} + (1 - r)(U_j^{\max} - U_j^{\min})\,. \qquad [3.60]$$

By definition, r may change from zero to unity depending on the steady-state concentration of limiting substrate (Eqns [3.56–3.57]). Under starvation $\bar{s} \rightarrow 0$, $r \rightarrow 0$, $P_k \rightarrow P_k^{\min}$, i.e. all P-components simultaneously come down to low limits (in reality, complete starvation could not be attained due to cell leakage and lysis), while under conditions of prolonged unlimited growth when $\bar{s} \gg K_s$, $r \rightarrow 1$, $P_k \rightarrow P_k^{\max}$, they attain maximal values.

The r-variable allows the derivation of a simple formula for any metabolic activity whose expression is associated with physiological adaptations:

$$v_k = k_{\text{cat}}\, P_k^{\max}\, r \frac{s}{K_{s_k} + s}\,, \qquad v_j = k_{\text{cat}}\, U_j^{\max}\, (1 - r) \frac{s}{K_{s_j} + s}\,. \qquad [3.61]$$

For example, the specific rate of substrate uptake q_s is given by

$$q_s = rQ_1 \frac{s}{K_1 + s} + (1 - r)\, Q_2 \frac{s}{K_2 + s}, \qquad [3.62]$$

where the first terms on the right-hand side stand for activity of a transport system with low affinity (P-component, expressed under conditions of intensive microbial growth), while the second describes uptake by a transport system with a high affinity (U-component, well expressed under severe growth limitation). A variable r allows the description of many other functions dependent on the cell's physiological state, e.g. specific turnover rate ($a = a_0 r$, a_0-constant), rate of synthesis of secondary metabolites ($q_{\mathrm{prod}} = (1 - r) Q_{\mathrm{prod}} S(s) x$, where $S(s)$ symbolizes a simple function of substrate concentration, say, in the form of Eqn. [1.58]), cell quota variations ($\sigma = \sigma_0 + r(\sigma_m - \sigma_0)$), etc.

Transient changes of cell composition

The steady-state value of r-variable, $\bar{r}$, in chemostat culture is at once related to the concentration of limiting substrate in the medium (Eqn. [3.57]). Under transient conditions, when substrate concentration rapidly changes from $\bar{s}_1$ ro s_2, the average r-value approaches a new steady-state value predicted by Eqn. [3.57]. However, the entire transition involves slow biosynthesis reactions, and so is far from instantaneous. Our aim now is the derivation of differential equations of the type $\mathrm{d}r/\mathrm{d}t = F(r,s)$.

To begin with we will consider the particular case of shift-up transients which is induced by an increase of substrate concentration to new higher level (denoted as s_2) as compared with the foregoing steady-state value ($s_2 > \bar{s}_1$). Let us assume (see Figure 3.45 for illustration) that throughout transients: (a) there is no mixing of *de novo* synthesized small cell increment Δx with the bulk of the cell biomass; (b) the growing increment is already in equilibrium with the new environment, i.e. it contains some representative P-component at an amount which follows from Eqn. [3.57], $P(s_2)$; and (c) the remainder of the cell mass retains the old content of P-constituent denoted as $P(\bar{s}_1)$. The increment of P per small time interval Δt is given by $\Delta p = p(s_2)\Delta x$, where p is P-concentration per unit of culture volume. Substitution of Δt with an infinitesimally small $\mathrm{d}t$ gives $\dot{p} = P(s_2)\dot{x}$. After differentiation of the product $p = Px$ we have:

$$\dot{p} = (Px)' = x\dot{P} + P\dot{x} = P(s_2)\dot{x},$$

wherefrom

$$\dot{P} = \mu[P(s_2) - P],$$

where P is the average concentration of P-component within the entire cell

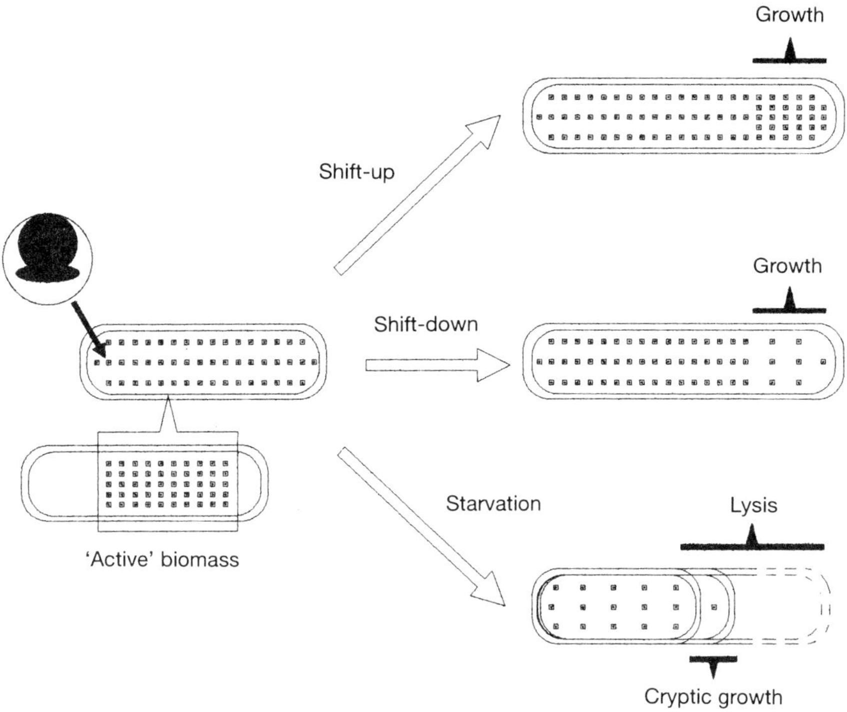

Figure 3.45 Cell composition changes during shift-up experiment (illustration of model derivation). Substrate concentration was shifted from low to high (shift-down), from high to low one (shift-up), and from a positive value to zero (starvation). The quantity of P-components is envisaged by the dot density, each dot schematically representing a ribosome particle. 'Growth' denotes small biomass increment Δx from the start of shift-up or shift-down. 'Cryptic growth' is a hidden increase of biomass at the expense of the lysed cell fraction. Bottom-left is an illustration of the term 'active biomass': suppose that ribosome particles from the whole cell are moved to such a confined cell volume that ribosome density attains its upper limit, corresponding to $r(s) = 1.0$. Then the mass of occupied biovolume is what we called 'active biomass', x_a

biomass. The equation for dr/dt follows immediately from relationship [3.56]:

$$\dot{r} = \mu(r(s) - r) = \mu\left(\frac{s^n}{K_r + s^n} - r\right). \quad [3.64]$$

Thus, the rate of change of r is proportional to the instantaneous growth rate and the 'driving forces' expressed as the difference between equilibrium and instantaneous r-values.

Similar considerations applied to the shift-down experiment result in the same equation: if $s_2 < \bar{s}_1$, then $[F(s) - r] < 0$ and the derivative dr/dt is negative, i.e. r-value throughout transient decreases.

In the case of the structured model we have to account for a number of secondary-rate effects neglected in the foregoing derivation:

$$\dot{r} = \beta(\mu + a)\,(1 - L)\left(\frac{s^{n}}{K_{r} + s^{n}} - r\right). \qquad [3.65]$$

Here, the following three modifications have been made: (a) apparent growth rate, μ, was substituted with the true rate, $\mu + a$, since modification of the vector $\mathbf{\Phi}$ takes place due to polymer turnover even in the absence of net biomass production; (b) the additional term $(1 - L)$ was incorporated since the biomass growth in only terms of pool increase does not affect macromolecular cell composition; and (c) the arbitrary constant β has been introduced for better flexibility of the approximation. Its interpretation is as follows. If $\beta = 1$, then any modification of the vector $\mathbf{\Phi}$ is performed synchronously with biomass growth. If $\beta > 1$ then the P biosynthesis *de novo* occurs both in 'daughter' increments and in the old 'mother' compartment (induction of biosynthesis), and finally, if $\beta < 1$, then modification of vector $\mathbf{\Phi}$ is retarded and after the onset of new growth conditions daughter cells proceed by inertia to reproduce old cell composition.

The special extreme case of a shift-down transient is cell starvation (Figure 3.45). Instead of normal net growth here we have cryptic growth fuelled by the consumption of the lysed cell compartment. Formally this complicated situation is well described by Eqn. [3.65], although for the sake of accuracy we should no longer tolerate the third assumption (which states that the intact cell compartment retains the old content of P-component $P(\bar{s}_1)$). An account of P-decay leads to following modification of Eqn. [3.64]:

$$\dot{r} = \beta(\mu + a)\,(1 - L)\left(\frac{s^{n}}{K_{r} + s^{n}} - r\right) - ar. \qquad [3.66]$$

Description of basic SCM

The complete model consists of the set of ordinary differential equations which were outlined separately for growth limitation by the source of carbon and energy (Table 3.9), and by conserved substrate exemplified by N source (Table 3.10). The last equation set also covers the situation when multiple (double) limitation may occur. Generally, the complexity of the SCM is not fixed and may vary depending on the research objectives. What we describe now is the so-called **basic** model. It contains all the essential features but is open to both reduction and augmentation. Reduction is justified for the modelling of a narrow range of growth conditions or fragmentary experimental data, while considerable enlargement of the model is required for the interpretation of scrupulous quantitative studies.

The basic model in Table 3.9 describes the conversion of organic substrates to biomass of heterotrophic microbes (growth of chemoautotrophic organisms requires only minor modifications, while phototrophs need to be considered specially). Differential equations give total account of fluxes included in the scheme [3.51] and follow the simple conservation conditions $x = x' + l$ or $X' + L = 1$ (see the Appendix for an explanation of the derivation technique). Since biomass and substrate (glucose) have different C contents ($\sigma_x \approx 0.46$ and $\sigma_s \approx 0.4$ respectively), to preserve C balance we have to introduce the conversion factors $k_1 \approx 1.15$ and $k_2 \approx 0.87$ in Eqns [A–C] in Table 3.9.

All kinetic terms were determined according to Eqns [3.61–3.62]. Leakage was assumed to be proportional to L. Total respiration (Eqn. [F]) was evaluated from C balance: $\Delta C - CO_2$ = C consumed – C incorporated into cells.

The next important issue is that of the energetic efficiency of microbial growth. It is known that 'heterotrophic microbial cells need to consume carbon substrate in order to grow, but the reverse is not true: cells do not have to grow to consume carbon substrate' (Tempest and Neijssel, 1984). Dissociation of catabolism from anabolism was accounted in the following particular forms.

1. **Turnover (endogenous decay)**. Cells have to replenish turned over macromolecular components, such as RNA, proteins and cell-wall polymers. SCM accounts turnover explicitly. The specific turnover rate a (scheme [3.51]) is directly related to the r-variable (since decay rates of P-components are higher than U-constituents) and is dependent on substrate concentration in the medium (Eqn. [E]). The latter dependence is based on observations that turnover is accelerated immediately after the onset of starvation conditions (Mandelstam, 1963; Marr *et al.*, 1963). However, we have not found turnover acceleration resulting from the shut-off of continuous culture (section 3.1). One tentative explanation of experimental facts includes the following mechanism. Macromolecular components (for example, enzymes) are not susceptible to decay if they are not idle, i.e they carry out some metabolic reactions via transient enzyme–substrate complexes, intracellular substrates being mainly components of monomeric pool L. Limitation of microbial culture by exosubstrate leads to a decrease in L and increases the probability of finding idle functional components which no longer possess immunity towards hydrolytic enzymes. These mechanistic considerations were translated into two interchangeable expressions (Eqn. [E]). One contains a threshold substrate concentration s^* which triggers the change from slow (a_1) to rapid

Table 3.9 SCM for description of microbial growth limited by the source of carbon and energy

Variable	Equation	
Basic model		
Limiting substrate ($mg\ l^{-1}$)	$\dot{s} = D(s_0 - s) - q_s x + k_1 v L x$, $q_s = rQ_s \dfrac{s}{K_s + s}$ (or Eqn. [3.62]) , $k_1 = \dfrac{\sigma_x}{\sigma_s}$	[A]
Monomeric intermediate ($g\ (g\ biomass)^{-1}$)	$\dot{L} = k_2 q_s - q_L + a(1 - L) - vL - \mu L$, $k_2 = \dfrac{\sigma_s}{\sigma_x}$ $q_L = q_L^{\blacklozenge} + q_L' = rQ_L \dfrac{L}{K_L + L} + Q_L' \dfrac{L}{K_L' + L}$	[B]
Biomass ($mg\ l^{-1}$)	$\dot{x} = (\mu - D)x$, $\mu = k_2 q_s - q_L - vL + Y(q_L - q_L' - m)$, $m = rQ_m \dfrac{L}{K_m + L}$	[C]
r-Variable (dimensionless)	$\dot{r} = \beta(\mu + a)(1 - L)\left(\dfrac{s}{K_r + s} - r\right)$	[D]
Turnover rate (h^{-1})	$a = \begin{cases} a_1 r & s \geqslant s^* \\ a_2 r\ (a_2 > a_1) & \text{at } s < s^* \end{cases}$, or $a = r\dfrac{a_0}{1 + a_3 s^n}$	[E]
Respiration ($mmol\ CO_2\ h^{-1}\ (g\ biomass)^{-1}$)	$q_{resp} = q_L - Y(q_L - q_L' - m) = Aq_s - B(\mu + vL)$	[F]
Reduced model		
Limiting substrate ($mg\ l^{-1}$)	$\dot{s} = D(s_0 - s) - q_s x - q_s' x + ax$, $q_s = \dfrac{rQ_s s}{K_s + s}$, $q_s' = \dfrac{Q_s' s}{K_s' + s}$	[G]
Biomass ($mg\ l^{-1}$)	$\dot{x} = (\mu - D)x$, $\mu = Y(q_s - m) - a$, $a = a_0 r$	[H]
r-Variable (dimensionless)	$\dot{r} = \beta(\mu + a)\left(\dfrac{s}{K_r + s} - r\right)$	[I]

Table 3.10 SCM simulation of growth under limitation by conserved substrates and under double limitation

Variable	Equation	
N-limited growth		
Limiting substrate (mg N l^{-1})	$\dot{s} = D(s_0 - s) - q_s x + vLx$	[A]
Intermediate (g N g^{-1} biomass)	$\dot{L} = q_s - q_L + a(\sigma - L) - vL - \mu L$	[B]
Biomass (mg l^{-1})	$\dot{x} = (\mu - D)x\ ,\quad \mu = \left[q_s + q_L\left(\frac{Y_N}{\sigma} - 1\right) - vL\right]\Big/\sigma$	[C]
r-Variable (dimensionless)	$\dot{r} = (\mu + a)\left(1 - \frac{L}{\sigma}\right)\left(\frac{s^n}{K_r + s^n} - r\right)$	[D]
Cell N content (g N g^{-1} biomass)	$\sigma = \sigma_0 + r(\sigma_m - \sigma_0)$	[E]
Biomass yield (g g^{-1} N)	$Y_N = \left[\sigma_0 + (\sigma_m - \sigma_0)\frac{s^n}{K_r + s^n}\right]^{-1}$	[F]
Account of C-substrate consumption and possible dual limitation		
C-substrate concentration, s^0 (mg l^{-1})	$\dot{s}^0 = D(s_0^0 - s^0) - q_s^0 x + k_1 v^0 L^0 x$	[G]
C intermediate, L^0 (g g^{-1} biomass)	$\dot{L}^0 = k_2 q_s^0 - q_L^0 + a(1 - L^0) - (\mu + v^0)L^0$	[H]
Biomass (mg l^{-1})	$\dot{x} = (\mu - D)x\ ,\quad \mu = \min\begin{cases} \left[q_s + q_L\left(\frac{Y_N}{\sigma} - 1\right) - vL\right]\Big/\sigma \\ k_2 q_s^0 - q_L^0 - v^0 L^0 + Y(q_L^0 - q_L^{0\prime} - m) \end{cases}$	[I]
r-Variable (dimensionless)	$\dot{r} = (\mu + a)(1 - L^0)(F(s, s^0) - r)\ ,\quad F(s, s^0) = \min\left(\frac{s^0}{K_r^0 + s^0}, \frac{s^n}{K_r + s^n}\right)$	[J]

(a_2) turnover. The other one is a continuous function describing the inhibition of decay by L.

Contrary to many kinetic models, the SCM does not identify turnover with respiration terms, i.e. products of decay are not CO_2 but monomeric compounds which may be reutilized.

2. **Maintenance requirements**. These are usually interpreted as energy expenditure for non-growth cell functions such as membrane energization, osmoregulation and cell motility. This term implies a direct use of energy source, so it was set to be similar to the total intracellular flux, q_L, as a function of r-variable and L concentration.

3. **Uncoupled (dissipative) respiration**. This plays a significant role under energy-sufficient growth conditions, especially during transients after relief of severe energy limitations. Terminal oxidases are divided into two groups in respect of both functional role and sensitivity to cyanide. CN-sensitive oxidases perform productive substrate oxidation coupled with ATP generation and should be attributed to P-type of cell components. CN-resistant oxidases carry out energy-spilling (dissipative) oxidative reactions not coupled with ATP generation. Their probable physiological function is the maintenance of the adenylate pool and cell protection from stresses caused by abrupt 'famine-to-feast' transitions (Tempest and Neijssel, 1984). CN-resistant oxidases are expected to be either independent of growth conditions (constitutive element of vector $\mathbf{\Phi}$) or negatively correlated with the r-variable (U-component). For the first and the second cases we have respectively:

$$q_L = rQ_L \frac{L}{K_L + L} + Q_L' \frac{L}{K_L' + L},$$

and

$$q_L = rQ_L \frac{L}{K_L + L} + (1 - r)\, Q_L' \frac{L}{K_L' + L}, \qquad [3.63]$$

where Q_L' stands for maximum dissipative catabolic activity.

Direct measurements of CN-resistant respiration revealed that it remains constant in the entire range of D in *Pseudomonas* chemostat culture (Figure 3.46). The ratio λ = (CN-resistant respiration)/(total substrate respiration) estimated near wash-out point was about 0.1. Approximately the same quantities were found from published data for other organisms (Akimenko *et al.*, 1983): average ± standard deviation for λ was 0.82 ± 0.11 and median 0.90 ($n \sim 100$, mainly bacteria and yeasts). Thus, CN-resistant respiration seems to belong to some 'fundamental' biological constant if it is expressed as a fraction of maximum potential

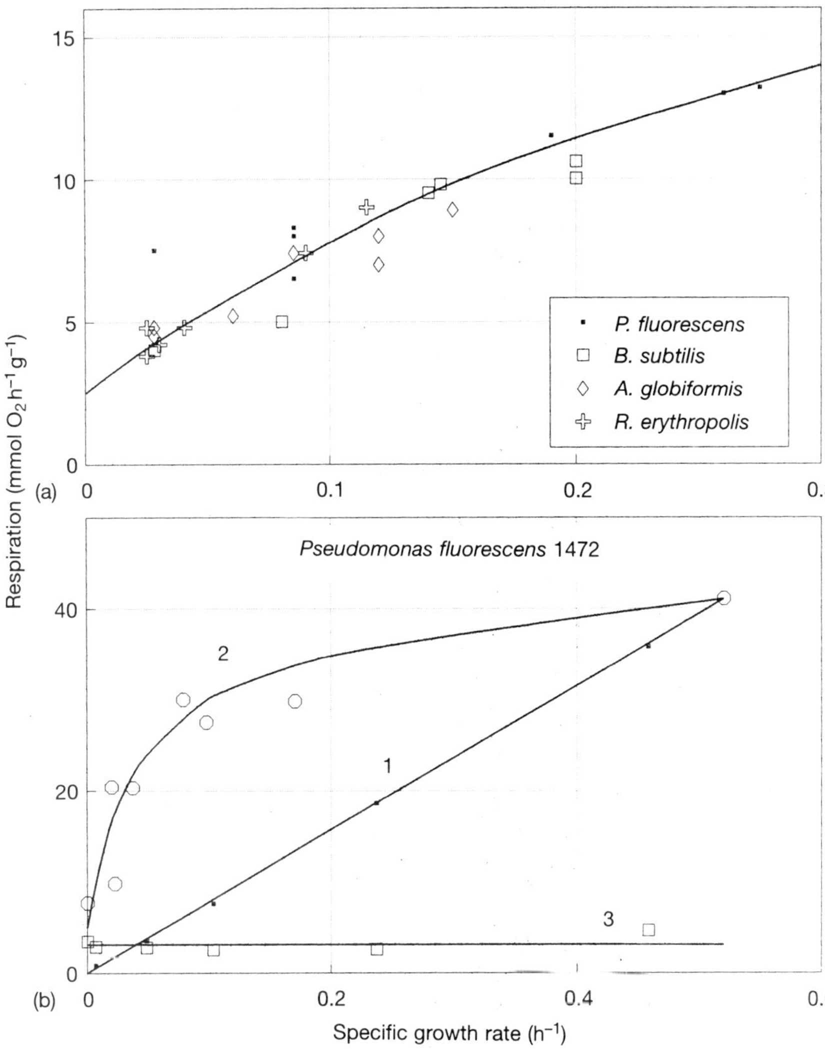

Figure 3.46 Synthetic chemostat model (SCM) simulation of *D*-dependent changes of respiration rates in steady-state chemostat culture (Panikov, 1991a). (a) N-limited growth of four different organisms. (b) Respiration rates of *P. fluorescens* measured by dynamic method directly in the fermenter of glucose-limited culture (curve 1), potential respiratory activity measured in culture subsamples with glucose amendment (curve 2), and with addition of glucose + KCN (curve 3). The curves are calculated according to the synthetic chemostat model (SCM).

catabolic activity. This implies that the actually measured relative contribution of uncoupled respiration should be higher under conditions of restricted growth, since at low s we have a lower r-value and low productive respiration activity, while CN-resistant oxidation potential is constant. From the two alternatives of Eqn. [3.63] the first postulate is the better option, so we have accommodated it in the basic SCM as Eqn. [B] in Table 3.9.

Apart from the basic model we derived a simplified unstructured version of SCM which includes only three variables (x, s and r). It is especially useful when there are no available data on intracellular pool sizes. According to this version, substrate is converted immediately to cell biomass, some fraction being oxidized in dissipative reactions. Since dissipative respiration and maintenance *sensu stricto* are practically indistinguishable, then for the sake of simplicity the maintenance term, m, was omitted. Turnover was simulated in the same way as in the complete model with only one exception: decay products instead of entering the L pool, were assumed to be released into the extracelluar environment for reutilization as exosubstrate. If this was not the case the conservation condition would be violated.

N-limited growth is simulated by the set of four differential equations (Table 3.10) which were derived from scheme [3.51] and the following conservation conditions:

N-balance

$$\sigma_N x = \sigma_N x' + l \quad \text{or} \quad \sigma_N = \sigma_N X' + L$$

mass balance

$$\frac{l}{\sigma_N} + x' = x \quad \text{or} \quad \frac{L}{\sigma_N} + X' = 1\,.$$

(Note that these conditions are valid only for the chosen dimensions L, mg N per g biomass, and X', g per g biomass.) As a result, the equation simulating r transient changes has been modified (compare Eqn. [3.65] and [D] in Table 3.10). Kinetic terms and turnover rate were as in the case of C-limited growth (Table 3.9).

The most important point to be discussed here is the expression for cell quota, σ, and biomass yield per unit of conserved substrate consumed, Y (Eqns [E] and [F], Table 3.10). Earlier we obtained Eqn. [3.28], which describes an inverse relationship between these two quantities ($\sigma = 1/Y$). SCM formulations gave basically different expressions. Why? Let us make clear the difference between σ and $1/Y$ in biological (physical) terms. Both quantities are of the same dimension, say mg N per g biomass. However, the entity $1/Y$ characterizes the **process** (the expenditure of conserved substrate to synthesized biomass unit), while σ is an index of cell composition (the content of intracellular N per biomass unit). Formally, $1/Y$ is equal to the σ-value of an infinitesimally small increment of cell biomass ($1/Y = ds/dx$), while σ is the averaged value for the entire cell biomass (e.g. in chemostat $\sigma = (s_0 - s)/x$). Cell quota is directly related to r and past history, while Y depends on the instantaneous limiting substrate concentration in the medium. Briefly speaking, σ is a slow, while $1/Y$ is a rapid variable, and this fact exerts the crucial consequences on the behaviour of the culture limited by conserved substrates (see below).

Taking into account dual limitation requires the enlargement of the model variables from four to six (second substrate and second monomeric intermediate are added). The specific growth rate is calculated as the minimum from two expressions calculated for the first and the second substrates (Eqn. [I], Table 3.1). A similar principle was chosen to define the functional relationship between r and substrate concentration (Eqn. [J], Table 3.1). It was further assumed that monomeric pools of N do not contribute separately to the total cell mass, since most of the real N-intermediates (amino acids, polypeptides, amides, nucleotides) except NH_4^+ do participate in the C-pool.

3.6.3 APPLICATION OF SCM FOR BETTER UNDERSTANDING OF MICROBIAL GROWTH

This section brings us back to experimental results. We will try to interpret these data which have not yet been explained within the framework of traditional kinetic models.

Steady-state microbial growth

We have already discussed most of the experimental observations on steady-state growth while deriving our basic model. In addition, most of the available facts have been accounted for by semiempiric models outlined in equations [3.17] (D-dependent variations of potential metabolic activities, affinity to substrate, maintenance requirements) and [3.29–3.35] (D-dependent variation of cell quotas σ_N and σ_P). So, we will confine ourselves to some brief comments.

1. ***D*-dependent variations of $\bar{x}$ and $\bar{s}$.** Even the reduced version of the SCM was able to reproduce accurately these variations in the entire D range, including extremely low values where the classic Monod–Herbert model fails. The model simulates perfectly well, maintaining an almost constant biomass yield at $D \to 0$ (Figure 3.7), almost linear growth in dialysis and fed-batch culture (Figures 3.14 and 3.15), a concave shape of the $\mu(s)$ curve at the concentration range 0.01–0.1 mg l^{-1} (Figure 3.16), an increase

with growth acceleration of apparent values of Q_{resp} and K_s (Figure 3.9) etc. Sensitivity analysis (see Appendix) revealed the importance of including the activity of the two transport systems (Eqn. [3.62]). On the other hand, the model was 'overburdened' with parameters responsible for energy dissipation (turnover, maintenance *sensu stricto*, and uncoupled respiration). It was sufficient to leave only one of listed terms to simulate practically all steady-state relationships. Differentiation of these biologically distinctive mechanisms require non-steady-state approaches.

2. **Consumption and oxidation of energy source under N-limitation**. This could not be described using non-structured models. Under these conditions respiration is no longer limited by respiratory substrates (O_2, glucose or intracellular intermediates). Nevertheless, specific respiratory activity, q_{resp}, monotonously declined at low D (Figure 3.46). Formally, this decline may be approximated by Pirt's equation with a μ-dependent maintenance coefficient (Pirt, 1982; Tempest and Neijssel, 1984). According to the SCM, the same decline is due to a decrease of the r-variable which is under the control of the N-source:

$$q_{resp} = rQ^{m}{}_{resp} + Q'{}_{resp}, \tag{3.64}$$

where Q'_{resp} is maximum dissipative (CN-resistant) catabolic activity. Under some conditions suppressive for biosynthesis of alternative oxidase (lack of biogenic elements), there is excessive excretion of metabolites to the environment (so-called overflow metabolism) which is simulated by SCM as intensive L leakage.

3. **Multiple limitation**. This has been described by the semiempiric model [3.36–3.38], but agreement was rather poor. The SCM provides a better fit and a clearer understanding of the entire phenomenon. The quantities of σ_N and σ_P are unique functions of the r-variable, and r in its turn is defined by the concentration of limiting substrate. In our experiment with step-by-step increase of input P concentration, σ_{0P}, under a constant N supply (Figure 3.30) we may identify exactly that particular σ_{0P} value which defines double limitation. At this point residual P concentration, $\hat{s}_P$, satisfies the condition $F(\hat{s}_P) = F(\hat{s}_N)$ or $\hat{s}_N/(K_N + \hat{s}_N) = \hat{s}_P/(K_P + \hat{s}_P)$ (Table 3.10, Eqn. [J]). Double-limited steady-state values of $\hat{s}$ are close to normal mono-limited residual substrate concentration, $\bar{s}$, the complete identity being fulfilled if $\sigma_N^0 : \sigma_N^m = \sigma_P^0 : \sigma_P^m$. Further step-by-step increase of s_{0P} proceeds under constant residual concentration of both limiting nutrients, differences in steady-states being attributed exceptionally to cell quotas σ_N and σ_P. This range of double limitation corresponds to movement of the phase point along line segments with equal r (between two arrow's heads):

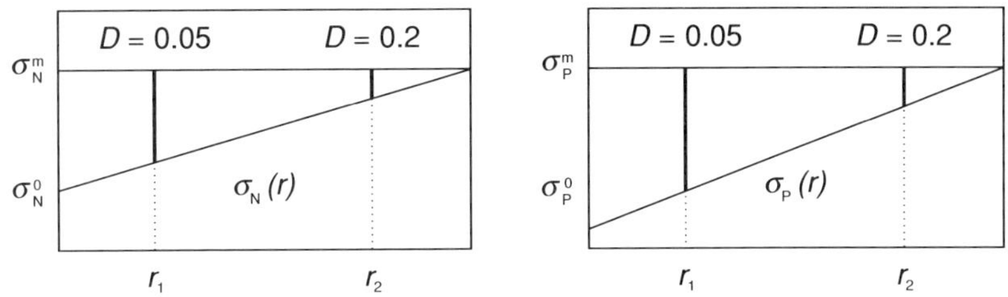

The lower the r-value, the longer is the corresponding segment, and therefore the broader the range of double limitation. It is interesting that the position of the phase point within the indicated region is beyond the pressure of any driving forces. Just imagine the ideal chemostat culture which 'does not remember' its past history, then all the steady-states between the arrows heads will be equiprobable (they are characterized by the same μ and r, and differ in respect to σ and $\bar{x}$). However, the behaviour of the real culture depends on its past history and particular trajectory of approaching steady-state. We have tried the following teleonomic rule which defines the choice of such a trajectory. In the range of double limitation, cells are steering themselves to maximize intracellular reserves of that nutrient which was deficient in the past history, but without the loss of the second limiting element. This strategy steers cells to occupy the upper arrow heads of both phase planes since this is allowed by the stepwise increase of input P concentration, s_{0P}. The subsequent increase in P input ensures an increase in $\bar{x}$ with a concomitant decrease of N quota down to the state of strict N-limitation. This strategy is easily translated into a computer simulation model. Calculation results are displayed in Figure 3.30.

Non-steady-state microbial growth in continuous culture

It is transient kinetics which should be considered the main descriptive objective of the SCM.

Our experimental studies of non-steady-state growth in continuous culture include the response of cells to glucose pulses (Figures 3.8 and 3.11) and the complete stopping of the medium flow (Figures 3.6 and 3.13). Integration of the differential equations of the basic SCM at a short-term temporal resolution gave an adequate simulation of the experimental curves: (a) an instant response to a small quantity of added glucose and a 1–2-min delay in response to heavier amendments (Figure 3.8); and (b) a steep decline in respiration rates after shutoff during 6–10 min with a subsequent levelling off of endogenous respiration (Figure 3.6). The only

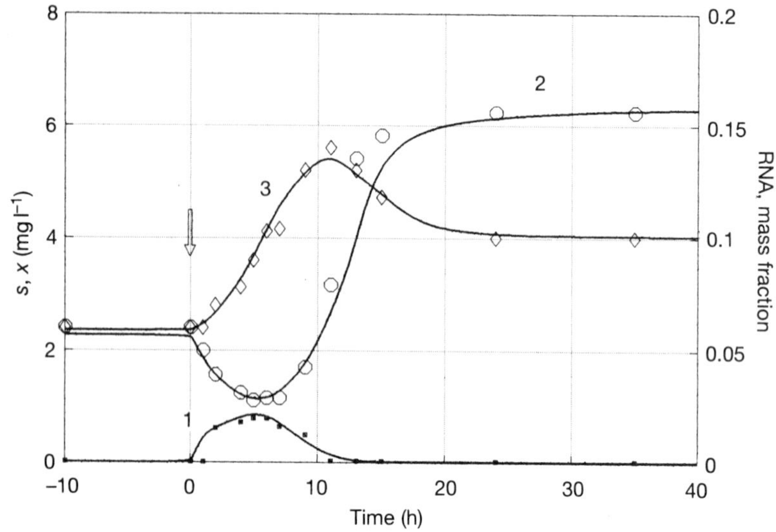

Figure 3.47 Synthetic chemostat model (SCM) simulation of transient growth induced by change of chemostat dilution rate. Original data of Tempest *et al.* (1967) for chemostat culture of *Aerobacter aerogenes* limited by glycerol. At $t = 0$ dilution rate was shifted from 0.004 to $0.24\,h^{-1}$. The dynamics of cell biomass (curve 2), residual glycerol concentration (curve 1) and cell RNA content (curve 3) was calculated according to reduced synthetic chemostat model (SCM) (Table 3.8) with the following parameters: $Q_s = 0.72$ g (g biomass h)$^{-1}$, $Y = 0.75$ g biomass (g substrate)$^{-1}$, $K_r = 17\,mg\,l^{-1}$, $K_s = 5\,mg\,l^{-1}$, $Q_s' = 5$ g (g biomass h)$^{-1}$, $K_s' = 1.75\,g\,l^{-1}$, $m = 9\,mg\,h^{-1}$ (g biomass h)$^{-1}$, $a_0 = 0$.

essential prerequisite of suitable agreements was the adjustment of model parameters to allow the pool sizes, L, to be a 'quick' variable with turnover time $L/q_L \sim 10^{-2}$–10^{-3} h. This condition was met in the case of *P. fluorescens* at $Q_L = 1.0\,h^{-1}$ and $K_L = 0.01$.

The second example of transient kinetics was a classic shift-up experiment with chemostat culture of *Aerobacter aerogenes* made by Tempest *et al.* (1967) (Figure 3.47). The SCM adequately describes the effects of 'overshoot' in respect of s and 'undershoot' in respect of x and RNA content during the transition from slow ($D = 0.004\,h^{-1}$) to intensive growth ($D = 0.24\,h^{-1}$). These effects are explained by inertia caused by slow biosynthetic reactions, primarily the synthesis of ribosomes. Needless to say, unstructured models are not able to simulate 'overshoots' and 'undershoots' at any numerical values of growth parameters.

All three examples belong to the relatively simple type of transient growth displayed by well-studied organisms with predictable growth patterns. More complicated behaviour of *Arthrobacter* will be considered below.

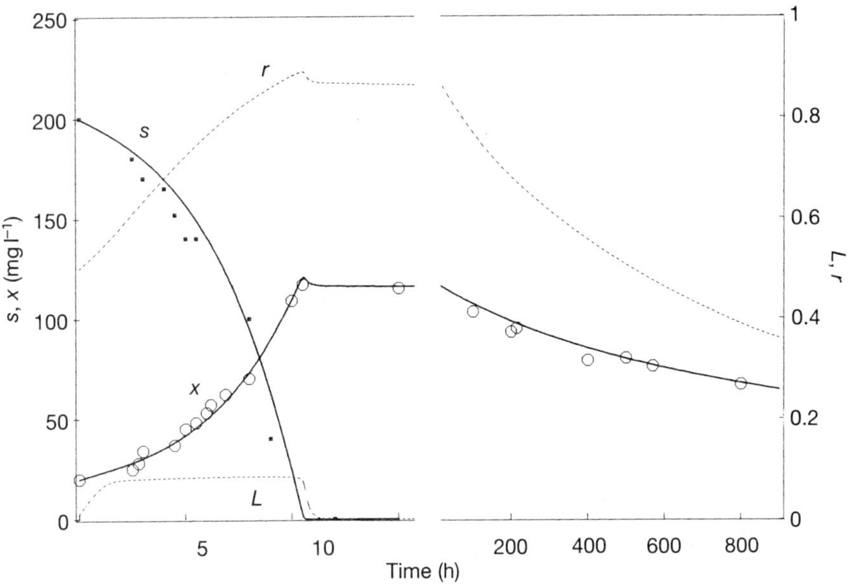

Figure 3.48 Synthetic chemostat model (SCM) simulation of complete dynamic curve of batch culture. Yeasts *D. formicarius* were grown on nutrient medium 6. Simulation was performed using the basic synthetic chemostat model (SCM) (Table 3.8) with the following parameters: $Q_s = 0.72$, $Q_L = 50$, $Q_m = 0.03$, $v = 0.25$, $a_1 = 0.03\ h^{-1}$, $K_s = 1.0$, $K_r = 0.2$, $s^* = 0.06\ mg\ l^{-1}$, $K_L = 1.0$, $K_m = 0.01$, $\beta = 1.0$ and $Y = 0.6\ g\ g^{-1}$.

Batch culture limited by the source of carbon and energy

Growth dynamics are fully simulated by the SCM – from inoculation and lag phase until complete decline, all intermediate phases being reproduced automatically (Figure 3.48). However, it is reasonable for the sake of clarity to split again the entire dynamics into two halves: (a) the period of unlimited growth (lag + exponential phases); and (b) the stage of substrate exhaustion and culture starvation and decline. With respect to these situations, even the reduced version of the SCM (Table 3.9, Eqns [G–I]) may be further simplified without loss of accuracy.

In the first case we may put $s \gg K_s$, $s \gg K_r$, and safely neglect substrate leakage ($vL \ll rQ_s$) as well as turnover and maintenance terms ($m \ll Q'_s$, $a \ll \mu$), uncoupled respiration remaining the only important component of dissipative catabolism. In addition, the coefficient β in Eqn. [I] was omitted, since in most simulations with the complete SCM it was identified as unity. Then, from the reduced SCM we have

$$\dot{s} = -(rQ_s + Q'_s)x\,,$$

$$\dot{x} = \mu_m rx, \quad \mu_m \approx YQ_s\,, \qquad [3.65]$$

$$\dot{r} = \mu_m r(1 - r) .$$

This equation set is integrated analytically with the initial conditions $x = x_0$, $s = s_0$ and $r = r_0$ at $t = 0$:

$$s = s_0 - Q'_s(1 - r_0)x_0 t - \frac{x_0 r_0 (Q_s + Q'_s)}{\mu_m}(e^{\mu_m t} - 1) ,$$

$$x = x_0(1 - r_0) + x_0 r_0 e^{\mu_m t} , \qquad [3.66]$$

$$r = r_0 \frac{e^{\mu_m t}}{1 - r_0 + r_0 e^{\mu_m t}} .$$

Contrary to a simple exponential equation this model takes into account the possibility of a latent phase, resulting from the low metabolic activity of inoculum. Accepting SCM jargon, we may define lag phase as the period required for microbial cells to increase their P-content to maximum values (corresponding to r increases from r_0 to $r = 1$). The higher the value of r_0, the smaller is the deviation from simple exponential growth. If $r_0 = 0$, then the inoculum is devoid of resuscitable biomass, and lag phase continues to infinity.

The product xr gives the so-called 'active biomass', x_a. It has the remarkable feature of displaying exponential growth at any r_0: $x_a = x_0 r_0 e^{\mu_m t}$. The meaning of the term 'active biomass' may be perceived from geometric considerations (Figure 3.45). Imagine that all available P-components (such as ribosome particles) from the whole cell are aggregated inside a confined biovolume to attain their highest possible level corresponding to $r = 1$. Then the mass of occupied biovolume is termed 'active biomass'. This notation will be used in our ecological studies.

Substrate concentration does not affect the growth rate, but the dynamics of organic substrate consumption and respiration is very informative since it closely follows the changes in both biomass density and cell physiological activity. This relationship will form the foundations of a new kinetic approach for microbial biomass determination and characterization in natural substrates like soils and waters (Chapter 6). Now we may state that equation set [3.65] demonstrates a fairly good agreement with the observed batch growth and respiration as dependent on the physiological state of inoculum (Figures 3.17 and 3.18).

We will consider the second stage of batch culture dynamics from the substrate depletion point, when limiting substrate in the medium and intracellular pool drop to starvation steady-state levels $\bar{s}$ and $\bar{L}$ respectively (the main source of S is cell leakage and lysis, while L is reimbursed with intracellular decay of macromolecular components). Since $\bar{s} \ll K_s$ and $L \ll K_L$, we may reduce all kinetic terms to first-order relationships. We may not neglect cell leakage and exometabolites excretion, but it would be reasonable to specify new features of the cell's extracellular products (denote them by h). They are transformed to refractive compounds (soluble humus-like substances and glycocalyces) which can no longer be reutilized by survivors. The rate of h production is assumed to be proportional to r and x.

Finally, we need to supplement the basic SCM by an equation describing the dynamics of dead cells. The probability of occasional loss of viability (lethal mutations, mistakes of autosynthesis) is fairly low ($\sim 10^{-5}$), but it should be taken into account for long-term starvation–survival dynamics. Even low frequencies produce, under non-growth conditions, significant amounts of non-viable cells (we have already denoted the necromass concentration as x_{necro}). The rate of x_{necro} formation is proportional to x and depends on the cell's physiological state. The latter dependence seems to be paradoxal in spite of the fact that it was discovered as early as 1924 by Sherman and Albus: the younger and the more active are cells, the more susceptible they are to death. In SCM terms this means a proportionality to r.

All postulates were translated into a set of differential equations which were fortunately solved analytically (Table 3.11). The model was used to simulate survival of several starving cultures, as an example we will consider the data on *Pseudomonas* (Figure 3.21). Altogether, six dynamic variables were fitted to the model simultaneously with a low sum of residuals. Thus, the model has proven to be adequate and we have the opportunity to discuss its biological interpretation.

First of all let us clarify the meaning of the three right-side terms of Eqn. [B] in Table 3.11 describing biomass dynamics. The first term, $k_d r$, is responsible for the occasional loss of viability. The numerical value of $k_d = 0.004\ h^{-1}$ is formally interpreted as the extinction frequency per unit of 'active biomass', rx. The second term, $a_0 r$, is the true decay rate caused by the turnover of the cell's macromolecules. The third term, $Y(q_s - m)$, is the rate of cryptic growth of survivors. It is cryptic growth which provides cell adaptation to starvation conditions, primarily the *de novo* synthesis of specific U-components responsible for the scavenging of trace amounts of released autolysis products and economizing the cell energetics. As an index of this adaptation we may use the dynamics of RNA content or potential respiration rate Q_{resp}. To assess the contribution of cryptic growth let us go back to the scheme used for the transient derivation of dr/dt (Figure 3.45). Then we have the following balance equation for a particular P-component – respiratory enzymes:

total activity		formed during cryptic growth		left in old compartment	
xQ_{resp}	$=$	$\theta x Q_{resp}$	$+$	$(1-\theta)x[Q^{min}_{resp} + (Q^{max}_{resp} - Q^{min}_{resp})r(s_1)]$,	[3.67]

where θ is the fraction of cryptically grown cells. Solution of this equation

Table 3.11 Reduced and supplemented SCM for description of death phase of depleted batch culture

Variable	Equation: Differential form		Equation: Result of integration
Substrate ($mg\ l^{-1}$)	$\dot{s} \approx 0$	[A]	$\bar{s} = \dfrac{a_0 r}{rQ_s/K_s + Q_s'/K_s'} \approx a_0 \dfrac{K_s}{Q_s}$
Biomass ($mg\ l^{-1}$)	$\dot{x} = -\eta r x$, $\eta = k_d + a_0 - Y\left(\dfrac{Q_s}{K_s}\bar{s} - m_0\right)$	[B]	$x = x_0(1 + r_0\zeta t)^{-\eta/\zeta}$
r-Variable (dimensionless)	$\dot{r} = \beta(\mu + a)(F(\bar{s}) - r) = \zeta r^2$, $\zeta \approx \beta\left[Y\left(\dfrac{Q_s}{K_s}\bar{s} - m_0\right) + a_0\right]$, $F(\bar{s}) \approx 0$	[C]	$r = r_0(1 + r_0\zeta t)^{-1}$
Exometabolites ($mg\ C\ l^{-1}$)	$\dot{h} = v' r x$	[D]	$h = \dfrac{v x_0}{\eta}\left[1 - (1 + r_0\zeta t)^{-\eta/\zeta}\right]$
Necromass ($mg\ C\ l^{-1}$)	$\dot{x}_{necro} = k_d r x$	[E]	$x_{necro} = \dfrac{k_d x_0}{\eta}\left[1 - (1 + r_0\zeta t)^{\eta/\zeta}\right]$
$CO_2 + HCO_3^-$ ($mg\ C\ l^{-1}$)	$\dot{p} = (\Omega r + \Omega')x \approx \Omega r x$, $\Omega = (1 - Y)\dfrac{Q_s}{K_s}\bar{s} + Y m_0$, $\Omega' = \dfrac{Q_s'}{K_s}\bar{s}$	[F]	$p = \dfrac{\Omega x_0}{\eta}\left[1 - (1 + r_0\zeta t)^{-\eta/\zeta}\right]$

for the 900th hour of starvation ($Q^{max}_{resp} = 11$, $Q^{min}_{resp} = 0.5$, $Q_{resp} = 1.5$ mmol h^{-1} per g biomass, $x = 18$, $x_{necro} = 85$, and $h = 63$ mg l^{-1}, Figure 3.21) produces a θ value of 0.905. Thus, cryptic growth of $18 \times 0.905 = 16.3$ mg l^{-1} of cells was provided by the reutilization of $200 - (18 - 16.3) - 85 - 63 = 50.3$ mg l^{-1} of viable cells (here 200 is bacterial biomass at onset of starvation, $x_{necro} = 85$ and $h = 63$). Yield of the cryptic growth, $Y_{cry} = 0.32$ g g^{-1}, is within common range observed for growth on organic substrates. Quantitative estimates made by Nion and Furusaka (1968) were by one order of magnitude lower when they tried to grow bacteria on medium with killed cells, but there is no contradiction with our results since growth on necromass is not equivalent to starvation survival.

Cryptic growth is extremely slow. To assess its specific rate, μ_{cry}, we measured the actual respiration rate, v_{resp}, in the starving culture and used the following expression:

$$\mu_{cry} = \frac{v_{resp}}{x_{via}}\left(\frac{Y_{cry}}{1 - Y_{cry}}\right).$$

Respiration activity gradually decreased throughout the starvation (Figure 3.21) as did the growth rate, e.g. at the 300th hour, μ_{cry} was 0.0087 d^{-1} which corresponds to a generation frequency of one division every 80 days!

The SCM gives quantitative interpretation to other characteristic features of starving microorganisms. In particular, the dependence of survival dynamics on preceding growth rate (Figure 3.23) has been fully explained by the effects of the initial physiological state expressed by the values of r_0.

Spore formation and maintenance state in *Bacillus* culture

There are two distinctive features of bacilli behaviour: (a) the ability to sporulate in batch culture after substrate depletion (a fact known from the last century); and (b) as discovered in the present study, the transition of vegetative cells to a non-growing maintenance state under chronic substrate limitation in dialysis and fed-batch culture. We feel that these two features have a common underlying mechanism. The most intriguing question is why bacilli lose their sporulation ability under some cultivation conditions. The answer was given by the SCM.

To describe spore formation the basic model was modified according to the following growth scheme:

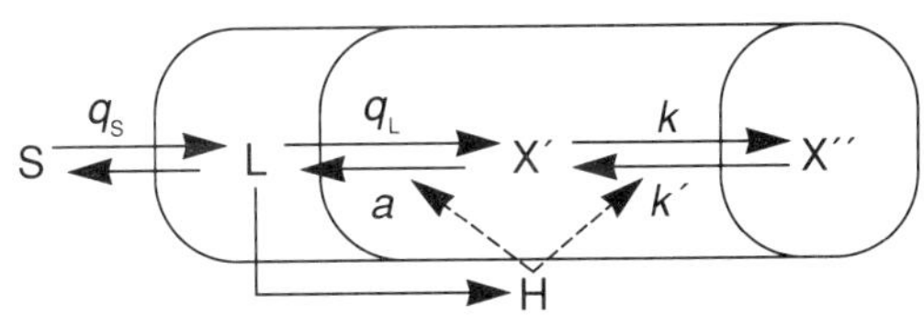

The major modification of the model is the introduction of the second macromolecular compartment, X″, which is a prospore or true spore. Since in reality the conversion 'prospore → spore' is accompanied by the complete loss of metabolic activity, we may identify spores as X″ compartment with an r-variable close to zero (say, $r < 10^{-5}$). The exchange between X′ and X″ compartments is under metabolic and environmental control. The second modification of the SCM is the formalization of the 'sensor model' for the initiation of sporulation (Jeong *et al.*, 1990), i.e. substrate depletion in the medium triggers the biosynthesis of metabolite H (in biochemical terms, it may be inosine, or highly phosphorylated nucleotides such as pppApp), which suppresses vegetative growth of bacilli and prompts spore formation.

The mechanism of spore formation is as follows. Under normal nutritional conditions exosubstrate S is converted to the biomass of vegetative cells X′, the flux from X′ to X″ being zero. The primary signal launching sporulation is substrate depletion down to the threshold level, s^*, which triggers accelerated X′ turnover and alters the direction of fluxes between compartments X′ and X″ in favour of prospore formation. If substrate is completely depleted as in batch culture ($s \sim 0$), then cryptic growth of prospore at the expense of degradation products of vegetative mother cells produces true spores with zero content of P-components ($r = 0$).

What happens in dialysis and fed-batch culture? Continuous supply of glucose prevents the formation of X″ compartment with $r = 0$ since always $s > 0$. However, the quasi-steady-state level of substrate concentration drops below threshold level, s^*, and precludes normal growth of X′ compartment due to: (a) the acceleration of turnover; (b) the sink of cell material to X″ compartment; and (c) the inhibitory action of metabolite H.

The role of inhibitor should be crucial. At this stage of model development we found that exclusion of this variable resulted in the general failure of simulation, i.e. it was impossible to keep the residual substrate concentration in continuous culture at a quasi-steady-state level below s^* and instead we obtained oscillations around s^* coupled with interrupted discontinuous growth. Various hypotheses have been tested, including the repression–derepression of transport systems with low and high affinity, competitive and non-competitive inhibitory effects of H on q_S, q_L, m, control of turnover by the L level, etc. All these assumptions have been found to be inconsistent with the observed growth pattern. The only positive result was generated with the following postulate. Autoinhibitor H decreases the efficiency of substrate conversion to biomass of vegetative cells, i.e. it reduces Y value. A specific experiment confirmed the presence of metabolic uncouplers in the dialysis culture of *B. subtilis*. In the presence of its sterile filtrate, bacteria displayed a significant decrease of yield with minor effect on growth rate (Evdokimova and Panikov, 1994).

The complete equation set of the modified SCM is presented in Table

3.12. The variable L was excluded because of the risk of overloading the model with parameters which are not supported by direct experimental determinations (there was no available data on the dynamics of the monomeric pool in *Bacillus* culture). Nonetheless, this reduction did not affect the essential points of the postulated mechanisms. In addition, we took into account the non-instantaneous development of spore refractivity by the introduction of a time delay, $\tau = 2\,h$, when the simulation was compared with the experimental data.

The simulation (Figures 3.14, 3.15 and 3.49) was successful from a purely statistical standpoint giving a low sum of residuals. In addition, it proved to have predictive abilities by generating relevant results which were not implied explicitly. Thus, the model adequately reproduced (a) spore germination in batch culture amended with substrate, and (b) a decline in cell biomass after stopping the flow in dialysis culture at maintenance state (see details in Evdokimova and Panikov, 1994).

In conclusion we may state that the common physiological mechanism underlying the peculiarity of bacilli behaviour under different cultivation conditions does exist. It is the combination of cell material segregation and metabolic control based on the uncoupling inhibitory action of a specific metabolite. Under plentiful supply of nutrients these peculiarities do not express themselves and growth kinetics is quite usual. Rapid depletion of substrate in batch culture results in normal sporulation which excellently preserves bacteria from extinction. However, any kind of 'chronic starvation' which is realized in dialysis or other continuously fed cultures are 'the traps' for bacilli populations, i.e. they are provoked into sporulation (by a decrease in s to the threshold level), but are not able to terminate it in a normal way.

Batch culture limited by conserved substrates

Here we will explain quantitatively the mysterious phenomenon frequently observed in an algae culture. That is, batch growth does not appear to cease after the depletion of limiting conserved substrate, but may even be accelerated. This effect is reproduced only by the complete basic SCM: conserved substrate exhaustion while the source of energy is in excess (light in the case of algal growth) triggers the intensive degradation of macromolecular components. Degradation products then enter the monomeric pool L and are reutilized with the formation of new biomass with much lower cell quota, σ, of deficient element. Thus, growth in the complete absence of conserved substrate in the extracelluar liquid is provided by a redistribution of the deficient elements between mother and daughter cells. But why should this be more intensive as compared with unlimited growth? To grasp the point we should bear in mind that the instantaneous specific growth rate depends on several variables (Eqn. [C], Table 3.10), including

Table 3.12 SCM for description of *Bacillus* growth and sporulation

Variable	Equation*
Limiting substrate ($mg\ l^{-1}$)	$\dot{s} = D(s_0 - s) - q_s x + ax$, $q_s = rQ_s \dfrac{s}{K_s + s}$
Vegetative cells (X′-compartment) ($mg\ l^{-1}$)	$\dot{x} = \mu x - ax - Dx$, $\mu = Y[q_s(1 - \theta) - m] - k + \dfrac{k'x'}{x} - a$, $m = m_0 r$, $a = a_0 r$, m_0, a_0 – constant
Spores or prospores (X″-compartment) ($mg\ l^{-1}$)	$\dot{x}' = \mu' x' - Dx'$, $\mu' = \dfrac{kx}{x'} - k'$
r-Variable for X′-compartment (dimensionless)	$\dot{r} = (\mu + a)\left(\dfrac{s}{K_r + s} - r\right)$
r-Variable for X″-compartment (dimensionless)	$\dot{r}' = \mu'\left(\dfrac{s}{K_r + s} - r'\right)$
Sporulation inducer ($mg\ l^{-1}$)	$\dot{H} = \theta q_s x - DH$, $Y = \dfrac{Y_0}{1 + \lambda H}$, Y_0, λ – constant
Sporulation control	If $s > s^*$, then $\theta = 0$, $a_0 = a_1$, $k = 0$, $k' = 0$ If $s \leqslant s^*$, then $\theta > 0$, $a_0 = a_2 > a_1$, $k > 0$, $k' = 0$

* D definition for different cultivation system:

Chemostat	D = constant, $0 < D < D_{washout}$
Simple batch culture	$D = 0$
Dialysis culture	Ds_0 = constant, $Ds = Dx = 0$
Fed batch culture	$D = \dfrac{F}{V(0) + Ft}$,

where $V(0)$ is volume of culture at time $t = 0$ and F is medium flow rate.

s, r, σ and Y. Recall also that biomass yield Y is no longer constant and displays a direct relationship with the 'quick' variable s (Eqn. [F], Table 3.10) and therefore it also is a quick variable. After substrate depletion ($s \approx 0$), Y values jump instantly to the upper limit $1/\sigma_0$ (although other variables, including the cell's average r and σ remain high). Thus a combination of fully economized biosynthetic efficiency with high residual metabolic activity may provide temporary growth acceleration at the onset of starvation. This effect should be most pronounced if the difference between σ_0 and σ_m is large, as in the case of conserved substrates such as phosphates and silicates. The agreements between observed dynamics of algae growth on phosphate and SCM simulation is demonstrated in Figure 3.31.

We have discussed the energy-sufficient phototrophic growth. However, in the case of heterotrophic organisms, starvation caused by conserved substrates results in an even more mysterious phenomenon related to 'substrate-accelerated death'. Its interpretation in terms of the SCM is as follows. Cells transferred to a medium devoid of some conserved substrate (in our experiments it was N, see Figure 3.32) maintain slow growth via the redistribution of N between mother and daughter cells. This growth is accompanied by intensive turnover of macromolecular constituents and a dramatic decrease of P-components, including all elements of protein-synthesizing machinery (ribosomes, tRNA, mRNA, etc.). After plating on recovery media these cells are principally viable but are so retarded (they have an extremely low r_0) that they fail to produce visible colony growth in customary incubation times. Let us assume the widely accepted routine of using 3 days of incubation. Then the critical level $r = r^*$ which provides 3 days of lag phase will vary as dependent on other parameters μ_m and β (calculated according to Eqn. [3.65]):

μ_m	β			μ_m	β		
	1.0	0.5	0.1		1.0	0.5	0.1
0.1	0.007	0.0116	0.0681	0.3	4×10^{-10}	9×10^{-6}	0.0094
0.2	6×10^{-6}	0.0003	0.0223	0.4	3×10^{-13}	2×10^{-7}	0.0043

Thus, under unfavourable combinations of μ_m and β, long-term lag phase may take place even at rather high r-values (up to 2–7% from maximum). Growth retardation is aggravated by inhibitory effects of catabolic products which are accumulated in the medium (decrease the effective value of μ_m and enlarge lag phase). Note that the value of $r^* \sim 10^{-4}$ should be close to that situation when an individual cell contains only one single ribosomal particle (assuming the normal concentration of ribosomes in *E. coli* at $r = 1.0$ to be in the range 30 000–40 000 per cell).

To test this hypothesis we have to calculate the index θ, the fraction of total cell population characterized by $r < r^*$. Exact solution is difficult since it requires the assessment of the population heterogeneity, so we will make only a rough approximation. Let us assume that the whole population consists of two types of cells. The first subpopulation, x_1, is fully active ($r = 1$), while the second part, x_2, contains almost dormant cells ($r = r^*$). The average for entire population, r, is defined by the mass balance $rx = x_1 r^* x_2$, then

$$\theta = \frac{x_1}{x} = \frac{r - r^*}{1 - r^*}. \qquad [3.68]$$

Calculation according to the basic SCM with inclusion of the relationship [3.68] gives tolerable agreement with the observed dynamics of CFU decline in N-starving culture with a simultaneous slow increase of total biomass (Figure 3.32).

Thus, substrate-accelerated death should be considered as an extremely severe metabolic limitation. This phenomenon is not exotic; it may frequently occur in many laboratory and natural circumstances (Chapter 6).

Acclimatization to new substrates

In section 3.4 we studied two categories of such acclimatization: (a) old substrate s_1 (glucose) is the intermediate of reactions involved in the transformation of the new substrate s_2 (starch); hence adaptation requires the induction (derepression) of the synthesis of only one enzyme or enzyme cluster (in our particular case, it was an amylase complex); (b) s_2 (phenol) is not a precursor of s_1 (glucose) and requires the induction of a basically different metabolic system, which competes with the old one.

In the first case we have the ordinary diauxie and derepression of enzyme synthesis which is simulated even by non-structured models (Table 3.3). The SCM reproduces the more delicate particularities of the entire phenomenon, e.g. the dependence of inducible enzyme production rate v_P on the physiological state of microbial cells:

$$v_P = \theta Y(q_L - m)x = Arx\ , \qquad [3.69]$$

$$A = \theta Y\bar{L}\left[\frac{Q_L}{K_L + \bar{L}} - \frac{Q_m}{K_m + \bar{L}}\right].$$

where θ is a valve-like regulator simulating the operon function similar to that used in the modelling of amylase biosynthesis (Table 3.3). Equation [3.69] predicts proportionality between induction rates and the preceding growth rate of the microbial culture. This relationship does occur in reality (Figure 3.50).

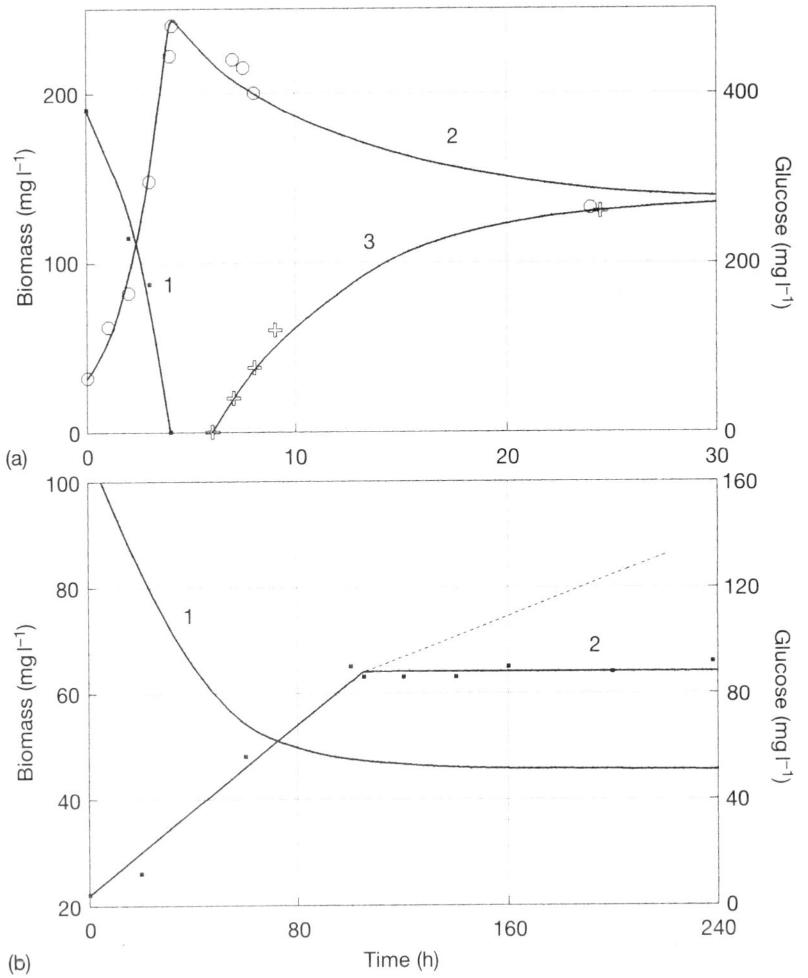

Figure 3.49 Growth and spore formation in *Bacillus* (a) batch culture and (b) dialysis culture: residual glucose (curve 1), total biomass (curve 2) and biomass of spores (curve 3) (Evdokimova and Panikov, 1994). The solid curves were calculated using the synthetic chemostat model (SCM) (Table 3.12); the dotted line is a simulation of biomass dynamics at $\theta = 0$ (neglecting the formation of autoinhibitory product).

In the second case basic SCM is amended with an additional variable of physiological state, ρ, which is a characteristic of the P-components of the alternative metabolic system (Table 3.13). The conserving condition is $r + \rho = 1$. Microbial cells are attributed to either the r- or ρ-state. For simplicity we assumed that monomeric intermediates belong to central metabolic pathways which are common for r- and ρ-states. The rules which

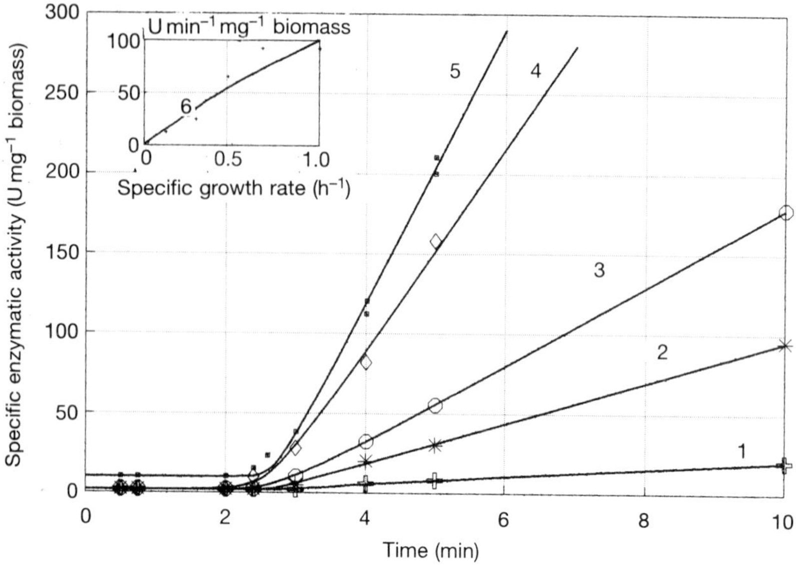

Figure 3.50 Synthetic chemostat model (SCM) simulation of dynamics of β-galactosidase formation on addition of inducer to the second fermenter of two-stage chemostat (Řičica *et al.*, 1967). Enzymatic activity is expressed as arbitrary units (U) mg^{-1} of biomass. Specific growth rate of microorganisms before inducer is added (h^{-1}): 0.02 (curve 1), 0.12 (curve 2), 0.29 (curve 3), 0.475 (curve 4), 0.675 (curve 5). Curve 6 shows the relationship between specific rates of enzyme production and microbial growth; the curve is calculated according to SCM, Table 3.9.

govern the metabolic strategy are formulated as follows. When microorganisms are in a *r*-state, they are allowed to start the consumption of s_2 only after the depletion of the preferential (for the *r*-system) substrate below the threshold level ($s_1 < s_1^*$), and synthesize the enzymes of alternative metabolic system after partial degradation of the first one below another threshold, at $r < r^*$. The exactly symmetrical conditions are set for the second ρ-state. Translation of these rules to the computer routine allowed the simulation of the main experimental observations: (a) long-term intermediate lag phase neccessary for the degradation of the old metabolic system; (b) the dependence of substrate preference on the past history (*r* or ρ state of inoculum); and (c) parallel consumption of both substrates in chemostat culture when simultaneously $r < r^*$, $\rho < \rho^*$, $s_1 < s_1^*$, and $s_2 < s_2^*$.

Cell cycle

Regulation of cell cycle in unicellular organisms is a specialized and vast field of microbiological research. In this book we do not intend to cover all

Table 3.13 SCM for description of microbial growth on a mixture of two alternative substrates

Variable	Equation
Substrate 1 (glucose) ($mg\ l^{-1}$)	$\dot{s}_1 = D(s_{01} - s_1) - q_1 x,$ $q_1 = rQ_1 \dfrac{s_1}{K_1 + s_1}$
Substrate 2 (phenol) ($mg\ l^{-1}$)	$\dot{s}_2 = D(s_{02} - s_2) - q_2 x,$ $q_2 = \rho Q_2 \dfrac{s_2}{K_2 + s_2}$
Intermediate (g (g biomass)$^{-1}$)	$\dot{L} = k_1 q_1 + k_2 q_2 - q_L + (a_1 r + a_2 \rho)(1 - L) - \mu L$
Variable of physiological state for glucose	$\dot{r} = \beta_1(\mu + a_1 r + a_2 \rho)(1 - L)\left(\dfrac{(1-\rho)s_1}{K_r + s_1} - r\right)$
Variable of physiological state for phenol	$\dot{\rho} = \beta_2(\mu + a_1 r + a_2 \rho)(1 - L)\left(\dfrac{(1-r)s_2}{K_\rho + s_2} - \rho\right)$
Biomass ($mg\ l^{-1}$)	$\dot{x} = (\mu - D)x\ ,$ $\mu = k_1 q_1 + k_2 q_2 - q_L + Y(q_L - m)$

modern aspects of this esoteric area. Our aim is quite modest, simply to apply the SCM to those situations when we can no longer ignore growth discontinuity resulting from cell division.

The interpretation of events associated with the bacterial cell cycle we will state here is in accordance with the most popular and influential model proposed by Helmstetter, Cooper and co-workers (Helmstetter *et al.*, 1968; Cooper, 1990). It postulates that the initiation of DNA replication is triggered by a threshold concentration of **initiator protein**, this protein being synthesized at a rate proportional to the total growth rate and requiring exactly one mass doubling time to reach its threshold concentration again. Replication time ('C' period) and subsquent cell division ('D' period) are constant and normally occupy, in the case of *E. coli,* about 40 and 20 min of the cell cycle respectively. The mass of the individual cell is growing exponentially with a specific rate dependent on cultivation conditions but not on the period of the cell cycle. It is this difference which yields the μ-dependent variations of cell sizes, since faster growing cells produce, in the same C + D = 60 min time, a bigger cell mass than slow-growing cells.

In SCM terms, the initiator protein belongs to constitutive cell components. The equations for its mass balance are different as they depend on the dimension of this variable:

per unit of cell biomass, g (g biomass)$^{-1}$		per one cell, g (l of cell volume)$^{-1}$	
$\dot{V} = \theta(\mu + a) - \mu V$	[3.70a]	$\dot{v} = \theta(\mu + a)x_{\text{cell}}$	[3.70b]

In the case of Eqn. [3.70a], we have the usual expression derived for any intracellular component (see Tables 3.6–3.7 and 3.9–3.10 for comparison). The V source is the biosynthesis rate, which is a uniform fraction of the total intracellular flux, while the sink is V dilution due to biomass growth. It is clear that the V quantity may quickly attain its steady-state value $\bar{V} = \theta(\mu + a)/\mu \approx \theta$. The notation of threshold concentration is not applicable to this variable, since it is kept constant throughout the cell cycle. Furthermore, we may not accept the widely used and tempting postulate that protein V is annihilated immediately after initiation of DNA replication, to be synthesized from zero at the start of a new division cycle, since it leads to the violation of the basic postulate on proportionality between its biosynthesis and total biomass growth.

In the case of Eqn. [3.70b], we have introduced the variable x_{cell}, the mass of single bacterial cell, which is intrinsically a transient entity, i.e. even during steady-state growth it changes continuously between lower and upper limits corresponding to masses of new-born cells and cells before division. If N is the total number of cells in the bacterial population, then the conventional variable x, the total biomass of bacterial culture, is the product $x = Nx_{\text{cell}}$.

For the sake of simplicity let us ignore the variations in cell diameter (implying a strict proportionality between cell length, l, and mass, x_{cell}) and divide the entire cell cycle into two parts: (a) the period of cell growth without chromosome replication from minimal length, l_0, of a new-born cell to some intermediate length, l^*, when the initiator attains its threshold concentration, v^*; and (b) the C+D period of chromosome replication and cell division which occupy the constant time τ, during which cell length attains the maximum length $2l_0$ and divides. Under steady-state conditions with constant μ, cell length grows exponentially ($dl/dt = \mu l$, $l = l_0 e^{\mu t}$), during the second part of the cycle it takes exactly τ min for the cell to elongate from l^* to $2l_0$, hence

$$2l_0 = l^* e^{\mu\tau} \quad \text{or} \quad l_0 = 0.5\, l^* e^{\mu\tau} \quad \text{and} \quad l_0 \approx 0.5\, l^* \mu\tau\,. \qquad [3.71]$$

Here we took into account the fact that $e^{\mu\tau} \approx \mu\tau$ for small $\mu\tau$ and came to an almost linear relationship between cell length at birth and specific growth rate, which was demonstrated for bacteria of various shapes (Figure 3.39). Note that the exact derivation leads to an exponential increase of cell length or mass, which was confirmed by accurate measurements made with a 'baby-machine' (Cooper, 1990). The linear approximation is used only for the mechanistic interpretation of the obtained empirical regression. The *y*-intercept of the regression lines in Figure 3.39 is $0.5\,l^*$, half the cell length when chromosome replication is initiated, and the slope is τ, the duration of C + D periods. It is interesting that τ was close to 60 min for bacilli, pseudomonas and nodule bacteria (0.75–1.0 h), as in the case of *E. coli*, while in cocci it was much shorter (0.41 h) and in *Arthrobacter* and *Rhodococcus* much longer (1.4–1.9 h). It is this quality which determines the specific morphology of coccoid and coryneform bacteria.

In order to simulate transient changes of cell sizes during the cell cycle, we introduced the auxiliary variable λ which simulates chromosome replication and cell division under the following conditions:

$$\dot{\lambda} = k, \quad \begin{array}{ll} \text{if } v \geqslant v^* \text{ then } k = k_1 > 0, \text{ else } k = 0 & \text{(initiation)} \\ \text{if } \lambda \geqslant \lambda^*, \text{ then } \lambda = \lambda/2,\ v = v/2,\ x_{\text{cell}} = x_{\text{cell}}/2 & \text{(division)} \end{array} \qquad [3.72]$$

For the sake of simplicity we again combined the C and D periods into one, formally assuming that physical division is performed instantly after termination of chromosome replication. Some of the simulation results are shown in Figure 3.51 to demonstrate the model sensitivity towards perturbations. Suppose that there are random variations in the content of initiator protein, which is to be expected for real bacteria. What consequences will this have on the division cycle? We can see from curve 1 that under a higher level of V as compared with standard, the model generates several abnormally frequent cell divisions with the subsequent reversal to normal steady-state multiplication pace. A symmetrical pattern was displayed under the reverse situation of V deficiency (curve 2), i.e. several divisions were delayed resulting in the production of abnormally long cells before steady-state was restored. Thus, the Helmstetter–Cooper model in the suggested mathematical shape does present the sufficient guarantee of cell cycle persistence to perturbations. The negative feedback mechanism which brings things back into line is based on the dynamic nature of variable V. It is characterized by a unique stable steady-state which is approached from different initial conditions. There was an instructive criticism of the initiator model stating that it 'does not include any mechanism adjusting the rate of initiator synthesis to the growth rate' (D'Ari *et al.*, 1990, p.10). The presented mathematical treatment does provide the required negative feedback. That is, V content is connected both positively

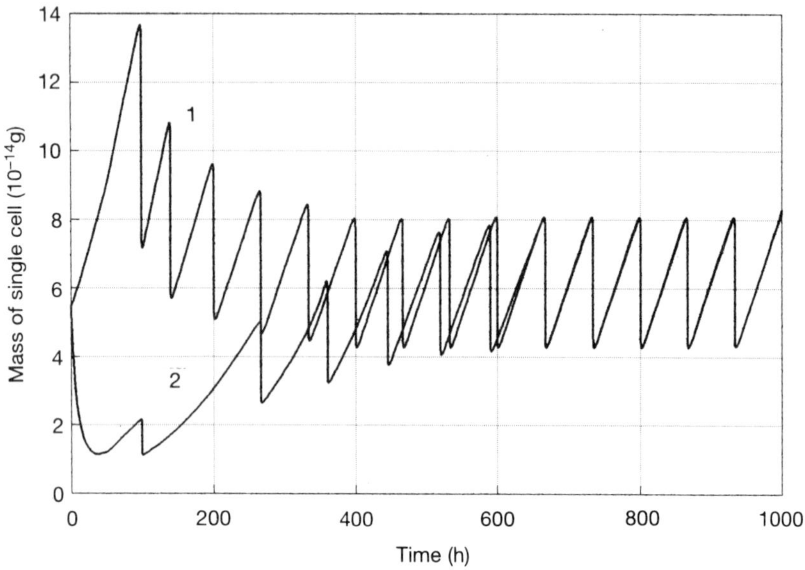

Figure 3.51 The test for stability of the Helmstetter–Cooper model of the cell cycle control by initiator protein. Synthetic chemostat model (SCM) supplemented with Eqns [3.70] and [3.72] simulates the elongation of single cell during steady-state growth. At $t = 0$ the intracellular concentration of initiator V was perturbed to either lower (curve 1) or to higher level (curve 2) as compared with normal steady-state value. As could be seen, the steady division frequency and cell sizes are restored in three to four generations.

and negatively with cell growth, negative effects coming from the initiator dilution due to biomass expansion, this being the method of V elimination in the possible event of oversynthesis. From this point of view, the cytoplasmic inhibitor model has no advantages over the Helmstetter–Cooper model. Another possible mechanism based on 'critical initiation mass' of the cell (Donachie, 1968) is equivalent to the discussed model since the time of threshold accumulation of the initiator corresponds to one unique mass of cell or cell length l^* which does not depend on the growth rate.

It may be easily shown that the described model adequately simulates the various morphological effects exhibited during non-steady-state growth, e.g. (a) more pronounced lag phase in batch culture when growth is surveyed by cell count rather than biomass measurements; (b) significant enlargement of cells during the transition from lag- to exponential growth phase; (c) the formation of dwarf cells in the starving population; and (d) the accumulation of 'division potential' (if division is blocked by inhibitor, then after the block is released all missing divisions take place in quick succession). By the introduction of the r-variable into Eqn. [3.72] we may account for the variation in C+D periods which takes place in very slow growing populations.

Unusual growth kinetics of *Arthrobacter*

Several hypotheses were tested to simulate all the essential features of *Arthrobacter*'s peculiar behaviour (multistability, false diauxie, sustained oscillations, etc.). These included: (a) delayed adaptation (low value of parameter β in Eqn. [3.65]); (b) substrate inhibition of growth (Eqn. [1.58]); (c) repression–derepression of high affinity transport system; and (d) 'internal' growth limitation by low content of respiratory enzymes (low upper limit Q_{resp}). However, all these assumptions were not adequate. The only acceptable results were provided by a significant enlargement of the basic SCM according to the following growth mechanism:

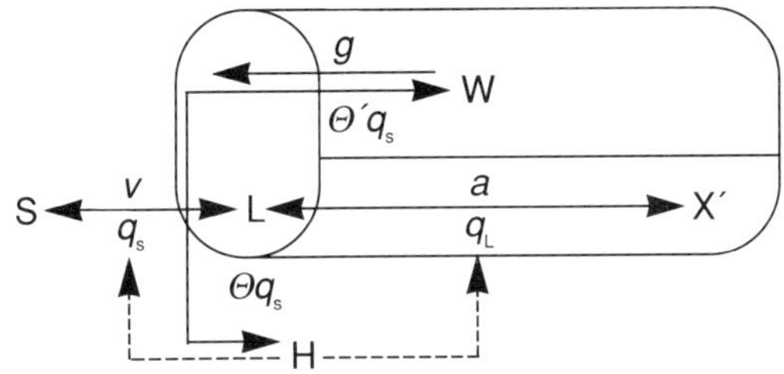

1. The monomeric pool consists of normal intermediates L and compounds H, which inhibit the transport of exosubstrate and L conversion to X′:

$$q_s = rQ_s \frac{s}{K_s(1 + \alpha_H H) + s}, \qquad q_L = rQ_L \frac{L}{(K_L + L)(1 + \beta_H H)}, \qquad [3.73]$$

where α_H and β_H are inhibition constants.

2. Autoinhibitor H originates from exosubstrate S and is inactivated by special protective enzymes:

$$H = \theta q_s - q_H - \mu H, \qquad q_H = rQ_H \frac{H}{K_H + H}. \qquad [3.74]$$

Since $\theta \ll 1$, then the contribution of H to the growth budget may be neglected.

3. H synthesis is governed by the following conditional rules: (a) H formation is initiated under high $L \geqslant L^* = cr$, where c is constant; (b) H production is stopped when its content achieves the upper threshold H^{max}; (c) recommencement of H formation requires the simultaneous condition $H \leqslant H^*$ (low threshold) and $L \leqslant L^*$.

4. *Arthrobacter* accumulates reserved carbohydrate W via the activity of an enzymatic system which belongs to the U-type of cell components:

$$\dot{W} = \theta'(1-r)q_s - gW - \mu W, \quad g = \text{constant}, \qquad [3.75]$$

where g is the specific rate of W conversion to L, and θ' is a valve-type regulator which is activated under the temporary relief from substrate limitation when $s \geqslant s^{**}$. Carbohydrate reservation competes with the primary metabolic pathway S → L → X′, so that W formation should be accounted by a modification of the mass balance Eqn. [B] (Table 3.9):

$$\dot{L} = [1-\theta'(1-r)]k_2 q_s - q_L + ar(1-L-W) + gW - L(\mu + v). \quad [3.76]$$

The modified SCM (Eqns [3.73–3.76]) reproduces all the essential features of *Arthrobacter* growth. Bistability in chemostat culture (Figure 3.4) is explained by the alternative possibility of having zero and non-zero steady-state H concentration. Hence the second type of steady-states with lower $\bar{x}$ and higher residual $\bar{s}$ may be interpreted as growth affected by both product autoinhibition and substrate limitation. Transient growth after a change of chemostat dilution rates is smooth only for small perturbations in a narrow range of low D (Figure 3.52, a), in this case $L < L^*$ and $H = 0$. In the range of medium D, perturbation induces temporary H formation resulting in long transient with the eventual recovery from inhibition (Figure 3.52, b). Finally at high D with a constant nutrient feeding rate, growth may be realized only as sustainable oscillations (Figure 3.52, c). The main cause of these oscillations is the periodic changes in inhibitor concentration: at $D > 0.25\ \text{h}^{-1}$ steady-state of the first type (when $H = 0$) is characterized by high $\bar{L} > L^*$ and hence, according to the conditions summarized above in point 3, autoinhibitor synthesis ceases at the upper threshold, H^{max}, but always resumes at the moment when H declines to H^*.

The first phase of false diauxie is completely explained by the filling with glucose of the storage compartment W, while the subsequent true growth is retarded by the accumulated autoinhibitors. Unpredictable and 'capricious' growth of *Arthrobacter* in batch culture is readily perceived as a consequence of unknown variations of the physiological state of inoculum (different values of r and H at $t = 0$).

We are far from being certain of the validity of the suggested mechanism. In fact, the agreement with experimental data is in no way proof of the premises that were translated into the simulation model. Nevertheless, we may consider the positive outcome of the simulation in two ways. Firstly, it prompts a more rigorous examination of the concept of autoinhibition. Obviously, we badly need to identify the chemical compound which plays the role of hypothetic metabolite H, the probable candidate being peroxides (Kuznetsov, 1981) or even 'normal' catabolites accumulated above the permissible level (Golovlev, 1983). The performed simulations

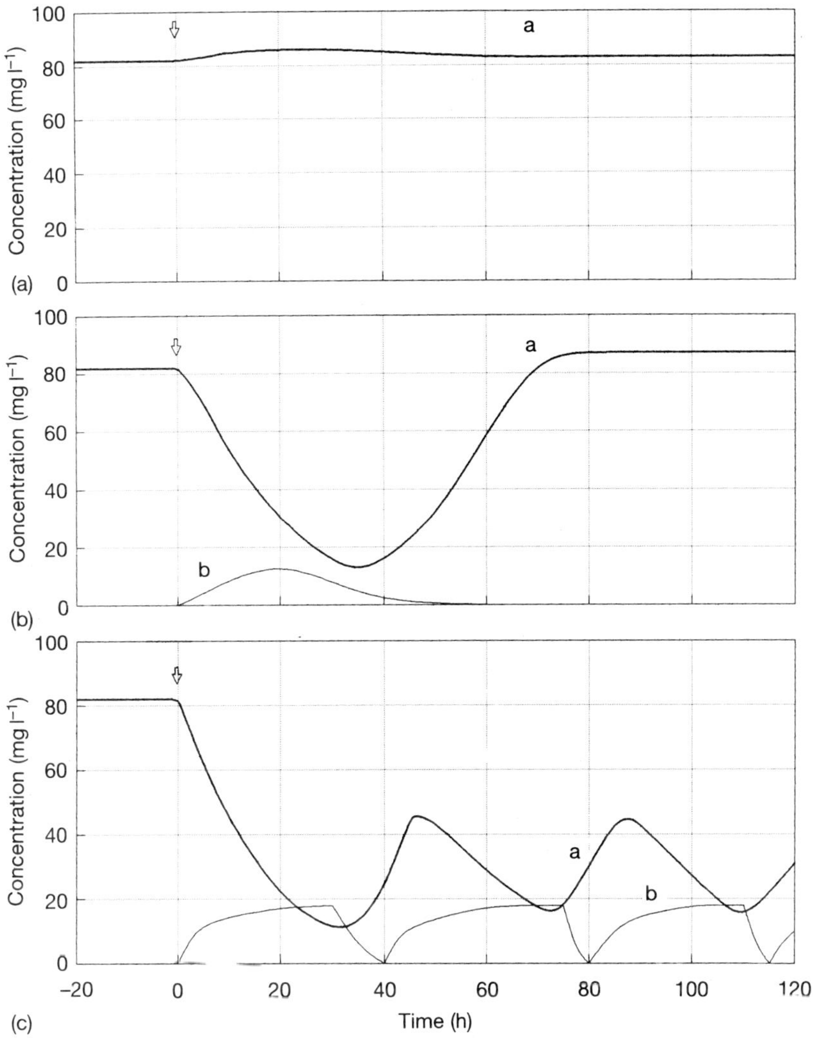

Figure 3.52 Simulation of transient growth kinetics of *A. globiformis* caused by shift of dilution rate in chemostat culture from $0.05\,h^{-1}$ to (a) 0.11, (b) 0.20, and (c) $0.27\,h^{-1}$ at $t = 0$. Calculation with the use of supplemented synthetic chemostat model (SCM) (Eqns [3.73–3.76]) of dynamics of biomass (curve a) and autoinhibitor (curve b). Parameters: $Q_s = 1.0$, $Q_L = 50$, $Q_H = 0.001$, $a = 0.003\,h^{-1}$; $K_s = 1.0$, $K_r = 0.9$, $s^* = 0.006\,mg\,l^{-1}$, $K_L = 0.8$, $K_m = 0.001$, $K_H = 20.0$, $\alpha_H = 12\,000$, $\beta_H = 120\,g\,g^{-1}$, $\theta = 0.008$, $Y = 0.55$, $\beta = 1.0$. (Dorofeyev *et al.*, 1992.)

suggested the improvement of the screening efficiency by the control of cultivation conditions.

Secondly, the simulations may be used for fermentation control aimed at avoiding the state of autoinhibition. This is especially important in the view of the great biotechnological significance of *Arthrobacter* and its relatives.

Thirdly, simulation is helpful for a better understanding of *Arthrobacter* population dynamics in a natural environment. For this purpose we will use the developed model as well as its reduced version (Chapter 5).

3.7 CONCLUSION

At last, we come to the end of this chapter. It turned out to be the longest one, and this is justified by the fact that homogenous growth is the primary focus of contemporary kinetic research. The main emphasis of our studies has been those aspects of microbial growth which have ecological perspectives, i.e. the diversity of organisms and cultivation conditions, very slow growth comparable with growth in natural habitats, starvation–survival dynamics, non-steady-state and unbalanced growth, multistability and multiple limitation, mixed substrates utilization, spore formation and cell cycles.

We will here remind the reader of some of the obtained experimental results:

1. A specially developed experimental technique allowed the direct measurement of variations in the turnover rates of macromolecular cell components and sizes of monomeric pools.
2. Growth parameters of continuous culture formerly supposed to be 'growth constants' (maintenance coefficient, saturation constant, yield factor, potential metabolic activity, cell content of biogenic elements) display regular variation coupled with the specific growth rate and the steady-state limiting substrate concentration.
3. Dynamics of batch growth in the lag phase is fully determined by the past history of the inoculum and in the exhausted batch culture degradation processes (cell death and lysis) are associated with biosynthetic ones (very slow but efficient cryptic growth, production of exometabolites).
4. Significant deviations from normal growth kinetics were observed in the continuous culture of bacilli (the occurrence of a maintenance state) as well as in batch and chemostat cultures of *Arthrobacter* (multistability, sustained oscillations, false diauxie).
5. Growth on polymeric and mixed substrates under N-limitation is affected by 'substrate-accelerated death' and irreversible repression.
6. Cell sizes of diverse organisms, including rods, cocci, branched and Y-shaped bacteria, uniformly increase with their growth rates.

As a first attempt we tried to fit the obtained experimental data to known unstructured kinetic models (Monod, Herbert, Pirt, Droop) or their modifications. However, successful results were acheived only on rare occasions. The main cause of this failure was the fact that almost all known models ignore the phenomenon of cell adaptations to the changeable environment. The problem of self-adjustment is not essential in chemical or enzymological kinetics, since normally abiotic catalysts and enzymes preserve their basic functional properties under any reaction conditions. However, in microbial kinetics the situation is quite different! Unstructured models actually identify microbial cells with these simple catalysts acknowledging their only ability to convert the nutrients into a uniform unchangeable biomass. Needless to say, the real microbial cell is a well-organized **combination** of numerous simple biocatalysts, the links between them being not absolutely rigid.

Adaptive changes of biomass quality are ambiguously expressed by the notation of 'physiological state'. To take up the quantitative route, we have addressed models which are based on an optimization paradigm. The assumption that microbial cells stir themselves to maximize their growth rate under the available nutrient supply did lead to realistic simulations. However, this did not lead to a better understanding of growth because optimization models turned out to possess almost the same degree of complexity as modelling archetypes. Finally, we developed the 'synthetic chemostat model', or SCM, which is a compromise between logical derivation and empirical generalizations. It formalizes the notion of physiological state by the introduction of only one non-conventional variable, r, which relates adaptive modifications of cell quality to the concentration of limiting substrate(s) in the medium. The benefit of using one extra variable was much greater than the cost of the slight model enlargement. The biological interpretation of the experimental data allowed the elucidation of a number of formerly mysterious growth phenomena which were discussed in section 3.6.3.

In the following chapters we will try to cope with more complicated matters (microbial colonies and biofilms, mixed cultures, microbial populations in soil, etc.) by splitting complex systems into several homogeneous units, each of which is capable of simulation by the SCM.

Heterogeneous microbial growth 4

Microbial growth associated with the formation of aggregates, colonies and biofilms is incomparably more variable and complicated than homogeneous growth. There are no uniform and comprehensive theories which explain the spatial organization of aggregated growth. Here we shall consider just a few specific cases amenable to quantitative analysis.

4.1 GROWTH OF COLONIES ON SOLID AGAR MEDIA

Plating on agar media in Petri dishes and streaking on slants are probably the most widely used cultivation procedures in the everyday routine of all microbiologists. In addition, the colony itself is a traditional fundamental subject in general microbiology. The problems of how it is organized, how it functions and develops have always attracted attention, being one of the most fascinating areas of microbiological research (see details in the exquisite reviews of Shapiro, 1984, 1987).

As shown by the preceding theoretical analysis (section 2.3), colony growth is essentially different for unicellular and mycelial organisms. The hyphae of fungi and actinomycetes are able to penetrate into the depth of the agar layer, so that mycelium expansion cannot be controlled by nutrient diffusion from outside the colony. If substrate concentration is not very high, then the mycelium follows a chemostat-type growth pattern, being (a) continuous, (b) steady-state and (c) limited by substrate availability. The role of the fermentation vessel of the conventional chemostat is played by the peripheral zone of the colony, the product bottle is substituted by the central part of the colony, while the operation of the pump is imitated by the chemotropic movement of hyphae tips along the substrate concentration gradient.

In the case of unicellular organisms, bacteria and yeasts, the closest homogeneous parallel is fed-batch or dialysis culture. These organisms are not able to penetrate into the agar matrix. Their spatial expansion is restricted to the agar surface, and a continuous nutrient supply is provided by the diffusional flux coming from the underlying agar layer.

Here we shall test these speculative considerations by an examination of authentic experimental data.

4.1.1 BACTERIA

Bacteria *Pseudomonas fluorescens* 312 and 1472 were grown on trypticase soy agar (TSA) and on a defined medium with 1.8% Difco agar, $5\,g\,l^{-1}$ pyruvate and mineral base as in the nutrient medium 1 (section 3.1). Inocula were obtained on the respective liquid media without agar in shake flasks.

Dynamics of colony radius and height

The agar surface in small plastic Petri dishes was spot inoculated with a diluted 24-h broth culture. The incubation was performed at 25°C in a constant temperature room. Colony diameters were measured with a calibrated, filar-micrometer eyepiece mounted on a microscope (Yenaval, Germany). The heights were estimated either with colony cross-sections cut from agar (Palumbo *et al.*, 1971) or by alternate focusing of the microscope on the colony top and nearby agar surface. The height was then calculated from the difference of microscrew precalibrated positions. The second approach requires more patience but allows the non-destructive measurement of the height with one and the same colony.

Some of the results obtained are shown in Figure 4.1 for the particular case of bacteria growing on TSA (Rodin and Panikov, 1995). Rodin followed the dynamics of colony development on a Petri dish from one single cell to the visual macrocolony. (Note that all published data on micro- and macrocolonies were obtained in separate experiments, the microcolonies being cultivated on slides or chambers which provide different growth conditions than a Petri dish.)

The radius of colonies increased exponentially for the first 10 h and then expanded linearly at least for several days. The notation of 'radius' is hardly applicable to the early stages, when the shape of a microcolony is rectangular rather than circular. For this particular stage, the measurements were made of longer R_1 and shorter R_2 dimensions, assuming the effective diameter to be the average value $(R_1+R_2)/2$.

The height of the microcolony was equal to cell's diameter during the first 5–7 h when bacteria grew as a monolayer. The second layer was then formed and colony height started to grow, at first exponentially and then linearly. It is interesting that height increased rapidly on TSA ($8.2\,\mu m\,h^{-1}$) and very slowly on the defined medium with pyruvate

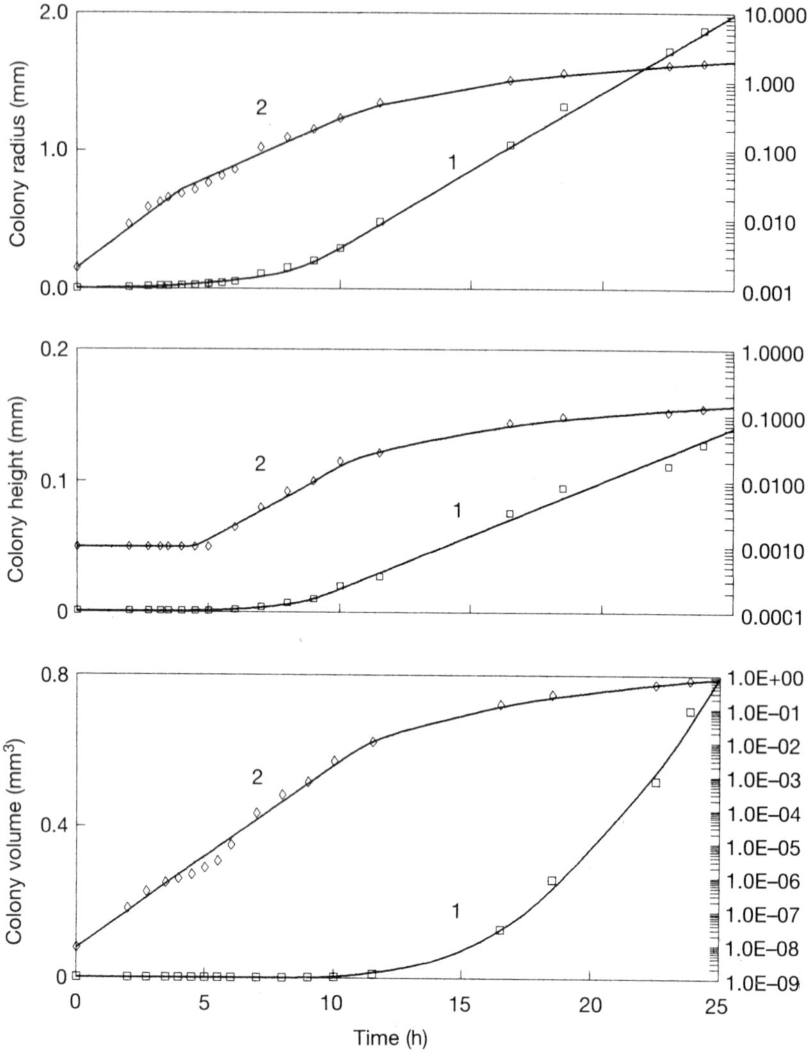

Figure 4.1 Dynamics of the development of a bacterial colony from single cell. Trypticase soy agar in a small Petri dish was spot-inoculated with *P. fluorescens* cells and incubated at 25°C. Changes in the diameter and height of a microcolony were followed by microscopy. Colony biovolume was calculated assuming it was a spherical segment (Eqn. [4.1]). Dynamic curves are presented in linear (curve 1) and log-scale (curve 2) forms. Simulations were performed using the model in Table 4.1. Explanations and parameters are given in the text. (Rodin and Panikov, 1995.)

($<1\ \mu m\ h^{-1}$). This disparity was responsible for the difference in colony shape: the TSA-colonies looked like spherical segments, while pyruvate-grown colonies were flat cylinders.

Bearing in mind this geometric resemblance, colony biovolume was calculated from two experimentally estimated quantities, radius R and height H as follows:

$$V_{\text{pyr}} = \pi R^2 H, \quad V_{\text{TSA}} = \frac{1}{6}\pi H(3R^2 + H^2) = \pi R^2 \hat{H}, \quad \hat{H} = \frac{1}{2}H, \quad [4.1]$$

where $\hat{H}$ is the average height, approximately equal to half of the H value measured at the centre of colony. As can be seen from Figure 4.1, the biovolume dynamics were exponential during the early stages and then increased parabolically.

The experimental data were in full support of Pirt's concept of colony growth (Pirt, 1975), implying the transition from non-limited exponential growth of the colony when it is quite small, to exponential growth of the colony's peripheral zone when any nutrient becomes limiting. Throughout the second growth phase, the colony increases its radius with constant rate $K_R = \mu w$ as long as the width of the peripheral zone, w, and the specific growth rate of cells within the peripheral zone, μ, are kept constant. However, the original model of Pirt requires some additional modifications to account for the height and volume dynamics.

The modified model (Table 4.1) is based on the following assumptions.

1. During the first phase of the colony development its total biomass and biovolume is growing exponentially with constant specific rate, μ_m (Eqns [a], [a]′).

2. The development of the microcolony is governed by several 'rules' illustrated by the following idealized scheme:

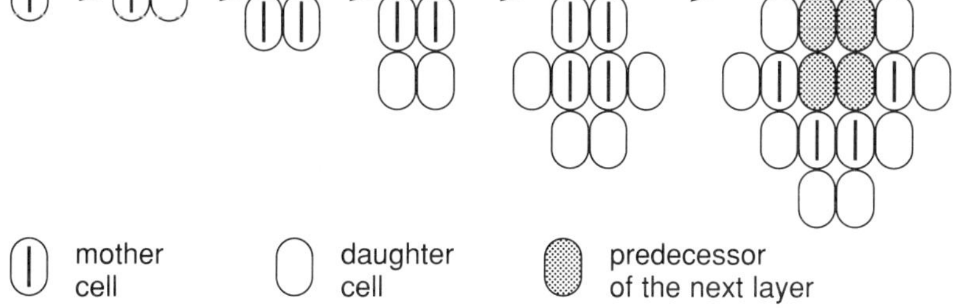

Table 4.1 Simulation of growth of a bacterial colony on agar medium

Growth phase	Differential form	Equation	Result of integration
Exponential growth of monolayer microcolony (0–τ h)	$\frac{d}{dt}V = \mu_m V$	[a]	$V = V(0)\, e^{\mu_m t}$
	$\frac{d}{dt}[\pi R^2 H(0)] = \mu_m \pi R^2 H(0)$	[b]	
	$\frac{dR}{dt} = \frac{1}{2}\mu_m R$	[c]	$R = R(0)\, e^{\mu_m t/2}$
	$\frac{dH}{dt} = 0$	[d]	$H = H(0)$

Table 4.1 Continued

Growth phase	Differential form	Equation	Result of integration
Exponential growth of multilayer colony ($\tau - T$ h)	$\frac{dV}{dt} = \mu_m V$	[a′]	$V = V(0)\, e^{\mu_m t}$
	$\frac{d}{dt}(R^2H) = \mu_m R^2 H = 2H\frac{dR}{dt} + R\frac{dH}{dt}$	[b′]	
	$\frac{dR}{dt} = \frac{1}{2}\theta\mu_m R$	[c′]	$R = R(\tau)\, e^{\theta\mu_m(t-\tau)/2}$
	$\frac{dH}{dt} = (1-\theta)\,\mu_m H$	[d′]	$H = H(0)\, e^{(1-\theta)\mu_m(t-\tau)}$
Linear growth ($t > T$ h)	$\frac{dV}{dt} = \frac{d}{dt}(\pi R^2 H) = 2\pi(K_R H + K_H R)$	[a″]	$V = \pi R^2 H = \pi K_R^2 K_H (t-T)^3$
	$\frac{dR}{dt} = K_R$, K_R – constant	[c″]	$R = K_R(t-T)$
	$\frac{dH}{dt} = K_H$, K_H – constant	[d″]	$H = K_H(t-T)$

where numerals indicate the number of generations. The cells in the microcolony behave as if they are 'striving' (a) to align to mother or sister cells, and (b) to attach to the agar surface. Following these 'rules', the microcolony is a characteristic four-cell array at $n = 2$ (Shapiro and Hsu, 1989), and then grows as a monolayer as long as there are vacant places within the horizontal plane on the agar surface (Eqns [b] and [c], where $H(0)$ is the diameter of a single cell). At $n = 5$, the eight cells in the centre of the microcolony have no option but to form the second layer. From this simplified viewpoint it follows that growth of a colony's height should be delayed, compared with the radius, by about five generations (τ h).

3. The allocation of cell mass between the horizontal (increase of R) and vertical (increase of H) directions during exponential growth of a multi-layer colony is defined by the coefficient θ (Eqn. [c′]). It is confined within the interval [0,1] and may be interpreted as the probability of daughter cells occupying a position in the same layer as the mother cell. Thus, for instance the value $\theta = 0.8$ corresponds to the growth pattern, when eight of every ten emerging cells shift cells that are adjacent and in the same horizontal row towards the periphery of the colony, thereby contributing to an increase in radius, while only two cells move to the next upper layer, enlarging the colony's height. Equations [b] and [b′] specify the conservation condition preserving the total exponential growth of cell mass composing the cylindrical colony.

4. When the flux of some nutrient from the agar becomes smaller than the rate of its consumption, then the growth of the cells in the centre of the colony stops and subsequent linear expansion of the colony in horizontal and vertical directions is provided by the growth of the peripheral zone (Eqns [c″] and [d″]). The simultaneous linear increase of R and H is equivalent to a parabolic enlargement of the total colony volume proportional to the cube of time (Eqn. [a″]).

Calculations using the described model produced excellent agreement with experimental data (Figure 4.1). The plot of $\ln R$ *versus* t consists of three domains divided by two inflexion points. The first inflexion corresponds to $\tau = 4.5$ h, which is close to the expected five generations of monolayer stage in the colony development. The second inflexion point, $T = 9$ h, marks the transition from exponential to linear growth phase. At this moment $R = 250\ \mu m$ and $K_r = 113\ \mu m\ h^{-1}$. If we assume that subsequently the width of the peripheral zone remains at 250 μm, then the specific growth rate in the colony periphery is $\mu = K_R/w = 113/2500 \approx 45\ h^{-1}$. It is assumed to be constant during the linear growth phase but lower than the upper limit, μ_m, due to substrate limitation. The

μ_m value displayed during the preceding exponential phase was 1.24 h^{-1} or 2.8 times higher.

It is interesting that maximal specific growth rate on agar was 15–30% less intensive than on the same medium under homogeneous conditions in shaker flasks. There is no mechanistic explanation for this fact, besides the vague declaration that nutrient consumption by the microcolony is controlled by substrate diffusion from the agar layer. Parameter θ, in the case of TSA, was equal to 0.60, indicating an almost equal probability of newborn cells contributing to the radius and height of the colony. On the defined medium, allocation was in favour of radial growth ($\theta < 0.1$). This difference may be teleonomically interpreted as the cell's attempt to leave the environment with the non-attractive 'menu' of one single C-compound.

The obvious drawback of the suggested model is that it does not express explicitly the driving forces of the transition from one phase to another. For this purpose we need to consider the available data on substrate consumption and other stoichiometric aspects of colony growth.

Stoichiometry of bacterial growth on nutrient agar

The agar column (30 × 30 mm) in a glass tube containing defined nutrient medium with 5 g l^{-1} of pyruvate was surface-inoculated by *P. fluorescens* 1472 to produce a lawn. During incubation at 25°C part of the replicate tubes were withdrawn for determination of the bacterial biomass dry weight, cell viability by the Postgate (1967) technique and residual substrate: the agar column was cut into three to eight sections, pyruvate extracted by hot phosphate buffer (70°C) and determined spectrophotometrically with lactate dehydrogenase. The parallel replicates of agar columns were incubated in air-tight jars to measure the respiration rate by gas chromatography (GC) as the rate of CO_2 accumulation in the headspace (Belova *et al.*, 1995). The use of a lawn instead of a colony was aimed at reducing the three-dimensional growth pattern of individual colonies to the simpler one-dimensional case when spatial distributions may be considered only along the column height.

Results are presented in Figure 4.2. During the first day of incubation, cell biomass grew exponentially with a specific rate comparable with that in homogeneous culture (0.25 h^{-1}). Growth then decelerated and the total biomass tended to a steady level. Stabilization was accompanied by accumulation of dead cells which accounted for almost half of the total population (35–50%).

The spatial distribution of residual carbon substrate was typical for the expected diffusional process, i.e. during the first day there was a dramatic decrease of pyruvate in the upper agar layer adjoining the biofilm; the depletion zone then gradually moved to the column bottom. The dynamic curves fit the standard diffusion equation

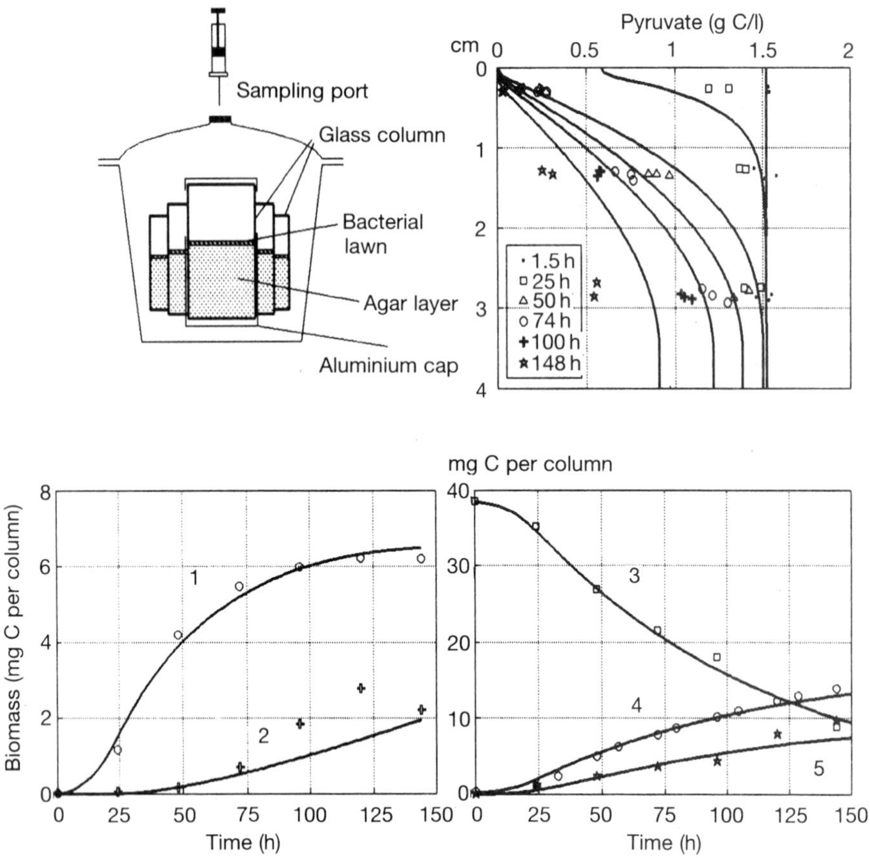

Figure 4.2 Growth dynamics of bacterial lawn. Bacteria *P. fluorescens* were grown on the surface of agar column with pyruvate–mineral medium. In the course of incubation the following parameters were recorded: (1) total biomass of bacteria; (2) necromass; (3) residual pyruvate in agar layer; (4) CO_2 in gas phase; and (5) extracellular products. The top panel shows experimantal design and dynamics of spatila distribution of pyruvate in agar (Belova, Dorofeyev and Panikov, 1995), the bottom panel presents experimental data and calculations according to Eqn. [4.2].

$$\frac{\partial s(z,t)}{\partial t} = D_s \frac{\partial^2 s}{\partial z^2},$$

with D_s about 0.05 $cm^2\,h^{-1}$. Integration over the entire profile gave the total substrate flux to the bacterial lawn, F_{dif}. It appeared to be more or less constant (0.2 mg of pyruvate C h^{-1}) throughout the 150-hour experiment, exhibiting an almost linear decline of residual substrate concentration (Figure 4.2, curve 1). (Deviations from linearity inherent to diffusional mass transfer were observed at the onset of the lawn growth (low substrate uptake rate) and before the complete substrate consumption. However, these deviations were not seen at used scale of display.)

The sum of all the measured components of the C-budget (residual substrate + biomass + CO_2) was less than the amount of added pyruvate C, the deficiency being attributed to some refractile exometabolites produced by the bacteria. The quantity of these products estimated from the C-balance increased in time from zero to 10 mg C per tube (Figure 4.2, curve 5).

It is appropriate at this point to state our conclusions about the mechanisms of lawn growth. As we expected, it was indeed close to that of dialysis culture, characterized by a continuous supply of limiting substrate at nearly a constant rate, $F_{\rm dif}$, without biomass wash-out. To account for the other effects, including the formation of end-products (h) and dead cells ($x_{\rm necro}$), as well as the differentiation of total biomass, x, into growing cells of peripheral zone, x', and the inert central part, x–x', we have to modify Eqn. [2.2] derived for the analogous homogeneous system as follows:

$$
\begin{aligned}
\dot{x} &= Yq_{\rm s}x' - ax\,,\\
\dot{x}' &= Yq_{\rm s}x'\,,\ \text{if } s \geqslant K_{\rm s},\ \text{else } \dot{x}' = 0\,,\\
\dot{x}_{\rm necro} &= ax\,,\\
\dot{s} &= F_{\rm dif} - q_{\rm s}x' - mx - q_{\rm h}x\,,\\
\dot{h} &= q_{\rm p}x\,,\\
\dot{p} &= (1 - Y)\,q_{\rm s}x' + mx\,, \qquad [4.2]
\end{aligned}
$$

where s is the substrate concentration within the biofilm, $q_{\rm h}$ is the specific rate of product formation, while other notations are as elsewhere. The segregation of the peripheral zone was postulated to be controlled by substrate concentration. If the substrate concentration drops below the saturation constant then there is no further increase in active biomass.

The kinetic parameters characterizing peripheral growth were found to be the same as in the homogeneous culture of *P. fluorescens*: $Y = 0.5$ g C g^{-1} C, m=12 mg pyruvate C h^{-1} per g biomass C, and $\mu_{\rm m} - Y(Q^{\max} - m) \geqslant 0.25\ {\rm h}^{-1}$. The biomass of the peripheral zone was about 2 mg C per tube, which is equivalent to a biofilm width of 6 μm, or less than 20 cells. However, this estimate should be interpreted with caution, since the actual spatial distribution of substrate and growth centres within the biofilm may be random or uniform as shown by direct autoradiographic studies (Reyrolle and Letellier, 1979).

Model [4.2] predicts exponential and non-differentiated growth ($x = x'$) during the early stages of lawn development with a subsequent transition to a decelerated growth phase, deceleration being caused by a rapid loss of the cell's viability. We recall that in dialysis culture the growth was linear for at least 3 months. So, the last and most intriguing question is, why are bacteria grown on agar more susceptible to death as compared with homogenous culture? The banal idea about the inhibitory action of metabolic products is not immediately acceptable, since the agar layer should serve as the same sort of sink for products as fresh medium in batch dialysis system (apparatus a, Figure 3.2). The difference may be only in quantitative rather than qualitative terms. Another explanation is that the survival of starving cells inside the biofilm (colony) is controlled by some density-dependent killing factors. The survival is better in dialysis culture, where population density is generally lower by several orders of magnitude.

4.1.2 FUNGI

The linear growth rate of colonies as dependent on substrate concentration

Fungi of the genera *Penicillium*, *Mucor*, *Trichoderma* and *Mortierella* (Table 3.1) were cultivated on synthetic medium 7 and various concentrations of glucose (0–800 mg l^{-1}). Upon inoculation by injection of spore suspension, one 'giant' colony per plate was grown. During incubation at 25°C, the colony diameter was measured with a calibrated eyepiece mounted on a stereomicroscope SMS-9 or simply by using a ruler when colonies were large.

After a lag phase of 10–40 h, which reflects the germination of spore inoculum, there was a short period of exponential growth; the radial growth rate, K_R, then became constant and remained so indefinitely.

Significant variations in K_R values were found among the various fungi species as well as for one particular organisms, dependent on glucose concentration (s_0) added to the agar medium (Figure 4.3). The latter dependence could be approximated by kinetic equation [1.58] containing a substrate inhibition term:

$$K_R = K_R^{max} \frac{s_0}{K_s + s_0 + s_0^2/K_{ss}}. \qquad [4.3]$$

According to this equation radial growth rate has a single maximum at $s_0 = \sqrt{K_s K_{ss}}$ (note that the value of K_R at this point is less than K_R^{max}). In *Trichoderma* and *Penicillium* colonies, substrate inhibition was practically absent, which formally corresponds to the case $s_0 \ll K_{ss}$). By contrast, the growth of *M. ramanniana* was under severe substrate inhibition, i.e. even the smallest amounts of added glucose (s_0 = 10–50 mg l^{-1}) retarded colony expansion as compared with the glucose-free agar medium.

In terms of enzyme kinetics, K_{ss} is the dissociation constant of a non-

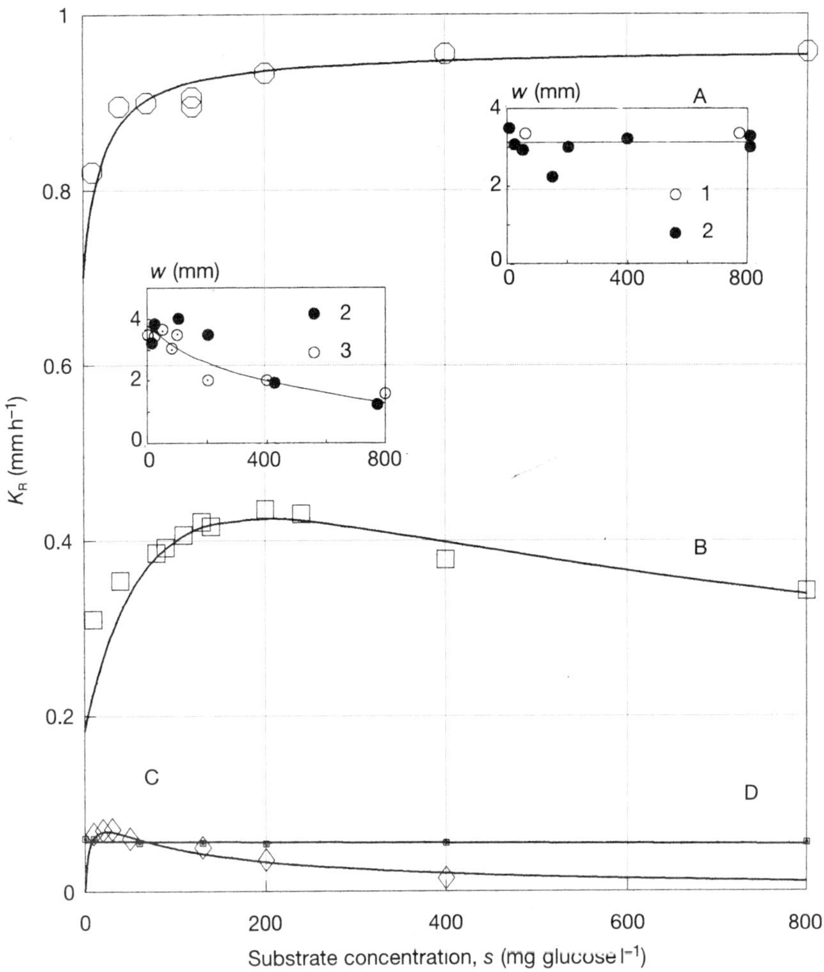

Figure 4.3 The effect of glucose concentration on the radial growth rate and width of peripheral zone (see inserts) of fungi colonies *T. koningii* (A), *M. plumbeus* (B), *M. ramanniana* (C), and *P. funiculosum* (D). The width of peripheral zone was estimated by dissection of the colony (1), by using cover glass (2), and by RNA determination (3). The curves are calculated according to Eqn. [4.3]; parameters are given in the text. (Panikov *et al.*, 1981.)

productive triple complex, ES_2. As far as the colony is concerned, the interpretation should be quite different and related to an *s*-dependent variation of the peripheral zone. Its width (w), as described in the original papers, may be assessed by the following techniques:

1. The circular colony is cut along a chord so as to cut off hyphal tips of graded lengths and colony growth is followed by photography, w being identified as the minimum length of hyphal tip necessary to sustain the same rate of R increase as that of an intact colony (Trinci, 1971).
2. A cover glass is pressed into the agar gel in front of the colony edge and the peripheral zone is visualized as mycelium overgrowth on the glass (Panikov *et al.*, 1981).
3. The colony cross-sections cut from the agar are treated with an RNA-specific dye containing pyronine, the peripheral zone being perceived as the region of elevated RNA content (Panikov *et al.*, 1981).

The first technique is the most direct way to measure the colony's elongation zone, but it is very laborious and requires great skill to produce accurate results. The second technique is the simplest but it fails in the case of slowly growing and fastidious fungal species. Finally, the third method was found to be the most versatile, accurate and reliable, although it is not direct since the RNA content is only correlated with the growth rate.

Even so, all three methods gave agreeable results (see inserts in Figure 4.3). With an increase of s_0 a hyperbolic decline in w was observed, which fairly closely resembled the descending part of the $K_R(s_0)$ curve. This leads to the following interpretation of Eqn. [4.3]. At low $s_0 \sim K_s$, the width w is roughly constant and K_R variation is caused exclusively by the effect of s_0 on μ. At high $s_0 \sim K_{ss}$, no inhibition of mycelial growth occurs (μ is close to μ_m) but the rate of colony spread over agar, $K_R = \mu w$, declines as a result of the reduction of the peripheral zone.

Dynamics of colony growth and respiration

The close association of mycelium with the agar prevents accurate biomass determinations by dry weights. So, growth dynamics were recorded indirectly, from the CO_2 dynamics. Fungi were grown on agar strips inside sealed test tubes (Figure 4.4a).

This ensured the unidirectional and uniform propagation of the colony, which eventually occupied the entire agar strip. Contrary to the circular colony in the Petri dish, the peripheral zone of such a rectangular colony remained to be of constant mass, simplifying the following calculations and assays. The increase of CO_2 content in the gas phase of test tubes during mycelial growth was determined by gas chromatography in air samples taken with a syringe through a rubber cap. The pressure reduction was compensated by the introduction of a similar volume of CO_2-free air. The parallel recording of the linear size of the colony allowed, under certain assumptions holding at $s_0 < 1\ \mathrm{g\,l^{-1}}$, the determination of all basic stoichiometric and kinetic parameters of the growth process.

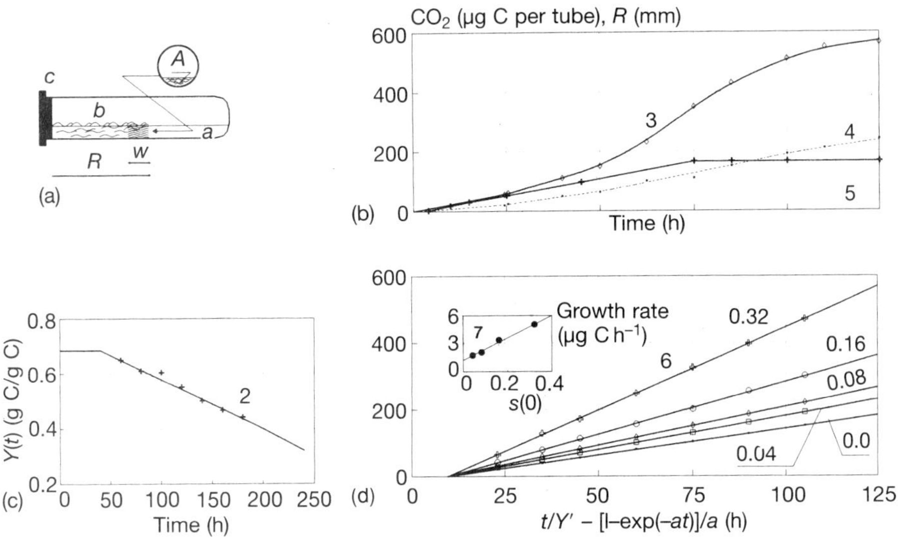

Figure 4.4 Kinetic analysis of colony respiration. (a) *T. koningii* was grown on the agar strip inside of test tube: *a* and *b* are parts of the agar strip respectively free and occupied by mycelium; *R*, *w* and *A* are the length, width of peripheral zone and cross-sectional area of the colony, respectively. (b) Growth was followed by recording CO_2 evolution (curves 3 and 4) and colony length (curve 5). Curves 3 and 4 are CO_2 formed during growth on agar with 400 mg l^{-1} glucose and unamended agar (control) respectively. (c) Results of kinetic analysis: dynamics of apparent yield (curve 2), linearization of CO_2 dynamics (curve 6), relationship between mycelium growth rate in peripheral zone and glucose concentration (curve 7).(Panikov *et al.*, 1983)

These assumptions are as follows: (a) glucose is completely consumed in those parts of the agar layer which are occupied by mycelium, the amount of substrate utilized (Δs) being calculated from the geometry of the growing colony as $-\Delta s = s_0 A K_R t$, where A is the cross-sectional area of the agar strip; (b) the contribution of non-utilizable exometabolites to the growth budget is insignificant; and (c) the mycelial growth and substrate respiration occur only in the peripheral zone of constant mass x', whereas biomass in the central part of the colony is decayed at an apparently constant specific rate a (in fact the true turnover rate less the rate of cryptic growth). Then the dynamics of the total biomass of the rectangular colony (x), the apparent biomass yield from the substrate ($Y(t)$, which, contrary to true yield, is a dependent variable), and the respiration rate are governed by the following equations:

$$\dot{x} = \mu x' - ax \, ,$$

$$x = \frac{\mu x'}{a}(1 - e^{-at}) \, , \qquad [4.4]$$

$$Y(t) = \frac{\Delta x(t)}{\Delta s(t)} = \frac{\mu x'(1 - e^{-at})}{as_0 A K_R t} \approx Y'\left(1 - \frac{at}{2}\right), \qquad [4.5]$$

$$\dot{p} = (1 - Y')q_s x' + ax = \frac{1 - Y'}{Y'}\mu x' + ax\,,$$

$$p = p_0 + \mu x'\left(\frac{t}{Y'} + \frac{e^{-at} - 1}{a}\right), \qquad [4.6]$$

where Y' is the biomass yield in the peripheral zone, which is close to true yield and could be assumed as a constant, and p_0 is the initial CO_2 concentration in the test tube. Approximation [4.5] was performed by numerator expansion into a Taylor series and neglecting all but the first two terms.

One example of the concurrent dynamic measurements of p_0 and R is shown in Figure 4.4 (curves 2–4). R was always a linear function of time, while the $p(t)$ curve was S-shaped, indicating a considerable contribution of the maintenance term ax to total respiration (Eqn. [4.6]). The CO_2 formation was related to s_0 in a linear rather than a proportional way, since even at $s_0 = 0$ a substantial CO_2 evolution was observed due to agar utilization. Subtraction of this background respiration allowed the calculation of CO_2 production, Δp, coupled with glucose uptake, Δs (estimated from the volume of occupied agar), and finally to assess apparent mycelium yield, $Y(t) = 1 - \Delta p/\Delta s$. As predicted by Eqn. [4.5], the value $Y(t)$ declined linearly with time, the slope being equal to specific decay rate a (Figure 4.4, curve 2). To examine further the adequacy of the model, CO_2 dynamics were plotted *versus* the composite argument

$$\frac{t}{Y'} - \frac{1 - \exp(-at)}{a}$$

of Eqn. [4.6] for a number of s_0. As can be seen from Figure 4.4, the curves 6 were perfectly linearized giving slopes equal to the mycelial growth rate in the peripheral zone of the colony ($\mu x'$).

The dependence of $\mu x'$ on s_0 was linear (Figure 4.4, insert 7), in agreement with data on radial growth measurements. Indeed, we have the general expression and its reduction to a linear relationship in the particular case $K_{ss} \gg s_0 \gg K_s$:

$$\mu x' = \dot{s}Y' = Y'AK_R s_0$$

$$= Y'AK_R^{max}\frac{s_0^2}{K_s + s_0 + s_0^2/K_{ss}} \approx Y'AK_R^{max}s_0\,. \qquad [4.7]$$

Since $x' = Y'Aws_0$, we may then calculate separately the mycelium biomass in the peripheral zone, x', and its specific growth rate, μ, as well as the maximum specific respiratory activity, $Q_{resp} = \mu/Y_{CO_2}$. Results of the identification procedure are summarized as follows:

Organism	w (mm)	μ (h^{-1})	Q_{resp} (mg C h^{-1} per g biomass)	Y' (g C biomass per g glu C)	a (d^{-1})	K_s/K_{ss} (mg l^{-1})	K_R^{max} (mm h^{-1})
T. koningii	3.0	0.25	63	0.64	0.13	20/∞	0.96
M. plumbeus	2.3	0.18	49	0.63	0.13	100/400	0.67

The growth parameters Q_{resp}, Y_p and a' are in a full agreement with determinations made for the same organism using continuous homogeneous cultures (Seto and Tazaki, 1973; Peitersen and Ross, 1979). It is worthwhile to note that the described relatively simple experiments with Petri dishes and test tubes are not only more economic than expensive fermentation equipment, but also provide a better extrapolation to *in situ* conditions of fungal growth.

4.2 MICROBIAL GROWTH IN A SYSTEM OF GLASS MICROBEADS

Glass microbeads are chemically inert, transparent, and uniform in shape and size. For this reason, they are convenient to use as a solid phase of porous media in microbiological research. An interesting and so far unexplained fact observed in experiments with microbeads is their fungistatic action, i.e. it is only bacteria that are capable of luxurious growth in a fine microbead environment, whereas fungi fail to show significant grow under such conditions (Parr *et al*, 1967).

We used 'Balatini' glass microbeads of three sizes: fine (150–250 μm), intermediate (600–800 μm) and large (1500–2000 μm in diameter). The maximum pore space was 30, 100 and 300 μm, respectively. Defined medium with glucose 7 ($s_0 = 0.2$–$10\ g\ l^{-1}$) was mixed with spore inoculum (to give finally 2×10^6 cells ml^{-1}) and placed in penicillin vials containing 3 g of washed sterile microbeads to fill half of the pore space. The vials were incubated at 25°C, and growth of microorganisms was followed by recording CO_2 evolution by gas chromatography.

Illustrations of the results are presented in Figure 4.5. In the medium with fine microbeads, only one organism, *Ulocladium consortiale*, exhibited feeble growth and then only at high s_0. The growth of other fungi was completely arrested. Microscopic observations revealed that their spores swelled but failed to develop growth tubes. Germination of spores of *U. consortiale* was also significantly suppressed in the medium with fine beads as compared with medium and large ones.

Let us now consider the possible causes of fungal growth suppression in fine porous media. It was argued (Parr *et al.*, 1967) that a key role is played by two limiting factors – the deficiency of oxygen and the non-availability of water bound to the glass interface. These hypotheses were tested in additional experiments involving 'chemical aeration' with H_2O_2 and variation of the water content. This, however, had no clear effect on the rate of fungal growth. Such an outcome could well be foreseen. Indeed, the oxygen supply proceeds via the liquid–air interface which is roughly the same in the case of fine and large beads. Moreover, even if gas mass transfer exchange was completely blocked, then the amount of dissolved oxygen in the medium and entrapped in air bubbles is sufficient to oxidize all added glucose. The lack of water is also unlikely because even in the case of fine microbeads water was still in a physically available state.

In our view, the most probable cause of the observed fungistatic effect is the high degree of dispersion of the distribution of limiting substrate among microsites. As a result each singular microhabitat does not possess sufficient substrate, even for the growth of one single hypha. Our hypothesis is illustrated in Figure 4.5a. The initial content of substrate in an almost cylindrical microcapillary pore is

$$\delta s = \frac{\pi D^2 l s_0}{4},$$

where D is pore diameter and l its length; apical growth of the hypha along the pore results in biomass formation

$$\delta x = \frac{\pi d^2 l \rho (1 - W)}{4},$$

where d is the hypha diameter, ρ is the cell density, and W the cell water content. Since substrate flux from adjacent microzones is blocked by air locks, we can write $\delta x = Y \delta s$. From this equation we may easily derive the threshold value of the hyphal diameter, d, of filamentous organisms growing in the microtunnel of diameter D:

$$d = D \sqrt{\frac{s_0 Y}{\rho (1 - W)}}.$$

Calculations showed that in fine microbeads at the tested glucose concentration, only those organisms which have hypha diameters 0.4–0.8 μm may develop best. This size range is attributed to actinomycete rather than fungi (hyphae diameters of fungi studied varied between 1.4 and 4 μm). Thus, fungi are able to grow only in intermediate and large-size microbeads media (with d equal to 1.4–2.8 and 4.2–8.5 μm, respectively), the probability of growth increasing with s_0. The latter prediction of the model was confirmed by an additional experiment with *Penicillium spinulosum*. In fine beads, a high glucose concentration did induce growth, although

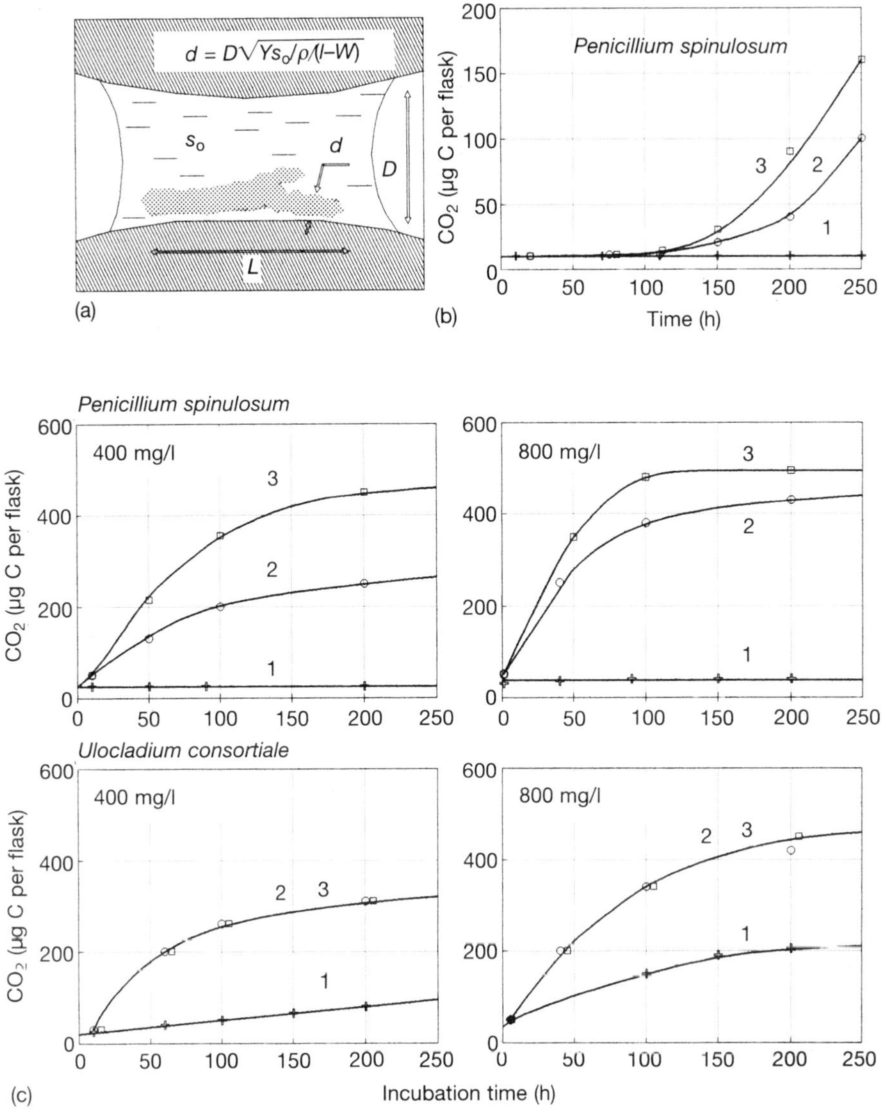

Figure 4.5 Growth and respiration of fungi as dependent on sizes of microbeads (Makarova *et al.*, 1987). (a) Schematic representation of microlocus with germinating spore and microvolume of nutrient solution containing glucose at concentration s_0; D and d are diameter of micropore and hyphae, respectively. (b) Respiration dynamics of *P. spinulosum* in fine microbeads at different glucose concentrations ($g\,l^{-1}$): 0.8 (curve 1), 3.0 (curve 2) and 10.0 (curve 3). (c) Respiration dynamics as dependent on microbead size: fine (curve 1), medium (curve 2) and large (curve 3).

only after a 100-hour lag period (Figure 4.5b). We failed, however, to obtain growth of *Z. heterogamus*, even at large initial glucose concentrations, s_0. Within the framework of our model, the successful development of *U. consortiale* in a microporous environment can be explained by large spores sizes with a high content of reserve substrates. Therefore, spores of this particular organism can germinate even in fairly small microsites with a low amount of substrate and then one branched hypha may exploit several microzones.

4.3 MICROBIAL GROWTH IN PACKED COLUMNS WITH INERT CARRIER

Methods

We used columns packed by sand, diameter 30 mm and height 100 mm, uniformly inoculated by yeasts *D. formicarius* at a ratio 10^6 cells g^{-1} of sand. The nutrient medium 6 (section 3.1), containing ammonium as a growth limiting substrate (s_0 4.7 mg N l^{-1}), was continuously delivered to the column top at a rate of 5 ml h^{-1} (1.4 cm h^{-1}), aeration being achieved by the continuous passage in the same direction of sterile moistened air (25–30 cm^3 min^{-1}). The microbial growth was followed by recording the dynamics of NH_4^+ and cell count in the column effluent. In addition, after termination of several identical experiments of different durations, the column was dismantled, and its contents split into 2-cm segments, which were examined by scanning electron microscopy and analysed for cell number and RNA content as an index of physiological state.

Results

These are shown in Figures 4.6 and 4.7. After a very short transient, the effluent substrate concentration reached a steady-state level. The cell concentration in the effluent exhibited chaotic fluctuations and was generally increasing with time, although most microorganisms were retained on sand grains. A characteristic hyperbolic profile distribution of cells along the column was established as early as on the fourth day, after which the steepness of the profile considerably increased.

Simulation

To derive a quantitative description of the observed dynamics, we used differential equations in partial derivatives of three variables – the limiting substrate concentration (s), the biomass of free cells (x), and biomass of immobilized cells (x_{im}) attached to solid phase. The immobilization was

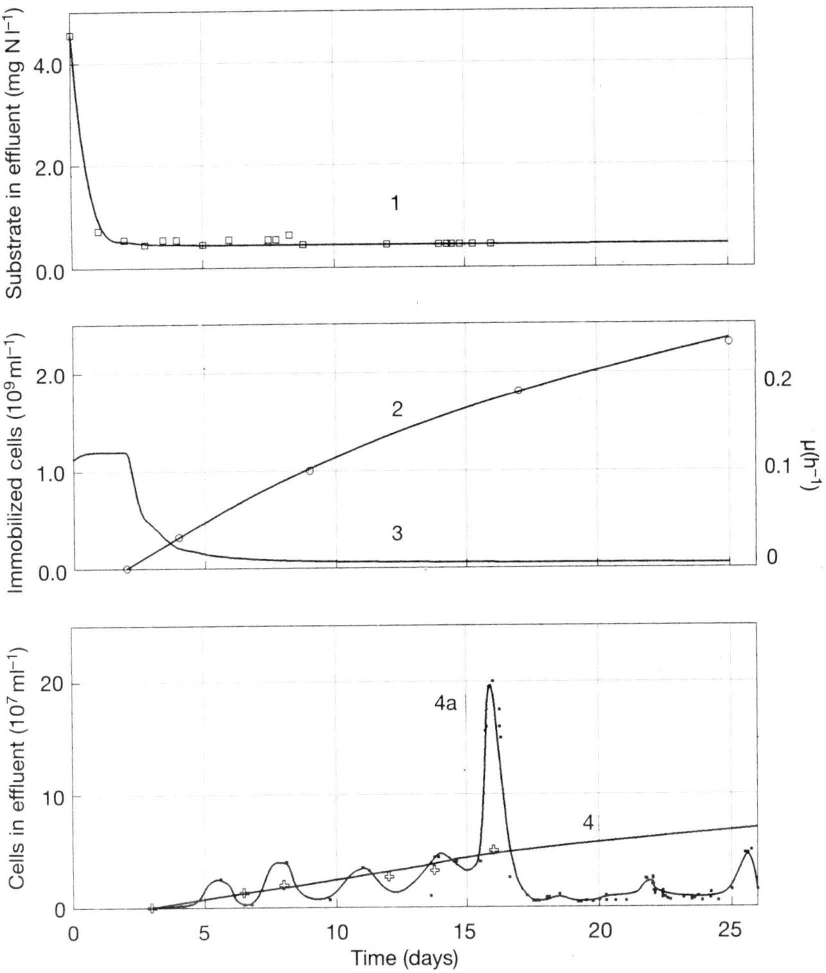

Figure 4.6 N-limited growth of *D. formicarius* in continuous-flow sand column: residual concentration of NH_4^+ in the column effluent (curve 1); average amount of cell immobilized on sand grains (curve 2) and its specific growth rate (curve 3); moving-average (curve 4) and instant (curve 4a) cell concentrations in the column effluent. Curves 2, 3 and 4 are calculated by using Eqn. [4.9]. (Panikov *et al.*, 1980.)

assumed to be reversible and governed by two rate constants, p_1 (the release of attached cells to solution), and p_2 (adhesion to the surface). Then

$$\frac{\partial s}{\partial t} = D_s \frac{\partial^2 s}{\partial z^2} - \frac{1}{Y}(\mu x + \mu_{im} x_{im}) + u \frac{\partial s}{\partial z},$$

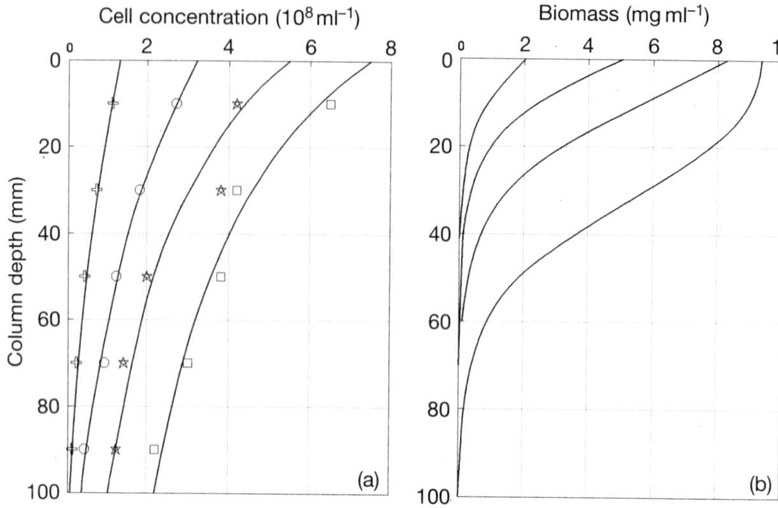

Figure 4.7 The profile distribution of cells in the column: (a) experimental data; (b) simulation by using Eqns [4.8]. The symbols denote days of growth: ✣, 4 days; ○, 9 days; ☆, 17 days; □, 30 days.

$$\frac{\partial x_{im}}{\partial t} = \mu_{im} x_{im} + p_2 x - p_1 x_{im}\,, \qquad [4.8]$$

$$\frac{\partial x}{\partial t} = D_x \frac{\partial^2 x}{\partial z^2} + \mu x + p_1 x_{im} - p_2 x + u \frac{\partial x}{\partial z}\,,$$

where z is spatial dimension, D_s and D_x are diffusion coefficients for substrate and microbial cells respectively, and u is the linear flow rate of substrate. Since limiting substrate belongs to 'conservative' type, then not only μ but also Y were allowed to be a function of s according to the SCM (Table 3.9), for simplicity we assumed $n = 1$.

The boundary conditions for the top, $z = 0$, and the bottom of column, $z = L$, were defined as follows:

$$\left.\frac{\partial s}{\partial z}\right|_{z=0} = u[s(0,t) - s_0] \quad \left.\frac{\partial s}{\partial z}\right|_{z=L} = 0\,,$$

$$\left.\frac{\partial x_{im}}{\partial z}\right|_{z=0} = 0 \quad \left.\frac{\partial x_{im}}{\partial z}\right|_{z=L} = 0\,, \qquad [4.8a]$$

$$\left.\frac{\partial x}{\partial z}\right|_{z=0} = \frac{u}{D_x} x(0,t) \quad \left.\frac{\partial x}{\partial z}\right|_{z=L} = 0\,.$$

In addition, we used a simplified model by accepting the average values of the variables s, x and x_{im} throughout the column and by neglecting diffusion, thus reducing the equations in partial derivatives to the ordinary differential equations:

$$\frac{ds}{dt} = D(s_0 - s) - \frac{1}{Y}(\mu x + \mu_{im} x_{im}),$$

$$\frac{dx_{im}}{dt} = \mu_{im} x_{im} + p_2 x - p_1 x_{im}, \qquad [4.9]$$

$$\frac{dx}{dt} = \mu x - Dx + p_1 x_{im} - p_2 x,$$

where D is the dilution rate defined as being similar to that used in a classical chemostat model. Strictly speaking, Eqn. [4.9] is a chemostat model accounting for partial biomass retention in the fermentation vessel.

The unabridged model [4.8] was solved numerically by a computer routine based on a finite-difference approach and Novikov's algorithm (see Appendix). Its homogeneous counterpart, [4.9], could be solved analytically dependent upon a quasi-steady-state approximation, $\dot{s} \sim 0$. The simulation results are shown in Figures 4.6 and 4.7 and compared with the experimental data. The complete model [4.8] simulated the observed dynamics of the spatial distribution of cells but only in a qualitative way. It failed to provide the exact quantitative fit to the experimental pattern using any combination of parameters. Surprisingly, the abridged model gave reasonable agreement with the experimental data if one is content with average values and not looking for details, such as chaotic fluctuations of cell concentration in the effluent. The adequate simulation of cell attachment to the sand grains was provided by a rather low ratio of the parameters $p_1/p_2 \approx 10^{-3}$.

Thus, the following discussion is only in terms of the average values within the framework of abridged model [4.9].

The main point of this discussion is the differentiation of the entire microbial population into free and attached cells. These cells differ with respect to metabolic activity and growth rates as indicated by RNA determinations, i.e. free cells were characterized by an RNA content close to the maximum (above 10% of dry weight), while immobilized cells tended to the lower limit of RNA content 4–6%. The direct assessment of the specific growth rate may be performed only for attached cells from the dynamics of x_{im}. The value of μ_{im} decreased over the first week of the experiment from 0.12 to 0.001 h^{-1} (Figure 4.6, curve 3).

The retardation of the growth of attached cells could not be explained by negative effects of the solid phase *per se*. Direct microscopy revealed that yeasts grew on the surface of sand in the form of microcolonies submerged in the extracellular gel, while multilayered accumulation of cells has never been observed and hence deficiency of oxygen is also not feasible. Much more important should be the deficiency of limiting substrate in some parts of the column completely devoid of substrate input. These microsites probably arise from random changes in the trajectories of the percolating nutrient solution, the plugging of microcapillaries by cells and other stochastic events. If this is true, then it is clear why model [4.8] failed to predict the experimental behaviour, since it completely neglects the micro-gradients and mentioned stochastic events.

Packed column and chemostat

It is a widespread opinion that a continuous-flow column is very close in nature to a chemostat culture. So, it is worth emphasizing the essential distinctions between these two cultivation techniques.

Chemostat	*Packed column*
The rapid approach of true steady-states with respect to all variables (x, s, r).	Quasi-steady-state is attained only with respect to the 'quick' variable s, but not for x or r.
The attained steady-state specific growth rate is equal to dilution rate, $\mu = D$.	For attached microorganisms, $\mu < D$ at all times, the difference increases steadily with time.
The value of μ does not depend on either s_0 or $\bar{x}$ and is a function of s alone.	The value of μ_{im} depends on s and on the amount of biomass accumulated in the bioreactor.
The value of μ is under full engineering control.	The possibilities for effective control are very limited.
The cell population is relatively homogeneous.	The population is highly heterogenous.

It is apparent from the above summary that packed columns are hardly the best choice as a means of controlled cultivation. The physiological state of attached cells is generally obscure, but is likely to be close to that which we normally observe in dialysis and long-term starved batch cultures.

4.4 CONCLUSION

We have considered only three particular cases of spatially ordered microbial growth. In fact, there should be infinite manifestations of inhomogeneous growth, simply because the notion of 'homogeneity' is an idealization and any real culture does exhibit some heterogeneity both in its composition and spatial organization.

The traditional approach to the quantitative analysis of spatially heterogenous biological systems is based on the mathematics of differential equations in partial derivatives. Some simple examples were surveyed in Chapters 1 and 2 (Eqns [1.82], [1.84], [2.7] and [2.8]) and used here in an attempt to describe the dynamics of a packed column. This effective mathematical technique was not used in our consideration of microbial colonies, and this gap has been filled elsewhere (for example, Georgiou and Shuler, 1986). However, we would not appraise partial derivatives as a best option. Although somewhat more consistent in terms of formal mathematics, these models may be less effective in simulation than simpler kinetic models based on ordinary differential equations. Frequently the latter are preferential since they provide a better match for the crude experimental data usually obtained with microbial cultures.

What in essence are these simple models? In fact, we have coped with heterogeneity by introducing new variables, in particular, by resolving the net biomass into x and x' (the biomass of the central and peripheral parts of the colony), x and x_{im}(the biomass of free and immobilized cells), or by assigning in heterogeneous porous media the small entities ∂s and ∂x, the infinitessimal amount of substrate and the increment of biomass within one of the countless microzones. These extra variables reflect the change in biomass 'quality' brought about by the spatial gradients of environmental factors, mainly of limiting substrate concentrations. The reduction of a heterogeneous system to a set of homogeneous subsystems provides a methodology to exploit all the advances of 'homogeneous' kinetics described earlier in Chapter 3.

Growth kinetics and the life strategy of microbial populations

5

Up until this page of the book, our main emphasis has been on the general kinetic principles common to all microbes. Now we will concentrate on specific features of individual organisms related to their ecology. Our premise is that microbial isolates, even those that have been cultured for many years under artificial laboratory conditions, should preserve some memory of the life strategy of their wild progenitors.

We have introduced the concept of life strategy (section 1.5). The key term 'strategy' is defined as a set of adaptations which provides the possibility for a given population to survive and maintain their stability in natural ecosystems. Usually strategy characterizes the so-called 'survival triad': (a) the ability to compete with other populations; (b) to recover after perturbations; and (c) to survive stresses. In this manner one may distinguish three types of natural selection:

1. K-selection operates in climax ecosystems under stable and predictable conditions without frequent perturbation and stresses. The habitats of this type are overcrowded, thus the main feature of K-selected species should be a high competitive ability ('lions' type of strategy). Their generation time is relatively long and they have few progeny, but nevertheless these species maintain high population densities.
2. r-Selection operates on the pioneer stages of succession initiated by some perturbation of a climax ecosystem, i.e. a sudden change of environmental conditions (not necessarily adverse), a flash of nutrients, a cataclysmic elimination of competitors. The main result of perturbation is a temporary relief from the pressure of severe competition for nutrient resources. r-Selected species survive in ephemeral, unpredictable habitats due to mobility and high reproduction rates (opportunistic or a 'jackal' type of strategy). They are not good competitors and are always ready to leave the resources once they become depleted or overcrowded.
3. L-selection operates under adverse environmental conditions caused

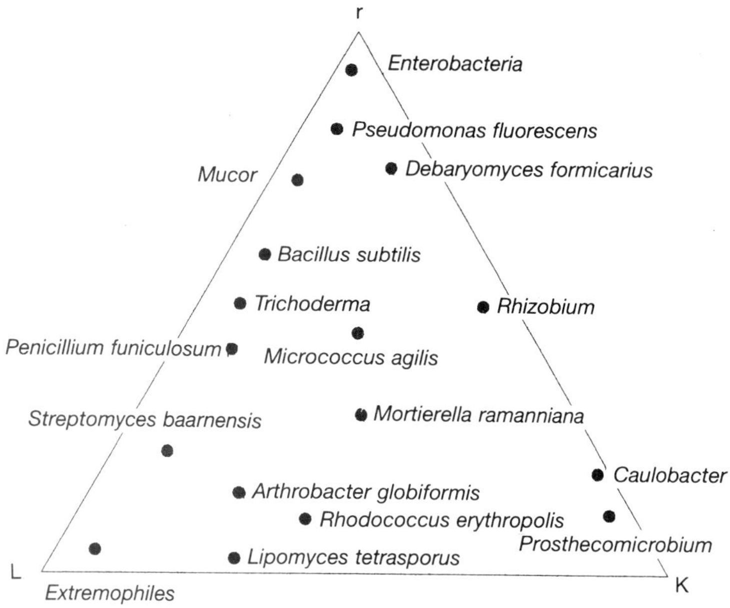

Figure 5.1 Hypothetical position of studied organisms in rKL-space.

by various stresses. Stress factors could be abiotic (non-optimal salt concentration, temperature, pH, water content, etc.) or biotic (antagonism, starvation caused by the depletion of substrate by more successful competitors). The products of L-selection are the patient species resistent to a particular stress factor ('camel' type of strategy). (Note the peculiarity of starvation as a specific kind of stress: it is a factor of L-selection and, on the other hand, is an immediate consequence of K-selection, being caused by severe competition. This instructive example illustrates the basic idea that there are no pure L-, K or r selections; instead we have three different circumstances of one and the same undivided process of natural selection.)

The concept of rKL-selection is not absolute, being meaningful only in the comparison of several organisms. However, to economize on words we will loosely refer to, for example, an r-strategist instead of the correct but awkward expression 'population superior to many others under conditions (2) of predominant r-selection'. The best way to identify the life strategy of some studied organisms would be to locate them in one common rKL-continuum. The more prosperous a particular species is under the conditions (1), (2) or (3), the closer it is placed to the K-, r- or L-pole of this continuum. An example of such an ordination is shown in Figure 5.1. Below, we justify this arrangement of microbial species by surveying the data on their growth kinetics in pure cultures, mixed culture and their population dynamics *in situ*.

5.1 GROWTH KINETICS OF PURE CULTURES

The experimental data on homogeneous and heterogeneous growth of various organisms have been demonstrated in Chapters 3 and 4. The natural situations (1), (2) and (3) were somewhat simulated by the use of several cultivation techniques under both steady-state and perturbed growth conditions. Thus, K-selection was imitated in steady-state or quasi-steady-state continuous cultures under deep limitation, i.e. in a chemostat running at very low *D*, or in dialysis and fed-batch cultures with slow feeding rates (Figures 3.4–3.16 and 3.26–3.30). The conditions for r-selection were simulated by different kinds of shift-up experiments accomplished by an abrupt increase of *D* in chemostat culture (Figures 3.47 and 3.52) or during the start-up of a simple batch cultivation with starving cells as the inoculum (Figures 3.17–3.19 and 3.31). The last situation, L-selection, is typified by the starvation stress which occurs when the cell is deprived of an energy source (Figures 3.21–3.25) and a source of conserved substrate (Figure 3.32).

Each of these growth strategies has been described by mathematical models, every organism being characterized by a unique set of growth parameters. Thus, we may assess the competitive advantages and disadvantages of the studied organisms under various circumstances by performing computer experiments, simulating competition between two, three or even more species for one common substrate. The details of these numerical experiments will be presented in section 5.3. Here, we shall consider the qualitative diversity of the growth regulation of the various organisms relevant to their life strategy.

The most detailed experimental studies were performed with three bacterial species, *P. fluorescens*, *A. globiformis* and *B. subtilis*. By a happy chance, these bacteria are located far from each other in the rKL-continuum allowing us to deduce the characteristic features of typical r-, K- and L-strategists.

Pseudomonas fluorescens

These bacteria bear many characteristic qualities attributed to r-selected organisms. After any kind of perturbation (shift of *D* in chemostat, amendment of starving culture with deficient nutrient, start-up of batch culture) they display an intensive and almost balanced growth with a smooth transition from one steady-state to another or from very slow multiplication to growth with maximum specific rate. On the other hand, under conditions of chronic starvation in dialysis and fed-batch culture they are inferior to *Arthrobacter* exhibiting at the same s, lower μ and Y values. The onset of complete starvation induces rapid autolysis and death.

Which particular growth parameters are responsible for the described type of dynamic behaviour? In SCM terms the most important are (a) low K_r (the steep dependence of r on s), and (b) high potential specific rates Q_s, Q_{resp}, Q_L and μ_m. Even substrate-limited slowly growing *Pseudomonas* cells do maintain rather high r values, and high instantaneous potential activities (expressed as the products rQ_s, rQ_{resp}, etc.). Following relief from substrate limitation their activity immediately jumps to this submaximal level and shortly afterwards attains maximal values at $r = 1$. However, the maintenance of excessive metabolic machinery requires an extra amount of ATP (the resynthesis of turned over proteins and RNAs, membrane energization, etc.). It is a heavy burden for the energy budget and thus substrate-limited growth of *Pseudomonas* is not efficient (low biomass yield, intensive endogenous respiration and a high maintenance ration).

Arthrobacter globiformis

These organisms are characterized by a high competitive ability under extreme substrate limitation (product of K-selection). They are superior to *Pseudomonas* in dialysis and fed-batch culture and display phenomenally long survival under complete starvation (although they are not as persistent as *Bacillus* spores). Short-term perturbations in steady-state substrate concentration caused by its increase from $\bar{s}$ to higher values $s > K_s$ is very beneficial (the first phase of false diauxie). However, a long-term excessive supply of nutrients is detrimental and results in slow feeble growth (the second phase of false diauxie), multistability, oscillations and long transients (chemostat at high D).

The SCM explains *Arthrobacter* behaviour by the following kinetic features: (a) a shallow $r(s)$ dependence (high K_r resulting in low r-values under growth limitation); (b) a high affinity to substrate (low K_s, high ratio Q_s/K_s); (c) the 'frugal ' method of reserve compound formation directly from transported limiting C-substrate (contrary to *Pseudomonas* which produces reserve C-compounds only under conditions of C-sufficient growth limited by N); (d) a moderate respiratory and biosynthetic activity (rather low Q_L and Q_{resp}); and (e) a loose metabolic control of intracellular fluxes (the formation of autoinhibitory metabolites, H, when monomeric intermediates exceed the threshold level, L^*).

The first feature, (a), is responsible for the small difference between actual and potential metabolic activity and hence for the rather low maintenance requirements of *Arthrobacter*. Points (a), (b) and (c) explain why these bacteria are competitive under oligotrophic conditions, while the combination of (c), (d) and (e) accounts for the unbalanced growth of these bacteria, i.e. false diauxie, retarded transients after a change in D and sustained oscillations in a chemostat.

Unbalanced growth is, by definition, the growth associated with changes of cell composition. In *Arthrobacter* culture these changes are due to the partial dissociation of transport from biosynthesis and respiration, as well as the substrate diversion from primary metabolic routes to the formation of reserved compounds. Contrary to *Pseudomonas*, unbalanced growth is a quite normal physiological feature of *Arthrobacter* from which it benefits more than it loses under natural conditions. For instance, the response of a natural *Arthrobacter* population in soil to the addition of a small amount of substrate is its instantaneous consumption without concomitant growth acceleration. Instead, cells refill their storage compartments and only afterwards accomplish delayed divisions. The advantage of such a strategy is obvious. Cells preserve their normal physiological state of retarded metabolic activity, associated with a high persistence to adverse factors and required for better survival under prevailing oligotrophic conditions. The disadvantage of this strategy for population dynamics is exhibited under eutrophic growth conditions when an excessive supply of nutrients persists for a long period. Then metabolic inhibition ($H > 0$) prevents rapid multiplication.

Anticipating the puzzled look of a reader experienced in bacteriology, I would like to finish this portrayal of *Arthrobacter* by giving an answer to the question, why has this organism been regarded as a 'lion'? In the fermenter or shaking flask it behaves like a stupid donkey, i.e. its slow growth is unpredictable, it is easily contaminated, and is obviously inferior to the growth of *Pseudomonas*. The answer is that this is the behaviour of an 'imprisoned lion' under artificial laboratory conditions. We should remember that *Arthrobacter* dominates in oligotrophic natural ecosystems (mainly mineral soils) where low nutrient content is the result of extremely tense competition for deficient nutrients between the members of the microbial community. However, the **domination** in a climax microbial ecosystem is equivalent to the statement that *Arthrobacter* is one of the **strongest competitors**. Strictly speaking, to be really dominant *in situ* *Arthrobacter* should also be persistent to stresses such as desiccation and freezing. Therefore, these bacteria combine the features of K- and L-strategists.

Bacillus

These bacteria are referred to as L-strategists. The persistence to stresses (starvation, drought, heat) is due to the well-known ability of these bacteria to form endospores. However, it is not immediately clear why bacilli are patients and have low competitive ability. Why does sporulation, seemingly so advantageous from an ecological standpoint, not allow the domination of bacilli in natural environments? Why do they go into a dormant state under conditions when other microbes preserve activity? Basically these questions were answered by experiments with dialysis culture. Bacilli belong to the category of weak competitors, since under severe limitation they lose the most important quality – the ability to multiply (the transition to maintenance state).

The specificity of bacilli dynamic behaviour is determined by the following kinetic properties: (a) a high threshold substrate concentration triggering the turnover of the cell's macromolecular components; (b) the segregation of cell material; and (c) a metabolic control based on the formation of inhibitors with an uncoupling function at low substrate concentration (contrary to *Arthrobacter* which produce metabolic inhibitor at high s).

At the same time under conditions of moderate substrate limitation and unlimited growth (high and intermediate D in chemostat, exponential phase of batch culture) bacilli exhibit r-type growth dynamics similar to that of *Pseudomonas*.

Other organisms

Two yeasts, *Debaryomyces formicarius* and *Lipomyces tetrasporus*, are the eukaryotic counterparts of *Pseudomonas* and *Arthrobacter* respectively. Similar to *Pseudomonas*, *D. formicarius* displays intensive balanced growth, smooth transient dynamics and low survival ability under starvation. It has a slightly lower respiratory and transport activity than *P. fluorescens*, but higher yields; thus the maximum specific growth rates are almost equal. *L. tetrasporus* is characterized by slow and efficient frugal growth, false diauxie and resistance to starvation similar to that of *Arthrobacter*, but it has a smaller potential metabolic activity. In addition, a specific feature of *Lipomyces* is the formation of lipid reserves under N-limitation.

Coryneform bacteria *Rhodococcus* and *Micrococcus* diverge from *Arthrobacter* in different directions (Figure 3.24). *Rhodococcus* is even closer to the K-pole, exhibiting a higher affinity to substrate, slower growth and better survival dynamics. *Micrococcus* is closer to the r-pole as it grows faster, has a higher maintenance requirement and a modest survival ability. Nonetheless, all members of this group display multistability in chemostat culture and false diauxie.

The position closest to the K-pole is occupied by *Caulobacter* and *Prosthecomicrobium*. In batch culture (Figure 3.25) they grow extremely slowly with a pronounced effect of false diauxie and have no declining phase at all (!) due to the ability to scavenge volatile organic substances from the laboratory air. In chemostat culture these bacteria exhibit the lowest measured values of K_s, below 0.01 mg l^{-1}, and zero maintenance requirements (Semenov and Vasilyeva, 1985, 1986). The formation of reserved compounds in the form of poly-β-oxybutyrate takes place even on diluted nutrient media.

Batch growth of *Rhizobium* appeared to be close to that of *Pseudomonas*, i.e. balanced and without false diauxie, but considerably slower. The survival under starvation was better than in *Pseudomonas* culture but inferior to coryneform bacteria. Hence these bacteria occupy the middle of rK-continuum.

Fungi *Penicillium funiculosum* tend to the L-pole in view of their ability to produce spores (although less resistant than endospores of *Bacillus*) and the susceptibility to substrate-accelerated death. However, in contrast to *Bacillus,* they sustain linear growth in dialysis culture, escaping the maintenance state (in fact this is an indication of the basic difference in the control of spore formation in bacilli and fungi). Hence, in natural habitats they should be more competitive than bacilli. *M. ramanninana* is the product of K-selection (high affinity to substrate, sensitivity to substrate inhibition and a slow growth rate). The position of 'saccharolytic' fungi *Mucor* and *Trichoderma* is clearly r-oriented as they display rapid growth and rather intensive endogenous respiration.

Detection and colonization tactics

The data on colony growth give us an opportunity for some speculations about this very important component of life strategy. There are two principal ways of active migration in search of a better microenvironment: chemotaxis (motile bacteria) and chemotropism (mycelial organisms). In a solid-state heterogenous environment (litter, soil, composts, subsurface) the second mode is much more advantageous since the water films available for chemotactic movement of bacteria are discontinuous. The high colonizing ability of fungi is due to the enormous width of the peripheral zone and the possibility of nutrients translocating along the hyphae.

A very rough measure of the colonizing efficiency of mycelial organisms is simply the linear extension rate of the colony, $K_R = \mu w$. It varies from as much as 0.5–2 mm h^{-1} (*Trichoderma* and *Mucor*) to as low as 5–20 μm h^{-1} (streptomycetes). However, this index is too simple. It would be more interesting to consider the functional relationship $K_R(s)$ which follows from Eqn. [4.3]. Altogether we may distinguish three particular tactics:

low K_R^{max} and low K_{ss} (*Mortierella*)	high K_R^{max} and high K_{ss} (*Trichoderma*)	high K_R^{max} and low K_{ss} (*Mucor*)
thorough search for short distances	superficial search for large distances	thorough search for large distances

Low K_{ss} is the indication of the retardation of hyphae extension as soon as hyphal tips enter favourable microsites with elevated substrate concentration ($s \sim K_{ss}$). Here, the mycelium is lingering on getting denser due to intensive branching to acquire complete substrate utilization. Fungi with a high K_{ss} have a greater chance of missing such a beneficial spot and hence their tactic has been defined as 'superficial' search.

Unicellular organisms are characterized by the following correlations: r-strategists generally are motile (*Pseudomonas*, *Rhizobium*), while K-strategists are immotile (*Arthrobacter*, *Rhodococcus*, *Prosthecomicrobium*) or have flagella for only a fraction of their life cycle (*Caulobacter*). It is clear that active migration should be attributed exclusively to an opportunistic type of behaviour since energy-consuming locomotion is justified if it eventually brings about explosive growth in some beneficial microhabitat. The motility of K-strategists, adapted to slow uniform growth and the uptake of substrate diffusing outside, is an absolute absurdity.

Chemotaxis is an obvious advantage over tropism in aquatic ecosystems or such homogeneous terrestrial macro- and microhabitats which contain long continuous water channels and films (i.e. rhizoplane, intestine of animals, submerged soil horizons). It is a striking fact that microbial communities of these homogeneous habitats are absolutely dominated by bacteria, while heterogenous environments (surface soils, decaying plant matter, etc.) are the realms of fungi.

Correlation between growth parameters and ecological features of studied organisms

This topic is too huge to be considered in detail, so here we shall present just a brief comparison of our results with traditional ecological studies of microbial distribution, seasonal dynamics and survival of released microbes, as well as various laboratory tests of stress-resistance and a spectrum of utilized substrates (Table 5.1). The comments pertinent to Table 5.1 are as follows:

1. Differentiation between the members of one strategic group. K-, r- and L-selected organisms vary broadly in properties other than their competitive ability, recoverability or stress-tolerance. So they should not be painted by one and the same brush. The most important secondary differentiation axes for our collection are: (a) spatial heterogeneity of occupied habitat (see above); (b) spectrum of used substrates; and (c) association with other organisms, plants and animals.
2. Distribution and dynamics. All K-selected organisms dominate poor habitats, while r- and L-selected species occupy rich habitats. However, this apparent incongruity results from a confusion related to epithets 'rich' and 'poor' habitats. In kinetic terms, oligotrophic environments

Table 5.1 Summary of ecological features of studied microorganisms

Organism	Typical habitat	Position in succession and population dynamics	Reaction to stresses	Diversity of utilized substrates
Arthrobacter globiformis	Domination in fallow, tundra and mountainous soils under sparse vegetation	Late stages, stable dynamics without oscillations	High resistance to starvation, desiccation and salts; low resistance to acidity	Wide spectrum of aliphatic and aromatic compounds; moderate depolymerase activity
Rhodococcus erythropolis	Litter–soil interface, soil polluted by oil	"	"	Wide spectrum, especially *n*-alkanes and fatty acids; no depolymerase activity
Micrococcus agilis	Lichen crusts especially on carbonaceous rocks	Dynamics dependent on lichen	High resistance to alkali; moderate to desiccation	Wide spectrum, requires some vitamins and amino acids
Pseudomonas fluorescens	Planted moist soils, rhizosphere, excrements and fresh litter	Early stages, frequent oscillations, dependent on plant state	Low resistance to starvation and desiccation	Wide spectrum, especially carbohydrates and organic acids; moderate depolymerase activity
Bacillus subtilis	Rich organic or manured soils, composts and plant litter	Occasional bursts of growth, mainly dormant state (endospores)	Low resistance of vegetative cells; supreme resistance of spores to all kinds of stresses	Wide spectrum of substrates; high depolymerase activity
Rhizobium (free form)	Soils and rhizosphere	Early stages, dynamics dependent on plant state	Moderate and low resistance to starvation and desiccation	Wide spectrum; low depolymerase activity (except for invasion to roots)

Table 5.1 *Continued*

Organism	Typical habitat	Position in succession and population dynamics	Reaction to stresses	Diversity of utilized substrates
Debaryomyces formicarius	Ant nests, particular 'fungi gardens'	Early stages, dynamics dependent on larvae state	Moderate salt tolerance; low persistance to starvation	Wide spectrum, especially carbohydrates, alcohols, organic acids and proteins
Lipomyces tetrasporus	Forest and steppe soils, including deep horizons	Late stages, dynamics is independent on vegetation cover	High osmo- and acidotolerance, resistant to starvation and desiccation	Wide spectrum, preference for carbohydrates including polysaccharides
Mucor, Zygorynchus	Plant litter, forest soils	Pioneer stages, sporadic growth	Moderate acidotolerance, spores resistant to many stresses	Preference for mono-saccharides; low or absent depolymerase activity
Trichoderma	"	Pioneer stages with more stable dynamics	Susceptible to fungistasis	Carbohydrates, including cellulose
Mortierella ramanniana	Mineral soil horizons	Late stages, stable dynamics	Sensitivity to soil compaction	Mono- and polysaccharides
Pencicillium funiculosum	Arable acid soils, amended with mineral fertilizers	Early stages	Resistance to acids and desiccation	Mono- and polysaccharides
Ulocladium consortiale	Perturbed soils	Pioneer stages, unstable dynamics	Resistance to soil compaction	Wide spectrum of carbohydrates

belong to climax ecosystems kept under steady-state conditions, while copiotrophic ones are associated with pioneer stages under transient conditions. To be really rich, natural habitats have to have frequent changes of nutrients sources (manuring of arable soils, unexpected emergence of animal faeces, permanently changed 'addresses' of plant root exudation, etc.). This is why copiotrophic organisms are necessarily r-selected (opportunistic) and display unstable dynamics, while K-selected organisms (strong competitors) are oligotrophic and stable. They are more ubiquitous but in rich habitats are for some time secluded by an abundant growth of r-strategists.

3. Utilization of polymeric and refractory substances. This feature should be considered as a manifestation of L-selection. At first glance this may appear absurd, but let us make the point clear. Firstly, the use of refractory compounds is one way to avoid direct competition for readily available substrates and hence it is not a K-feature. Secondly, growth on these compounds is very slow and energetically expensive, so it is not an r-property. Thus the ability of microorganisms to use *n*-alkanes (*Rhodococcus*), and strong depolymerase activity (*Bacillus* and many fungi) are manifestations of endurance rather than competitive strength. From this point of view it is clear that the teleonomic meaning of the close correlation between the following seemingly disparate features as: (a) spore formation; (b) depolymerase activity: and (c) the capability to produce antibiotics. All these features are attributed to bacilli, streptomycetes and slowly growing fungi such as *Penicillium funiculosum, Penicillium daleae*, etc. All of these are weak competitors in an open contest for available substrate, and instead evolve in a 'cowardly' fashion, struggling for existence by sporulation, the use of second-rate nutrients and the use of toxic metabolites to repulse stronger rivals.

5.2 MIXED MICROBIAL CULTURES

Speculative discussions on life strategy now need to be verified by direct experimental examination of microbial competition for a common substrate when populations are grown in one fermentation vessel. We chose two bacterial species which have already been discussed most intensively, *P. fluorescens* 312 and *A. globiformis* C-4. Batch and dialysis cultures were inoculated with identical amounts of these bacteria. The growth dynamics were followed by the determination of total biomass dry weight and a microscopic count of the ratio between *Pseudomonas* and *Arthrobacter* cells, their biomasses being calculated from proportions.

Batch culture

The bacterium *P. fluorescens* multiplied more quickly than *A. globiformis* during the exponential phase of unlimited growth (Figure 5.2). In the

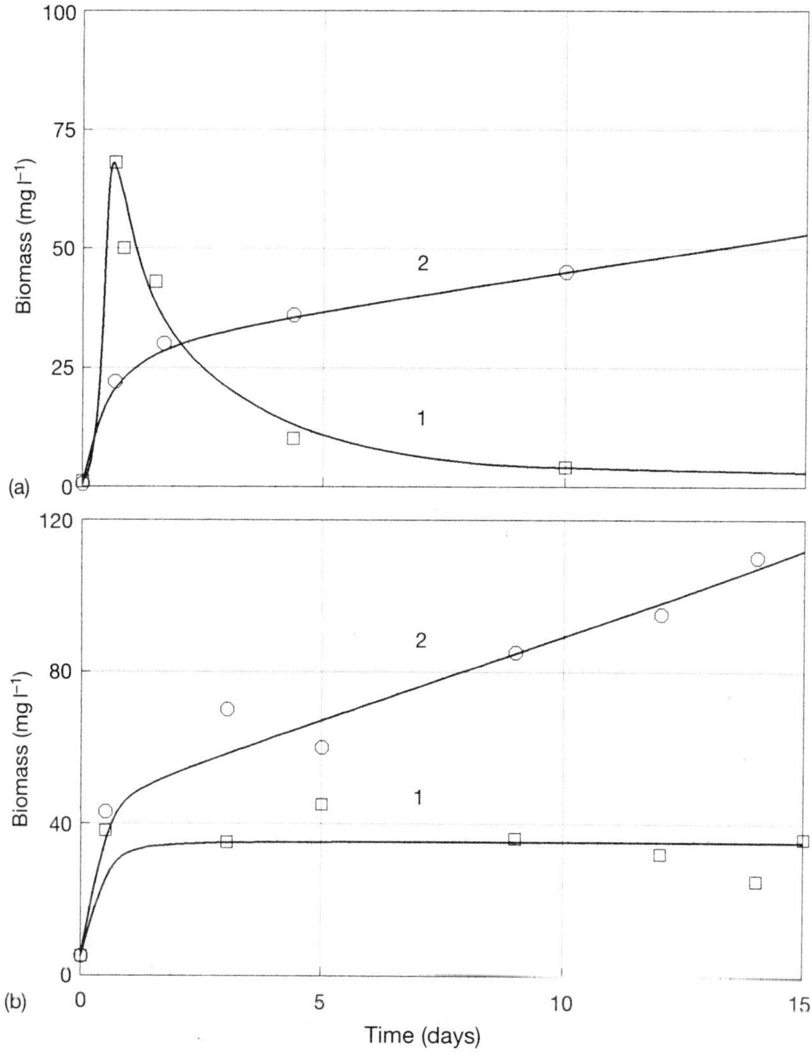

Figure 5.2 Dynamics of competition between *P. fluorescens* (curve 1) and *A. globiformis* (curve 2) in (a) mixed batch culture and (b) dialysis culture. The curves are simulated by synthetic chemostat model (SCM). (Panikov *et al.*, 1989b.)

starvation phase the r-selected bacteria quickly declined, and *Arthrobacter* became the dominant component feeding on the lysed cell material of *Pseudomonas*. The apparent yield of such secondary growth, calculated as the ratio (*Arthrobacter* cell mass increase)/(*Pseudomonas* cell mass decrease), was rather high, $0.33\ g\ g^{-1}$.

Dialysis culture

Here, the success of *Arthrobacter* was evident from the very start of experiment. Its biomass increased in a linear fashion similar to monoculture (compare with Figure 3.14). The competition pressure did affect the growth efficiency. The apparent growth yield was only 0.04 instead of $0.23\,g\,g^{-1}$ in monoculture, this difference being attributed to glucose consumption by the competing population. The cell density of *Pseudomonas* remained constant during experiment at one and the same level, i.e. a small biomass gain was exactly compensated by a decay.

The obtained data on mixed cultures are in full agreement with the SCM. The only modification of the original model (Table 3.9) concerns the equation for limiting substrate which is now common for two competing species:

$$\dot{s} = F - q_{s_1}(s)x_1 - q_{s_2}(s)x_2 + k_1 v_1 L_1 x_1 + k_{1_2} v_2 L_2 x_2\,, \qquad [5.1]$$

where the first term, F, is the substrate input rate ($F = 0$ in batch culture), the next two terms are substrate consumption rates and the last term represents product leakage from the cells. Subscripts 1 and 2 denote all parameters and variables related to the first and the second species respectively.

It is important that the outcome of competition may be explained by the differences in the kinetic properties of the two species without model augmentation by any other specially designed postulates concerning interactions, such as production of antibiotics or lytic enzymes by one of the competitors. The accelerated death of *Pseudomonas* in batch culture is accounted for by the fact that *Arthrobacter* is a vigorous sink of leakage products, preventing their reutilization which occurs in *Pseudomonas* monoculture. The same reason (substrate sink, although less vigorous) diminishes the growth of *Arthrobacter* in dialysis culture.

Thus, performed experiments support the basic idea that competition is mainly governed by hereditary preserved growth properties of individual microbial species. These properties are the same in pure and mixed cultures. We do not completely exclude antagonism as one of the operating mechanisms, simply because antibiotics have been detected in mixed cultures of *Arthrobacter* and *Pseudomonas* when very rich media were used (Sorokina *et al.*, 1984). The idea that competitive interactions do involve 'biochemical aggression' in the form of the production of antibiotics, killing factors, etc., is deeply rooted in contemporary biological mentality and is based on vast experimental data with batch growth on rich media. However, it seems to be of secondary significance in the substrate-limited conditions dominating in nature. Much more important are the self-dependent features of competitors, i.e. an affinity to limiting substrate, the ability to persist stresses, to recover from perturbations and to detect favourable habitats in mosaic environments, etc.

5.3 POPULATION DYNAMICS IN NATURAL HABITATS

This section deals with the computer simulation of population dynamics of microorganisms with different life strategies. Results of simulations then will be compared with observations made *in situ* by traditional ecological techniques.

The simulations were of two types. The first were computer exercises designed to expose the general principles governing microbial dynamics in imaginary 'soil' under assumed environmental conditions. Another simulation was related to the real and vital ecological problem of global climate change caused by the dramatic increase of greenhouse gases in the Earth's atmosphere (Fantechi *et al.*, 1991). We attempted to simulate the present-day seasonal dynamics of a microbial community in the arctic coastal tundra (Barrow, Alaska) and to predict its reaction to anticipated climatic changes. The choice of arctic soils was determined by the fact that they are the most vulnerable to global warming. In addition, primitive northern ecosystems seem to be simpler for quantitative interpretation than temperate ones. The particular Barrow site is the most intensely studied spot in the Arctic (Brown *et al.*, 1980).

5.3.1 MATHEMATICAL MODEL

The flow diagram of the model is shown in Figure 5.3 and the equations are summarized in Table 5.2.

The microbial community

The community is represented by three typical heterotrophic organisms competing for one common organic substrate, S. (We focus on heterotrophic organisms, as in tundra they absolutely dominate over sparse chemolythotrophic populations. The abundant photosynthetic microorganisms (algae and cyanobacteria) are also beyond the scope of simulation, being tacitly referred to as a functional component of a plant community.) *Arthrobacter* and *Pseudomonas* are in fact the most abundant bacteria in many real tundra soils (Parinkina, 1989). However, we will sometimes use these names in a wider context implying the whole strategic groups rather than the respective singular genera. Thus, the variable X_1 may be attributed not only to the principally dominant *A. globiformis* but also to the other less abundant tundra K-selected species, e.g. *Rhodococcus* sp., *M. ramanniana* and prostekate bacteria. The state variable X_2 represents the principal component *Pseudomonas* and secondary r-selected species – *Achromobacter*, yeasts and 'saccharolytic' fungi. The L-strategist *Bacillus* (producing two subpopulations – vegetative cells, X_3, and spores, X_4) was

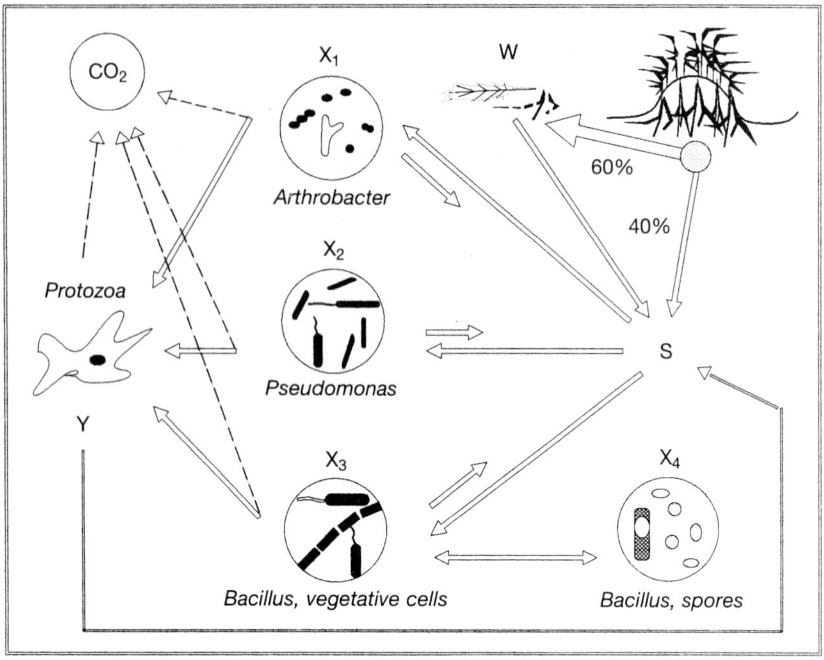

Figure 5.3 Flow diagram of fluxes described by simulation model.

included in the model to secure its entirety, although it is rarely observed in tundra (the reasons will be explained below).

The growth of *Pseudomonas* was described by the basic reduced SCM (Table 3.9). The complete models developed for pure cultures of *Arthrobacter* and *Bacillus* (Eqns [3.73–3.76] and Table 3.12) turned out to be unmanageable when combined and incorporated into the larger version. The compromise was found in the use of simplified submodels (Eqns [a1–a6] and [b1–b11], Table 5.2). These were much simpler and simulated the major experimentally observed features, although they failed in some details.

Sources and sinks of microbial biomass

The elimination of microbial cells was assumed to be caused by predation. It was simulated explicitly by the introduction of the variable y, protozoan biomass. The grazing obeys a second-order kinetic reaction and was supposed to be arrested if prey biomass declines below the threshold level, x^*. The elimination of the protozoans themselves (dominated by amoebae) was postulated to be by their autolysis, which occurs at a rate proportional to y^2.

Nutrient substrates for microbial growth are provided by the plant community. However, instead of phytomass as an explicit state variable, we used the value of gross primary production (GPP) which is a direct function of temperature. The input of organic matter to soil was assumed to be accomplished via (a) root exudation (40% of GPP) and (b) plant litter formation (60% of GPP). Root exudates consist of readily available compounds which enter the soil pool of S, while plant litter is composed of stable macromolecular compounds, W (pectin, lignocellulose, resins, etc.), which need prior extracellular transformation. Decomposition of W into S was assumed to follow first-order kinetics with the rate constant k_d dependent only on the temperature.

The effects of temperature

The dependence of the rate constant, k, of a chemical reaction on the absolute temperature, T, is accounted for by the Arrhenius equation derived from thermodynamic considerations:

$$\frac{d(\ln k)}{dT} = \frac{E_a}{RT^2} \quad \text{or} \quad k = A \exp\frac{-E_a}{RT}, \qquad [5.2]$$

where R is the gas constant, E_a is activation energy, and A is the integration constant standing for the frequency of collisions of reacting molecules. In a narrow T range around 273 °K, Eqn. [5.2] may be approximated by an exponential function of Celsius temperature, T_c:

$$\frac{d(\ln k)}{dT} \approx \lambda \quad \text{or} \quad k \approx B \exp(\lambda T_c), \qquad [5.3]$$

where $\lambda = E_a/RT^2$ is taken as an empirical constant related to the widely used coefficient $Q_{10} = \exp(10 \cdot \lambda)$, and B is the reaction rate at $T_c = 0$.

The biological reactions catalysed by enzymes deviate from the relationships predicted by Eqns [5.2–5.3] at high temperature due to the thermal denaturation of enzymes. If we assume that denaturation is reversible,

$$\begin{array}{ccc} \text{active enzyme} & K_T & \text{inactive enzyme} \\ E & \longleftrightarrow & E' \end{array}$$

then the amount of active enzyme may be found from the equilibrium constant, K_T, and the van't Hoff equation (Cornish-Bowden, 1976):

$$E = \frac{E'}{K_T} = \frac{E_0}{1 + K_T} = \frac{E_0}{1 + \exp(\Delta S^0/R - \Delta H^0/RT)}, \qquad [5.4]$$

Table 5.2 Simulation of microbial population dynamics in tundra soil

State variable	Mathematical expression	
General		
Air temperature	$T = 30 \sin^4 [(t - 10)\pi/360] - A, \quad A = 14 \ldots 24°C$	[a]
Microbial temperature factor	$T_M = \begin{cases} 0.5\, e^{0.166T}/[1 + e^{67.5(1-280/T)}] \\ 0.1\, e^{0.166T}/[1 + e^{67.5(1-290/T)}] \end{cases}$	[b] [b′]
Plant temperature factor	$T_P = \left[\operatorname{arctg}(1000\pi T) + \frac{\pi}{2} \right] e^{0.03T}$	[c]
Available substrate	$\frac{ds}{dt} = 0.4T_p \cdot \text{GPP} + T_M[k_d w + a_y y + a_1 r_1 x_1 + a_2 r_2 x_2 + a_3 r_3 x_3 - (q_{s1} + q'_{s1})x_1 - (q_{s2} + q'_{s2})x_2 - (q_{s3} + q'_{s3})x_3]$	[d]
Biomass of *Protozoa*	$\frac{dy}{dt} = Y_y(\sum \text{GRA}_i x_i) - T_M a_y y^2$	[e]
Plant litter	$\frac{dw}{dt} = 0.6T_p \cdot \text{GPP} - T_M k_d w$	[f]
Soil respiration	$\frac{dp}{dt} = \sum [(1 - Y_i)q_{si} + q'_{si} + T_M Y_i m_i r_i]x_i + (1 - Y_y)\text{GRA}_i x_i, \quad i = 1, \ldots, 4$	[g]
Arthrobacter		
Biomass	$\frac{dx_1}{dt} = (\mu_1 - \text{GRA1})x_1$	[a1]
Substrate uptake	$q_{s1} = \frac{T_M r_1 Q_1 s}{(K_i + s)(1 + bH)}$	[a2]
Waste respiration	$q'_{s1} = \frac{0.025 T_M Q_1 s}{K_1 + s}$	[a3]
Specific growth rate	$\mu_1 = Y_1(q_{s1} - T_M m_1 r_1) - T_M a_1 r_1$	[a4]
Grazing rate	$\text{GRA1} = T_M k_y y,\ k_y = 0.009 \text{ if } x_1 > x_1^* \text{ else } k_y = 0$	[a5]
r-Variable	$\frac{dr_1}{dt} = (\mu_1 + T_M a_1 r_1)\left[\frac{s}{K_{r1} + s} - r_1 \right]$	[a6]
Inhibitory by-product	$\frac{dH}{dt} = k_H q_{s1} - \mu_1 H$	[a7]

Table 5.2 *Continued*

State variable	Mathematical expression	
Pseudomonas		
Biomass	$\frac{dx_2}{dt} = (\mu_2 - GRA2)x_2$	[p1]
Substrate uptake	$q_{s2} = \frac{T_M r_2 Q_2 s}{K_2 + s}$	[p2]
Waste respiration	$q'_{s2} = \frac{0.1 T_M Q_2 s}{K_2 + s}$	[p3]
Specific growth rate	$\mu_2 = Y_2(q_{s2} - T_M m_2 r_2) - T_M a_2 r_2$	[p4]
Grazing rate	$GRA2 = T_M k_y y,\ k_y = 0.009$ if $x_2 > x_2^*$ else $k_y = 0$	[p5]
r-Variable	$\frac{dr_2}{dt} = (\mu_2 + T_M a_2 r_2)\left[\frac{s}{K_{r2} + s} - r_2\right]$	[p6]
Bacillus		
Vegetative cells	$\frac{dx_3}{dt} = (\mu_3 - GRA3)x_3$	[b1]
Substrate uptake	$q_{s3} = \frac{T_M r_3 Q_3 s}{K_3 + s}$	[b2]
Waste respiration	$q'_{s3} = \frac{0.1 T_M Q_3 s}{K_3 + s}$	[b3]
Specific growth rate	$\mu_3 = Y_3(q_{s3} - T_M m_3 r_3) - T_M a_3 r_3$	[b4]
Grazing rate	$GRA3 = T_M k_y y,\ k_y = 0.009$ if $x_3 > x_3^*$ else $k_y = 0$	[b5]
r-Variable	$\frac{dr_3}{dt} = (\mu_3 + T_M a_3 r_3)\left[\frac{s}{K_{r3} + s} - r_3\right]$	[b6]
Bacillus spores		
Flux from active cells to spore	$TR_1 = \frac{100 T_M x_3}{x_3 + x_4}$ if $s > s^*$ else $TR_1 = 0$	[b7]
Flux from spore to active cells	$TR_2 = \frac{100 T_M x_4}{x_3 + x_4}$ if $s \leqslant s^*$ else $TR_2 = 0$	[b8]
Cell turnover rate	$a_3 = 0.1 T_M$ if $s \leqslant s^*$ else $a_3 = 0$	[b9]
Spores biomass	$\frac{dx_4}{dt} = TR_1 + TR_2 - GRA4 \cdot x_4$	[b10]
Grazing rate	$GRA4 = T_M k_y y,\ k_y = 0.009$ if $x_4 > x_4^*$ else $k_y = 0$	[b11]

where E_0 is the total amount of enzyme, ΔS^0 is entropy and H^0 is the enthalpy of the denaturation reaction.

The combining of Eqn. [5.3] for an active enzyme with Eqn. [5.4] provides the temperature dependence of the reaction rate over the entire T range:

$$k = \frac{B\exp(\lambda T_c)}{1 + \exp[\xi(1 + \psi/T]}, \qquad [5.5]$$

where $\xi = \Delta S^0/R$ and $\psi = \Delta H^0/\Delta S^0$. Equation [5.5] was used to characterize the dependence of any elementary microbial reaction (substrate uptake, maintenance catabolism, turnover, etc.), as well as plant litter decomposition and grazing on ambient temperature. Parameter ξ defines the steepness of thermoinactivation, while ψ is numerically equal to that temperature (in K) at which activity is suppressed by 50% (formally half of the enzyme molecules are denaturated). The ψ-value is close to the optimum temperature of the particular reaction.

5.3.2 IDENTIFICATION OF THE MODEL'S PARAMETERS

The average sizes of primary production, $260\,g\ C\,m^{-2}\,year^{-1}$, as well as dependence of GPP on temperature were taken from Tiezen (1974). The rapid decline of GPP at $T_c < 0$ was approximated by the empirical arctg function (Eqn. [c], Table 5.2). The annual temperature dynamics of Barrow was taken as the average over 1930–1960 (French, 1974), and was approximated by a sine function (Eqn. [a], Table 5.2). The rates of plant litter decomposition and average litter reserves ($1500\,g\ C\,m^{-2}$) were assessed from the data of Flanagan and Veum (1974).

The parameters of the microbial populations were identified from data on batch and continuous growth of *A. globiformis* VCM-658, *P. fluorescens* VCM-1472 and *B. subtilis* VCM-428. The use of simplified models for *Arthrobacter* and *Bacillus* slightly worsened the agreement with experimental observations. In the case of *Arthrobacter* the model did not reproduce oscillations in chemostat culture and false diauxie. In the case of *Bacillus* the model did not simulate the abrupt cessation of growth in dialysis culture, producing instead a gradual transition to a maintenance state. To summarize, as described by the simplified models, *Arthrobacter* lost some of its competitive ability, while *Bacillus* became a little stronger.

Bearing in mind the importance of temperature characteristics and their possible changes during the laboratory cultivation of isolates, we undertook a special short-term experiment to identify the parameters of Eqn. [5.5] directly in soil. The respiration of fresh soil samples was continuously measured as a CO_2 evolution rate by infra-red gas analysis (IRGA) while the incubation temperature was automatically raised from −3°C to 35°C

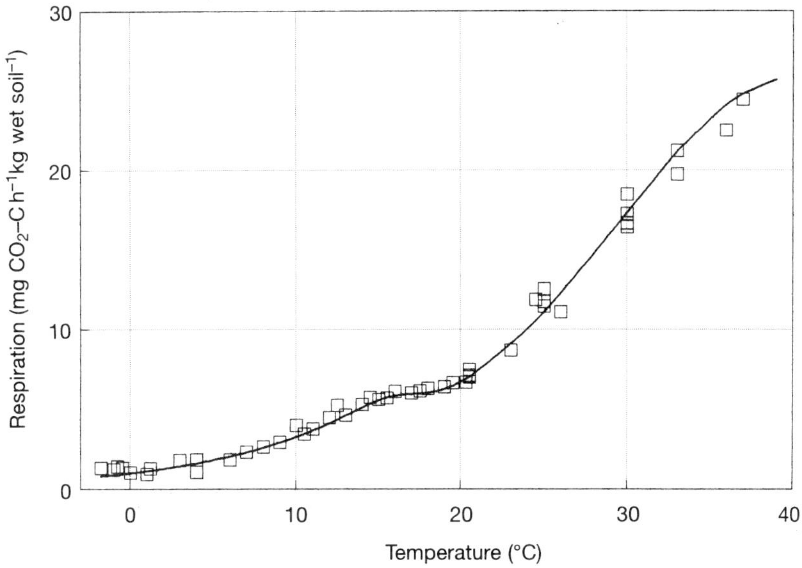

Figure 5.4 The effect of temperature on soil respiration. The respiration of fresh samples of tussock peat bog (Vasyugan wetland, West Siberia) was continuously measured as a CO_2 evolution rate by infra-red gas analysis (IRGA) during a slow automatic raising of incubation temperature from -3°C to 37°C with gradient 2°C h^{-1}. Obtained respiration rates were plotted *versus* respective instant temperature and fitted by Eqn. [5.6]; for parameters, see Table 5.2.

(gradient 2°C h^{-1}). The resulting experimental curves were generally in agreement with the predictions of Eqn. [5.5], but some northern and boreal soils displayed minor deviations manifested as inflection points at 5–16°C (Figure 5.4).

The most reasonable explanation is the heterogeneity of the soil microbial community, in particular the coexistence of mesophilic and psychrophilic microbial populations. The expansion of Eqn. [5.5] into the simplest two-term series

$$q_{\text{resp}}(T) = B_1 \frac{\exp(\lambda_1 T_c)}{1 + \exp[\xi_1(1 + \psi_1/T)]} + B_2 \frac{\exp(\lambda_2 T_c)}{1 + \exp[\xi_2(1 + \psi_2/T)]} \qquad [5.6]$$

allows a more accurate fit to the experimental curve. The two sets of parameters with subscripts 1 and 2 refer to psychrophilic and mesophilic components of the soil microbial community. In the samples taken from the West-Siberian wetland, the contribution of psychrophilic activity with an optimum temperature as low as 13.7°C, was 33%. It is curious that in some soils we have a clear-cut demarcation line between two distinct clusters rather than a continuum of temperature characteristics of the entire microbial community (otherwise there would be no inflection point on the experimental curve). The absence of a continuum may be indicative of abrupt (without 'halftones') changes in the thermal regime of northern ecosystems.

5.3.3 SIMULATIONS OF MICROBIAL POPULATION DYNAMICS IN TUNDRA

Present-day seasonal dynamics

In the course of the simulation all variables quickly converged to quasi-steady-state with reproducible seasonal patterns (Figure 5.5). In the cold climatic conditions normal for Barrow (average T_c in January −24°C, and in July +5°C) this pattern was characterized by the following features: (a) the microbial community was dominated by *Arthrobacter* and *Pseudomonas*, while bacilli displayed only sporadic small increases in activity during spring; (b) the dynamics of *Arthrobacter* was stable throughout the year; and (c) the *Pseudomonas* population varied within a wide range attaining two maxima, in late spring and autumn. These results were fully supported by experimental data for arctic tundra in North America and Russia (Nelson and Visser, 1978; Parinkina, 1989), which report the spring and autumn maxima of bacterial abundance, but the most striking agreement is in respect to the *Bacillus* population dynamics. That is, almost all authors declare a higher probability of detecting bacilli only at spring-thaw time.

It is important to note that we did not introduce any special temperature requirements for these bacteria. Their competitive exclusion in arctic soils was brought about by one factor – the trophic status of the environment. Variation of input parameters revealed that only strong eutrophication (GPP > 5000 g m^{-2}) led to a competitive advantage of *Bacillus* over other species. In accordance with this conclusion are the empirical data that bacilli are abundant in very rich and highly productive soils (alluvial, meadow, chernozemic and manured soils).

The prediction of the effects of climatic changes

What may happen after the anticipated climate warming? According to various scenarios mean air temperature in winter will increase in the Arctic by the middle of the next century by as much as 4–14°C (MacCracken and Luther, 1985). To simulate these events, parameter A (Eqn. [a], Table 5.2) was incrementally changed from −24 to −19 and −14°C. It was surprising that the resulting changes in the community structure were modest even at the maximum shift, $\Delta A = 10$°C. There was some increase in the biomass of *Arthrobacter*, a slight decrease of the average *Pseudomonas* populations under the pressure of predation, and (most significant) an acceleration of *Bacillus* growth, i.e. their spring bursts became higher by one order of magnitude (Figure 5.6).

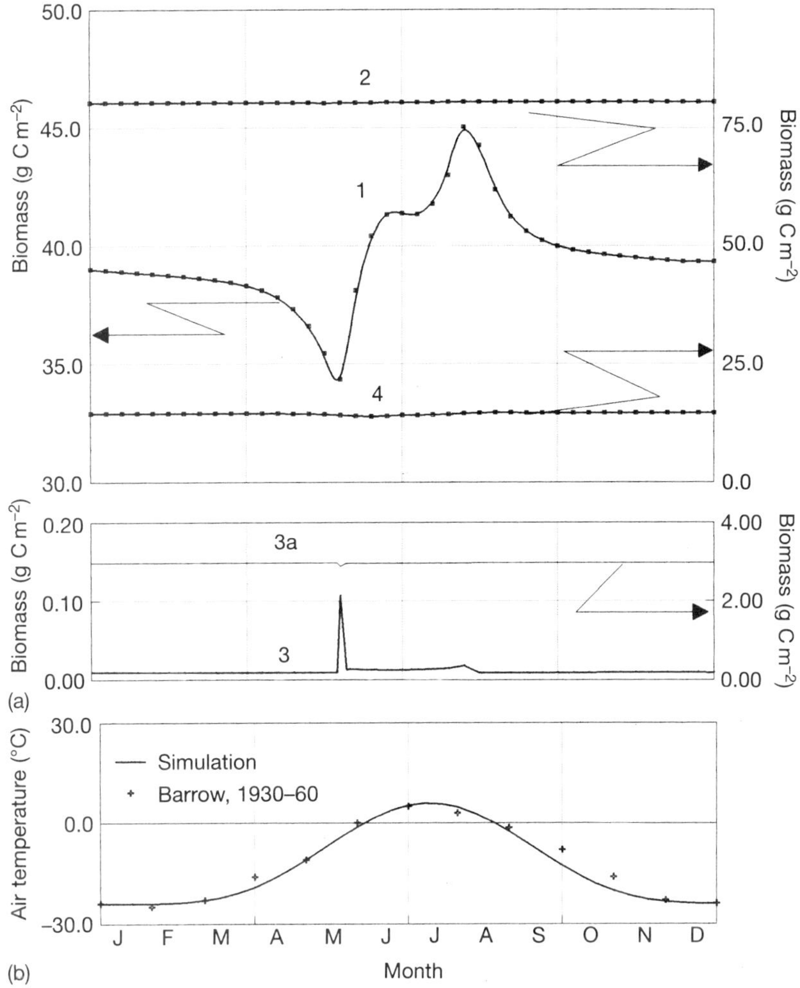

Figure 5.5 (a) Simulated present-day seasonal dynamics of microbial community of tundra soil: *Pseudomonas* (curve 1), *Arthrobacter* (curve 2), *Bacillus*, vegetative cells (curve 3) and spores (curve 3a), protozoa (curve 4). (b) Simulated dynamics of air temperature fitted to observation data averaged for 30 years. Displayed simulation results were obtained with the model shown in Table 5.2.

The main effect of rising temperature on the carbon balance of the ecosystem was the considerable activation of organic matter decomposition (Figure 5.7). Total mineralization (CO_2 evolution) became higher than primary production and as a result, the stores of plant litter dramatically decreased (15% over the first year). A new steady-state value of plant litter was attained after 20–30 years (not shown) at a level of about 740 g C m^{-2} or 50% of the initial stores.

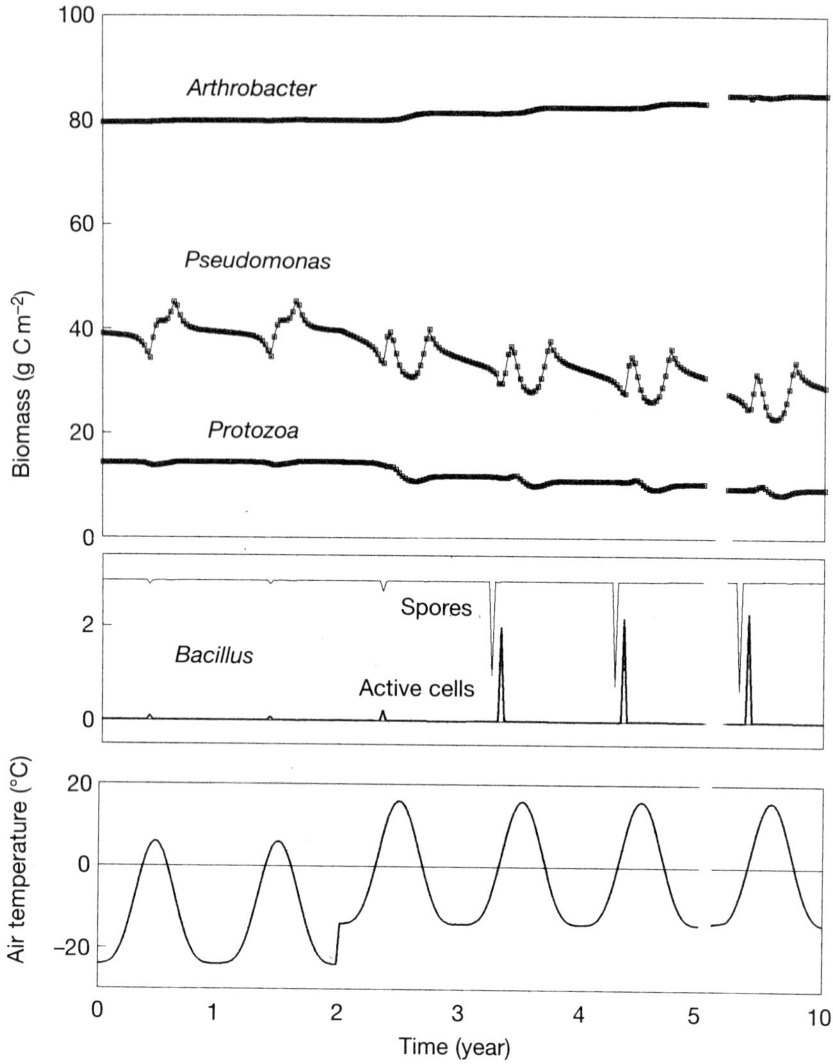

Figure 5.6 Simulation of tundra ecosystem response to elevated temperature: dynamics of bacterial biomass. Assumed changes in the climatic and ecosystem parameters: increase of the January temperature from −24 to −14°C, the number of frost-free days from 81 to 135, and index of degree-days from 293 to 1365. Displayed simulation results were obtained with the model shown in Table 5.2.

Separate simulations were used to follow the competition between psychrophilic and mesophilic subpopulations of *Pseudomonas* (Figure 5.8). The psychrophilic subpopulation (assumed to have an optimum temperature 10°C, Eqn. [b], Table 5.2) absolutely dominated the bacterial

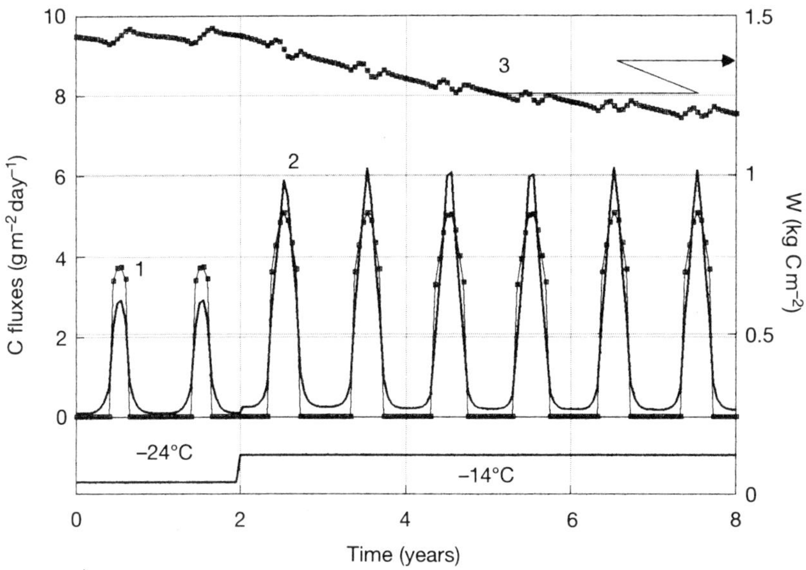

Figure 5.7 Simulation of tundra ecosystem response to elevated temperature: dynamics of the main sources and sinks of organic carbon. Soil respiration (curve 1), GPP (curve 2), plant litter (curve 3). For changes in the climatic parameters, see Figure 5.6. Displayed simulation results were obtained with the model shown in Table 5.2.

community under cold conditions, and an increase in T_c by 5°C did not affect their domination. Further warming by another 5°C resulted in a rapid 50% substitution of psychrophiles by mesophiles over 2 years. Under the new quasi-steady-state, the two subpopulations demonstrated a stable coexistence due to the regular seasonal changes of temperature, i.e. the first maximum in spring was due to the rapid growth of psychrophiles, while the second autumn maximum was attributed to the mesophiles.

Thus, simulations displayed a high degree of stability of the microbial community with respect to the anticipated warming of the climate. Significant changes in the functional structure occurred only after a huge increase in temperature, i.e. 10°C. Activation of *Bacillus* growth may be interpreted as a relief from the pressure of L-selection, the main factors being an increase of substrate availability due to increased mineralization activity and elevated primary production. The reliability of this prediction is unclear. The only optimistic sign is the fact that the model adequately describes contemporary seasonal dynamics.

5.3.4 SIMULATION OF POPULATION DYNAMICS IN SOIL AND THE RHIZOSPHERE

To gain a better insight into population dynamics we performed several simulations at different soil microsites, in particular in the rhizosphere and

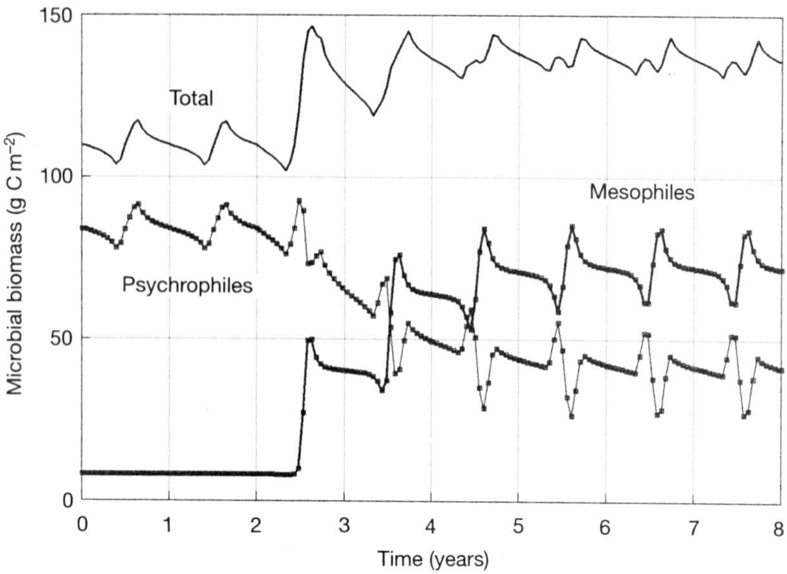

Figure 5.8 Simulation of competition between psychrophilic and mesophilic subpopulations of *Pseudomonas* in tundra soil after temperature shift-up. For changes in the climatic parameters, see Figure 5.6. Temperature characteristics of psychrophiles and mesophiles are described respectively by Eqns [b] and [b′] (Table 5.2). Displayed simulation results were obtained with the model shown in Table 5.2.

the surrounding non-rooted soil (Figure 5.9). In the rhizosphere we assumed the occurrence of diurnal cycles of root exudation according to the simple function $20 \times \sin^2(\pi t)$ ($g\,C\,m^{-2}\,d^{-1}$), which implies a release of organic substances in the light part of the day and its complete cessation overnight. The competing organisms started at equal population densities of 80 g $C\,m^{-2}$. Exclusion of *Bacillus* occurred fairly quickly, the main elimination factor being predation accompanied by the arrest of activity by nocturnal starvation. *Arthrobacter* exhibited a slow elimination, but for another reason, i.e. the formation of autoinhibitory by-products during intensive daytime exudation. The *Pseudomonas* population was the real master of the situation, displaying a steady growth at the highest density.

It is interesting that the simulated diurnal fluctuations in nutrient supply induced oscillations in the biomasses of *Pseudomonas* and *Bacillus*, but not of *Arthrobacter*. This is in accord with observations that the bacteria of the rhizosphere display regular and wide oscillations (Evdokimova, 1976). However, the agreement between simulation and observation is only qualitative, since *in situ* the amplitude of the fluctuation should be higher. The probable reason for the failure of the quantitative agreement is the fact that the spatial factor has been ignored, i.e. in the real rhizosphere

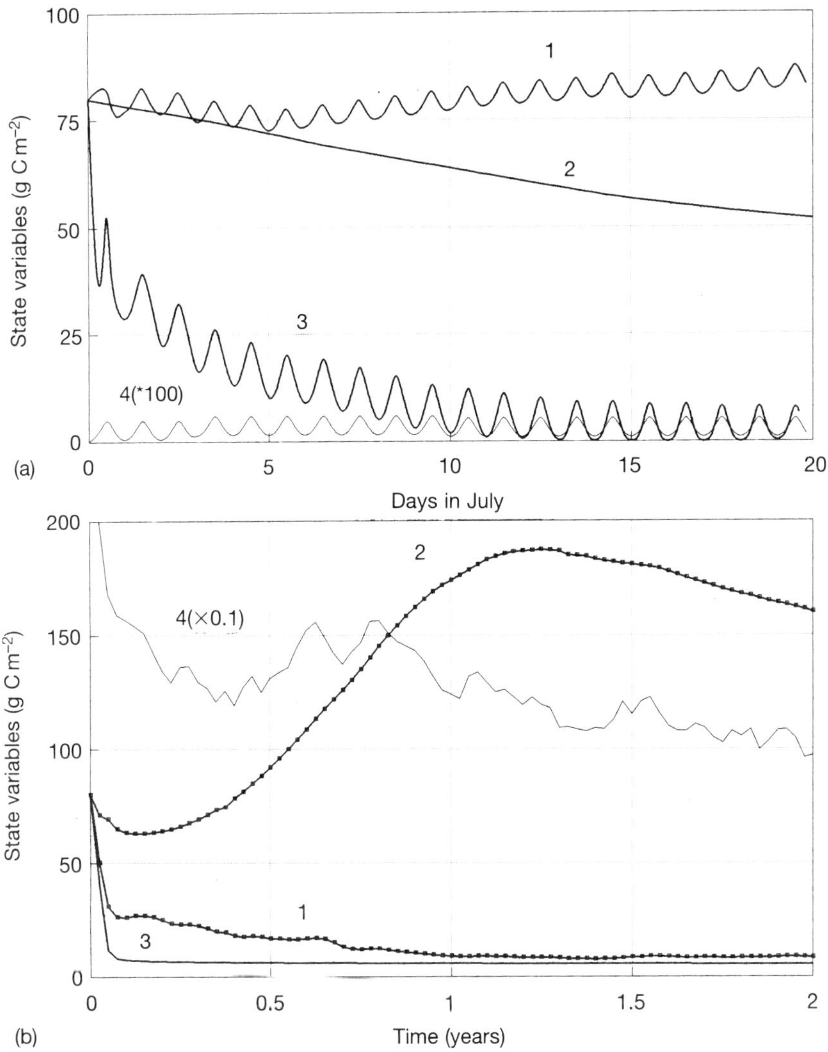

Figure 5.9 Simulation of competition between *Pseudomonas* (curve 1), *Arthrobacter* (curve 2) and *Bacillus* (curve 3) in (a) rhizosphere and (b) unplanted soil, curve 4 is available substrate. In the rhizosphere, substrate input was assumed to follow a diurnal pattern according to function $A \cdot \sin^2(\pi t)$, where A is constant and t is time, in days. In unplanted soil, input of organic substrate was assumed in the form of plant litter at 1% of the flux used in simulation with tundra soil. Displayed simulation results were obtained with the model shown in Table 5.2.

diurnal dynamic fluctuations of root exudation are accompanied by the spatial change of exudation loci originating from root elongation and growth. Thus, pioneer stages of succession are initiated in the rhizosphere every morning. It is this feature which makes the plant surface the real 'oasis' of r-selected soil bacteria.

Another situation occurs in the non-rooted parts of the soil, for which GPP was set equal to zero in Eqn. [d], but not in Eqn. [f] (Table 5.2), since there was supposed to be some low input of fresh plant litter. In this type of environment the only source of immediately available substrate is the slow depolymerization of plant litter producing monomeric products. Under these conditions the domination of *Arthrobacter* is unquestionable (Figure 5.7). *Bacillus* is again rapidly eliminated and preserved exclusively in the form of endospores, while the *Pseudomonas* population declines to a very low level, suffering from chronic starvation.

5.4 CONCLUSION

The concept of life strategy has been used to explain the origin of the diversity of growth characteristics among various microorganisms. We have concentrated on three types of life strategy (there could be more) which have evolved under the pressure of different circumstances occurring in natural selection: r-selection (pioneer stages of succession, the state of temporary ecological 'vacuum'); K-selection (climax stages at the most intensive competition); and L-selection (in adverse environments).

r-Selected organisms include opportunistic populations of well-known bacteria (*Pseudomonas*, and the members of the *Enterobacteriaceae* family intensively studied elsewhere), as well as yeasts and fungi (*Debaryomyces*, *Mucor*, *Zygorynchus*, *Trichoderma*). They are characterized by rapid and balanced growth in rich media, but do not survive prolonged starvation. K-selected organisms (*Arthrobacter*, *Rhodococcus*, *Micrococcus, Lipomyces*, *Caulobacter*, *Prosthecomicrobium*) grow slowly but have a high affinity to substrate and a frugal energetic metabolism which provides their competitive advantage in overcrowded environments with a limited supply of nutrients. L-selected organisms, typified here by *Bacillus*, are weak competitors and survive only because of the ability to sporulate. This group may be appended with other spore-forming bacteria (*Clostridium*) and various inhabitants of extreme environments (halo-, acido-, alkalo-, psychro- and thermophilic microorganisms, species resistant to drought, irradiation, toxic substances, etc.).

The experiments with mixed cultures showed that it is sufficient to know the growth parameters of several individual organisms in order to predict the outcome of their competition for a common substrate. Therefore the data on growth kinetics may be used to predict the dynamics of microbial succession *in situ*. We used this rationale to simulate the microbial population dynamics in arctic tundra induced by the anticipated global climatic changes. The simulation model appears to be realistic in that it accounts for the observed temperature conditions of tundra, portrays the most important natural sources of organic substrates (plant root exudation, litter formation and decay), and describes the elimination of microbial biomass via grazing. We have found that simulation results are in a fairly good qualitative agreement with observations on the microbial community in tundra soils.

However, we still have no solid experimental data to support our ecological paradigm in quantitative terms. There is serious scepticism over the extrapolation of laboratory growth data to the natural environment. Therefore, our final consideration was a kinetic study of microbial growth *in situ*.

Microbial growth in soil 6

This chapter deals mainly with laboratory microcosms, outdoor ecosystems being briefly described only in the later sections to show that they are fundamentally the same. The advantage of the microcosm is absolutely clear: it provides full control of environmental conditions (moisture, temperature, illumination), and the feasibility of accurate dynamic measurements. At the same time, carefully manipulated microcosms may preserve undistorted the composition and structure of the soil microbial community. In the context of the present study, it is especially important that in microcosms we are able study those naturally occurring interactions between microbial populations and other biotic components (plants and animals) which provide the specific environment and the source of nutrients, and exert native elimination control. Moreover, we are able to study these interactions *in situ*.

The microcosms exploited in the present study (Figure 6.1) were actually fresh soil samples or soil substitutes which, in different experiments, were either amended with various substrates (defined and complex, mineral and organic, volatile, soluble or solid), or used in combinations, i.e. soil + plants or soil + introduced invertebrates. The microcosms were incubated under a constant moisture regime of 50% of water-holding capacity (WHC) and at a temperature of 25°C (if not otherwise specified). Microbial dynamics and the activity of associated higher organisms were followed preferably by instrumental procedures. In addition, all the major components of a microcosm's mass (carbon) balance were considered, so that we were able to question the accuracy of measurements using the criteria of the conservation law. The main goal was to estimate true rates of microbial growth and extinction and understand the mechanisms responsible for the observed dynamics.

Accurate measurements, essential in kinetic studies, are hindered by the characteristic features of soil, i.e. heterogeneity, adsorption, opacity, etc. Hence we paid great attention to the refinement of research techniques.

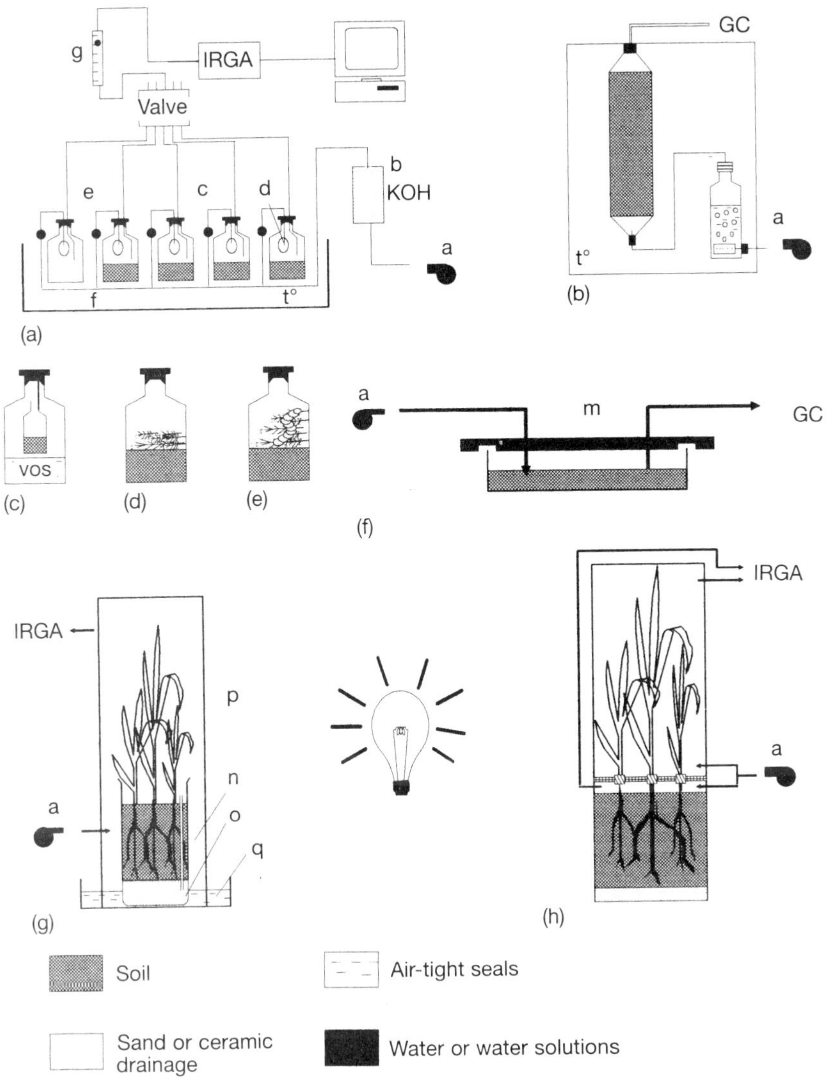

Figure 6.1 Laboratory microcosms and experimental designs. (a) Soil amended with non-volatile substrates with hardware providing automatic recording of CO_2 evolution rates, including a, air pump; b, CO_2 absorber; c, manifold of bottles containing soil; d, air-humidifier; e, flowstats; f, temperature bath; g, rotameter. (b) Column packed with soil continuously ventilated with air saturated by volatile organic substances (VOS). (c) Soil incubated in air saturated with vapour of VOS, liquid phase containing VOS of various concentration. (d) Soil incubated with plant litter. (e) The same with introduction of invertebrates. (f) Soil with crust of algae–bacterial community in 10-cm Petri dish with m, transparent Plexiglas lid. (g) Planted soil in: n, beaker with o, sand drainage under p, Plexiglass cover and q, water seal. (h) Planted soil in a two-compartment chamber providing independent measurements of soil+root respiration and CO_2 exchange rates between shoots and atmosphere. GC, gas chromatography; IRGA, infra-red gas analyzer.

6.1 THE INITIAL RESPONSE OF A MICROBIAL COMMUNITY TO SOIL AMENDMENTS: A KINETIC METHOD FOR THE DETERMINATION OF SOIL MICROBIAL MASS

Let us define 'initial response' as all short-term events in amended soil (substrate consumption, activation and growth of microorganisms) which are not yet associated with a sizeable alteration in the composition of the microbial community. There are two complementary aspects of these studies: (a) to describe the dynamics of growth and substrate transformation given the initial conditions, i.e. the initial values of microbial biomass, x_0, and the index of physiological state, r_0 (the direct problem); and (b) to find x_0 and r_0 from the dynamic curve of substrate transformation (the reverse problem). The second problem is matching the task of developing the kinetic method of microbial biomass determination.

The principle of kinetic methods

To determine microbial biomass, short-term enrichment experiments are performed, adding to the soil sample the particular substrate, following its uptake, and then fitting the obtained curve to an appropriate kinetic model. The desired entities x_0 and r_0 are found as the two parameters of the fitted model.

The model's adequacy is very important. It has to be robust and adequate and to be based on justified assumptions. To meet this requirement we put the following restrictions on the experimental set-up: (a) the added substrate must be distributed uniformly throughout the soil (this is a realistic requirement in the case of soluble or volatile chemicals); (b) the soil amendment should be large enough to provide zero-order kinetics of substrate transformation ($s_0 \gg K_s$), but well below the inhibitory level; and (c) environmental parameters must be physiologically suitable and kept constant. If all these conditions are satisfied, then the growth of responsive microbial populations should be rather simple: uniform (almost homogeneous) and unlimited. Moreover, for the time interval restricted by our definition of initial response, the growth should be unaffected by interactions such as competition or grazing (we will show below that the response of protozoan grazers is indeed delayed since they require a significant build-up of the prey population above a high threshold level). Hence, a multispecies soil community behaves, during a short-term enrichment experiment, as a single population. This gives us some freedom in the choice of simulation model. In particular we may rely on the double-reduced SCM used in section 3.6.2 for the simulation of lag and exponential phases of batch culture. Equations [3.65] and [3.66] are reformulated to express explicitly the relationship between the rate of uptake of added substrate, $v(t)$, and the required parameters x_0 and r_0:

$$v(t) = -\dot{s} = rxQ + xQ' = A + Be^{C},$$

$$A = Q'x_0(1 - r_0), \quad B = r_0x_0(Q + Q'), \quad C = \mu_m = YQ, \qquad [6.1]$$

where Q and Q' are maximum rates of coupled and uncoupled consumption of substrate respectively. (It was assumed that added substrate is the source of energy; in the case of conserved substrate we have simply $Q' = 0$.) In section 3.6.2 we have shown that the ratio $Q/(Q+Q') = \lambda$ varies among diverse microorganisms in a rather narrow range so it may be accepted as a basic stoichiometric constant ($\lambda \approx 0.9$). If not, this ratio may be estimated directly as a CN-sensitive fraction of the total soil activity. The biomass yield, Y, also should be either selected from reference data or determined from specially designed experiments for every particular group of soil microorganisms (see below). One should remember that Y is the 'true' yield which is related to the apparent experimental value Y^{app} as follows: $Y = Y^{app}/\lambda$.

Returning to Eqn. [6.1], the composite parameters A, B and C are easily estimated by non-linear regression. We used a special procedure based on general linear regression which insures the global minimization of residuals (see Appendix). Then,

$$Q = \frac{C}{Y}, \quad r_0 = \frac{B}{B + A/(1 - \lambda)}, \quad x_0 = \frac{\lambda B}{r_0 Q} = \frac{\lambda YB}{r_0 C}. \qquad [6.2]$$

Even more simple is the expression for the so-called active biomass (see Figure 3.45 for geometric interpretation):

$$x_{act} = \frac{\lambda YB}{C} = \frac{BY^{app}}{C}, \qquad [6.3]$$

i.e. active biomass is directly derived from experimentally measured quantities and is not affected by assumptions about the stoichiometric constants Y and λ. Now we shall consider the experimental data.

Aerobic microorganisms utilizing glucose

The response of soil microorganisms was followed by recording respiration (CO_2 evolution) (Figure 6.2). The biomass yield per unit of CO_2 formed, Y_{CO2}, was assessed from the C-balance in independent experiments. When CO_2 production, Δp, was measured parallel to residual glucose, s, then $Y_{CO_2} = \Delta x/\Delta p = (\Delta s - \Delta p)/\Delta p$, where $\Delta s = s_0 - s$. Prokaryotic (mainly bacteria) and eukaryotic (mainly fungi) microorganisms were differentiated by the use of actidione and chloramphenicol, being selective inhibitors of protein biosynthesis (Anderson and Domsch, 1973). Other inhibitors, KCN and benzhydroxamates (BH), were applied for the selec-

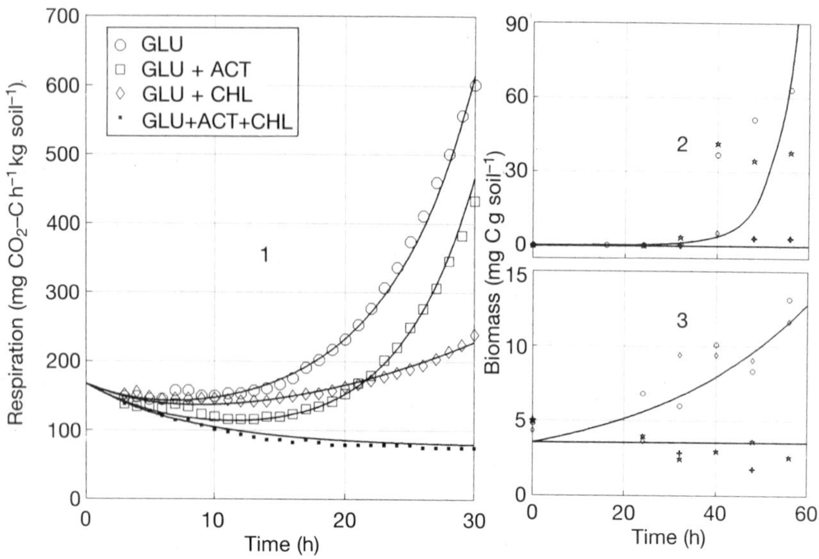

Figure 6.2 Dynamics of microbial growth and respiration in soil amended with glucose and specific inhibitors (kinetic method for differentiation of bacteria and fungi) (Panikov *et al.*, 1992). Peat soil sampled from layer 10–20 cm (ombrotrophic sphagnum bog, Tver) was mixed with glucose (GLU), 10 mg g^{-1}, $(NH_4)_2SO_4$, 1.0 mg g^{-1}, KH_2PO_4, 1.0 mg g^{-1}, as well as actidione (ACT) and/or chloramphenicol (CHL), 0.1 mg g^{-1}, and incubated at 25°C and moisture 50% of water-holding capacity (WHC) under continuous flow of CO_2-free air (0.3 l min^{-1}). Recorded dynamic variables: (curve 1) soil respiration by infra-red analyser (Infralyt-4); (curve 2) bacterial population by plating on MPA, the colony-forming unit (CFU) number being converted to biomass according to Eqn. [3.1] (assumed values $W = 0.4$, $\rho = 1.1$ g ml^{-1}, $\sigma_x = 0.5$; size distribution of CFU was measured directly in soil with scanning electron microscope); (curve 3) biomass of fungi assessed by direct soil microscopy from measurements of hyphae length and width. Curves 1–3 were calculated according to the following modification of Eqn. [6.1]:

G	$v = Ae^{-kt} + B_f\exp(\mu_f t) + B_b\exp(\mu_b t)$	G+C	$v = Ae^{-kt} + B_f\exp(\mu_f t) + B_b$
G+A	$v = Ae^{-kt} + B_f + B_b\exp(\mu_b t)$	G+C+A	$v = Ae^{-kt} + B_f + B_b$

where G, A and C stand respectively for glucose, actidione and chloramphenicol treatments of soil, subscripts f and b are labels of respectively fungi and bacteria, and k is decay rate.

tive suppression of total coupled respiration and uncoupled respiration of eukaryotic organisms respectively (see details in Panikov *et al.*, 1992).

The following results were obtained. The double-reduced SCM fitted well to experimental curves. Independent determinations of uncoupled respiration, $A = Q'x_0(1 - r_0)$, as a regression parameter and from the direct measurement of CN-resistant respiration were consistent. For the entire soil community, the parameter λ was close to the same value of 0.9 that was found in pure cultures. Fungi and bacteria contribute roughly the same share to the total uncoupled soil respiration as estimated by

KCN + BH inhibitors. The physiological state of the microbial population before soil amendment, r_0, varied in a wide range from as low as 0.0025 (bacteria in sphagnum peat) to 1.0 (bacteria in the rhizosphere). In most samples taken from planted soils r_0 was higher (0.5–1.0) than in bare soils (0.001–0.5). Thus, the physiological state of soil microorganisms *in situ* is quite comparable with that of laboratory chemostat or dialysis cultures and is more vigorous than in starving batch culture.

The relevance of the inhibition technique was confirmed by direct microscopy, i.e. the observed dynamics of bacterial and fungal growth was closely followed by the model predictions (Figure 6.2, curves 2 and 3). In general bacteria displayed a higher maximum growth rate, a lower actual activity and lower biomass 'standing crops', e.g. in a peat soil we found for bacteria $\mu_b = 0.194\ h^{-1}$, $r_0 = 0.0025$ and $x_0 = 464\ \mu g\ C\ g^{-1}$ soil, while for fungi $\mu_f = 0.028\ h^{-1}$, $r_0 = 0.87$ and $x_0 = 2195\ \mu g\ C\ g^{-1}$ soil. In mineral soils the growth rates μ_b and μ_f were roughly the same and the biomass was lower, e.g. the grey forest soil contained 60–100 and 305–790 $\mu g\ C\ g^{-1}$ of bacterial and fungal biomass repectively. In rhizosphere soil the relationship was reversed with bacteria accounting for about 60–75% of the total biomass. Thus, a soil microbial community developing on glucose is dominated by fungi, while bacteria are represented mainly by r-selected populations displaying explosive growth after soil amendment.

We should remember, however, that glucose is a rather selective substrate, being used by many saprotrophic microorganisms but not by the majority of them. In particular, many K-selected bacteria do not respond to glucose amendment. Compared with a microscopic counting method, the glucose-kinetic method accounts for no more than 20–50% of microbial biomass (although direct microscopy is very bad yardstick with all its bias and uncertainty). A more complete account of aerobic chemoorganotrophic microorganisms was achieved by the use of substrates less selective than glucose, e.g. sterilized soil. This amendment contains the full array of organic substances originating from killed microbial cells and partly decomposed soil humus.

Phototrophic soil microorganisms

These organisms are closely associated in the soil with heterotrophic satellite bacteria forming an algal–bacterial community (ABC), the main source of nutrients for bacteria being exometabolites excreted by the algae. We induced the laboratory growth of ABCs by cultivating a soil amended with axenic algae culture and exposing the soil surface to continuous artificial illumination. To assess the true rates of CO_2 photoassimilation and the sizes of fluxes of exometabolic products, we used a non-steady-state approach. The rates of CO_2 exchange under alternate light–dark cycles (Figure 6.3) were recorded.

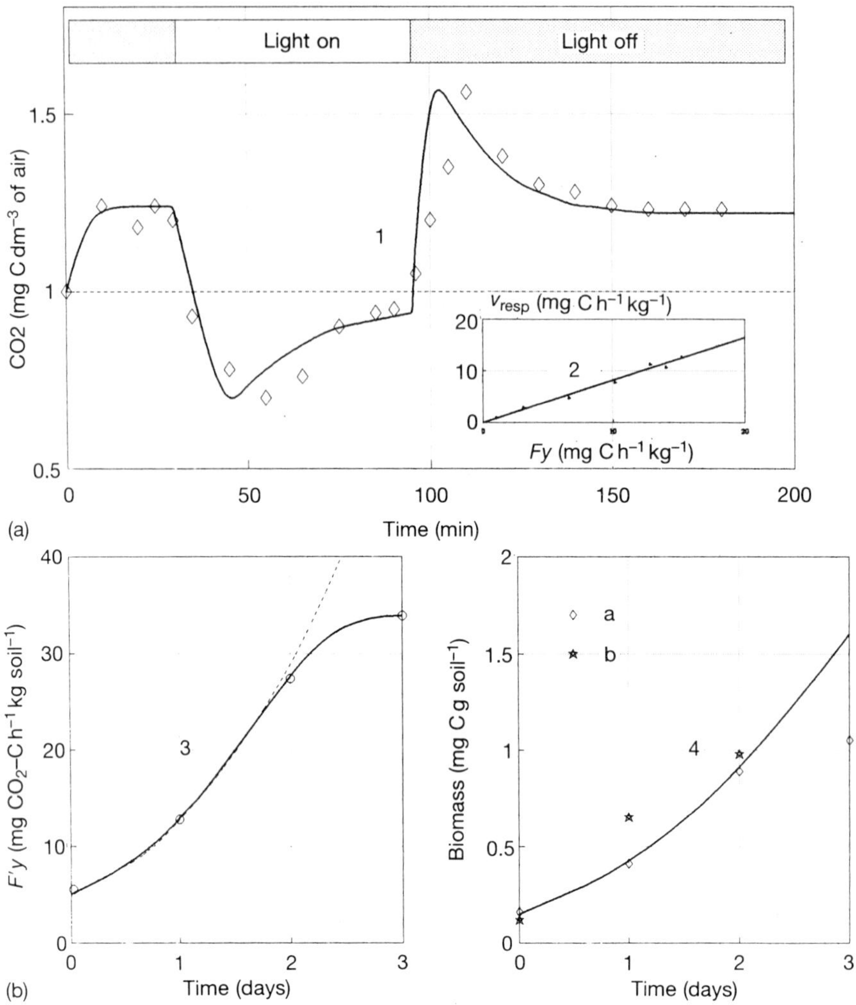

Figure 6.3 Kinetics of algae crust growth over the soil surface under artificial illumination (Panikov *et al.*, 1989a; Dedysh *et al.*, 1991). Soil (south chernozem under alfalfa, 0–10 cm, Odessa) was inoculated with washed cells of axenic culture *Bractococcus minor* 22 (10^7 g^{-1}) and incubated under continuous light, 10 000 lux at 25°C. (a) Light-to-dark transient response of algae–bacterial community (ABC). The soil was placed inside the chamber with continuous air flow (Figure 6.1 (f)) and incubated under alternate light–dark conditions. Dynamics of CO_2 in output air was recorded by gas chromatography. Curve 1 was fitted by Eqns [C] and [D] (Table 6.1) using the following parameters: $D = 0.33$ min^{-1}, $k = 0.07$ min^{-1}, $Y = 0.05$, $\theta = 0.55$, $\phi = 0.08$, $Fy = 0.35$ mg C min^{-1} dm^{-3}, $\beta = 0.82$, and $r_0 = 0.15$. Curve 2 is the relationship between bacterial respiration, $v_{resp} = \phi+(1-Y)\theta ks$ and gross photosynthesis, Fy, obtained from several transient experiments. (b) Growth response of ABC to continuous soil illumination: curve 3, dynamics of apparent CO_2 photoassimiation rates $(dp/dt)_d - (dp/dt)_{ill}$; curve 4, algae biomass determined by kinetic method (4a) and by direct count with epifluorescent microscopy (4b). The dashed line is calculated according to Eqn. [6.4]; the solid line according to Eqn. [6.5].

The mathematical model approximating the response of an ABC to intermittent soil illumination is based on the reduced SCM and contains a variable r for algae (Table 6.1). The main postulate is that exometabolite production is coupled with photosynthesis (no excretion in darkness) and is proportional to gross CO_2 photoassimilation (proportionality coefficient θ). The model does simulate the observed quick kinetics with a characteristic delayed response in the CO_2 exchange rate (Figure 6.3, curve 1). There was **undershoot** for the first 20 minutes after the onset of illumination (photoassimilation starts immediately, but heterotrophic respiration is low since there are, as yet, no exometabolic products), and **overshoot** occurred for the first 5–10 minutes after the light was turned off (the instantaneous stopping of CO_2 photoassimilation while bacteria continue, for some time, to intensively oxidize accumulated exometabolites).

The amplitudes of overshoot and undershoot are functionally related to the C-fluxes of the studied system. Parameter identification allowed the estimation of the true rate of algal photosynthesis, Fy, which was almost two times higher than the apparent rate, $F'y$, due to a hidden fraction of productivity (the formation of exometabolites, θ, varies from 0.4 to 0.6). The identified apparent yield, Y, of bacteria was very low, probably due to intensive grazing in the 'algosphere'. The respiration of heterotrophic microbes was closely correlated with Fy (Figure 6.3, curve 2) supporting the suggested mechanism.

For biomass determination by the kinetic method it is sufficient to measure the apparent photoassimilation rate, $F'y$, which is equivalent to our usual notation of specific growth rate. Soil was incubated for several days under continuous light. At least once a day, the rates of CO_2 evolution from illuminated soil, $(\mathrm{d}p/\mathrm{d}t)_{\mathrm{ill}}$, and during short-term dark incubation, $(\mathrm{d}p/\mathrm{d}t)_{\mathrm{d}}$, were measured. The difference between these two measurements is close to $F'y$ (Eqn. [E], Table 6.1):

$$\left(\frac{\mathrm{d}p}{\mathrm{d}t}\right)_{\mathrm{d}} - \left(\frac{\mathrm{d}p}{\mathrm{d}t}\right)_{\mathrm{ill}} = Fy[1 - \theta(1 - Y)] + (1 - Y')q_{\varepsilon}'y \approx F'y\,. \qquad [6.4]$$

During such experiments (Figure 6.3, curves 3 and 4) the photosynthetic activity displayed exponential growth for the first 2 days, after which algal growth was restricted by grazers, numerous protozoa and collembola. The calculation of biomass was performed in two ways: (a) according to the formulae [6.1–6.3] by taking $Y = 1\,\mathrm{g\,C\,g^{-1}\,C}$, $r_0 = 1$ and $v(t) = F'y$; and (b) by the numerical integration of Eqn. [6.4]. The second approach, in essence, is based on the fact that CO_2 is conserved substrate and hence an increment in the biomass, Δy, may be derived by the summation of the net CO_2 carbon uptake during all the illumnated periods of incubation. The initial biomass, $y(0)$, may be roughly found from the simple proportion:

$$y(0) : \Delta y = F'(0) : \Delta F'\,, \qquad [6.5]$$

Table 6.1 Mathematical simulation of algae-bacterial community (ABC)

Simulated process and time scale	Equations for ABC as dependent on illumination			
	Light on		Light off	
C fluxes: y, algae x, bacteria s, algal metabolites p, CO_2	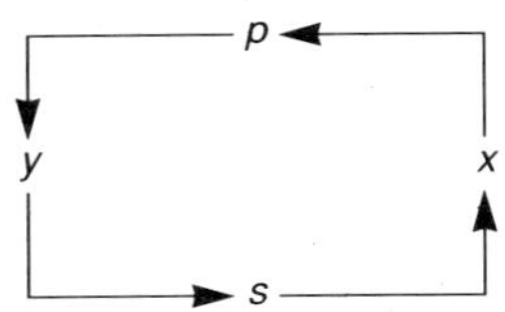		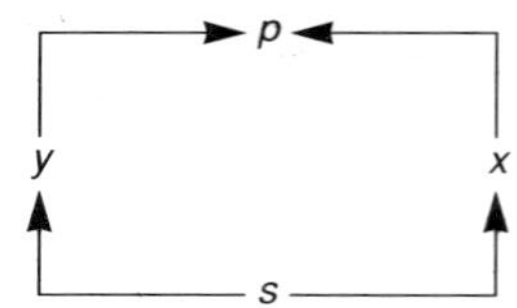	
The total mass-balance of the ABC	$\dot{y} = F(1-\theta)y$ $\dot{s} - \theta Fy - q_s x,\ q_s = \frac{Qs}{K_s + s}$ $\dot{x} = Yq_s x$ $\dot{p} = D(p_0 - p) + (1 - Y)q_s x + \phi - Fy$	[A]	$\dot{y} = Y'q'_s y,\ q_s = \frac{Q's}{K'_s + s}$ $\dot{s} = -q_s x - q'_s y$ $\dot{x} = Yq_s x$ $\dot{p} = D(p_0 - p) + (1 - Y)q_s x + (1 - Y')q'_s y + \phi$	[B]
Instant response (quick reactions)	$\dot{s} = r\theta Fy - ks$ $\dot{p} = D(p_0 - p) + (1 - Y)ks + \phi - rFy$ $\dot{r} = \beta(1-\theta)F(1-r)$	[C]	$\dot{s} = -\ ks$ $\dot{p} = D(p_0 - p) + (1 - Y)ks + \phi$ $\dot{r} = 0$	[D]
Long-term dynamics under quasi-steady-state ($\dot{s} = 0,\ s = \bar{s}$)	$\dot{y} = F(1-\theta)y = -F'y$ $\dot{x} = Yq_s x = Y\theta Fy$ $\dot{p} = D(p_0 - p) + \phi - F'y - Y\theta Fy$	[E]	$\dot{y}' = Y'q'_s y$ $\dot{x} = 0$ $\dot{p} = D(p_0 - p) + \phi + (1 - Y')q'_s y$	[F]

F and F' are respectively true and apparent specific rates of CO_2 photoassimilation; θ, fractional contribution of exometabolites to gross photosynthesis; ϕ, basal soil respiration; D, ventilation rate; Y, q_s and K_s are parameters attributed to satellite bacteria; Y', q'_s and K'_s are respective parameters of dark metabolic activity of algae.

where $F'(0)$ is the photosynthesis rate at $t = 0$. The two methods of calculation produced compatible results for the time interval before the onset of intensive grazing (not allowed for in the model of initial response). Direct microscopy also supported the obtained data generating a similar dynamic pattern of algal growth (compare a and b in Figure 6.3).

Microbovore protozoa

Grazing by soil animals is the most essential sink for microbial biomass. The growth kinetics of the most active predators, soil protozoa, was studied firstly under refined homogeneous conditions (Figure 6.4). The aim of experiments with the synthetic binary system *Tetrahymena–Pseudomonas* was to identify the principal growth features of protozoa and formulate the model (Table 6.2). The major findings are discussed below:

1. *Tetrahymena* displayed a rather high growth efficiency: 40–52% of the consumed substrate-C was incorporated into cell mass, 30–46% was respired and 14–18% was released as excrement.
2. The grazing rate was dependent on the physiological state of both the prey and the predator. Bacteria from an exponential growth phase were grazed quicker than bacteria taken from a long-starving culture. These features were accounted for by the introduction of r-variables for both organisms.
3. The protozoan growth rate was suppressed by a high concentration of bacteria, probably due to the toxic effect of some bacterial metabolites. On the other hand, grazing was precluded at low prey concentration. Both these effects were approximated by Eqn. [G] (Table 6.2) containing the substrate inhibition term and threshold prey level, x^*.
4. The maintenance catabolism of infusoria includes intensive mechanical work (active locomotion, osmoregulation involving the functioning of the contractile vacuole) and turnover of macromolecular components. Both these components were expressed by a first-order rate constant, a_y, directly related to the r-variable. It simulates the higher motility of grazers under abundant nutrient conditions and the formation of almost dormant retarded cells under starvation. It is interesting that during starvation, protozoan biomass rapidly declined, while cell numbers remained constant for at least 1 week. This is a characteristic feature of a starving multicellular organism and basically differs from that of bacteria which display a parallel decrease of both cell mass and numbers.

In the soil the response of protozoa (mainly naked amoebae) to a release of bacteria was adequately approximated by the same set of equations (Figure 6.5), although with different parameters. Amoebae grew more slowly and less effectively than *Tetrahymena*. The threshold prey concentration was much higher, indicating the beneficial effects for the prey and

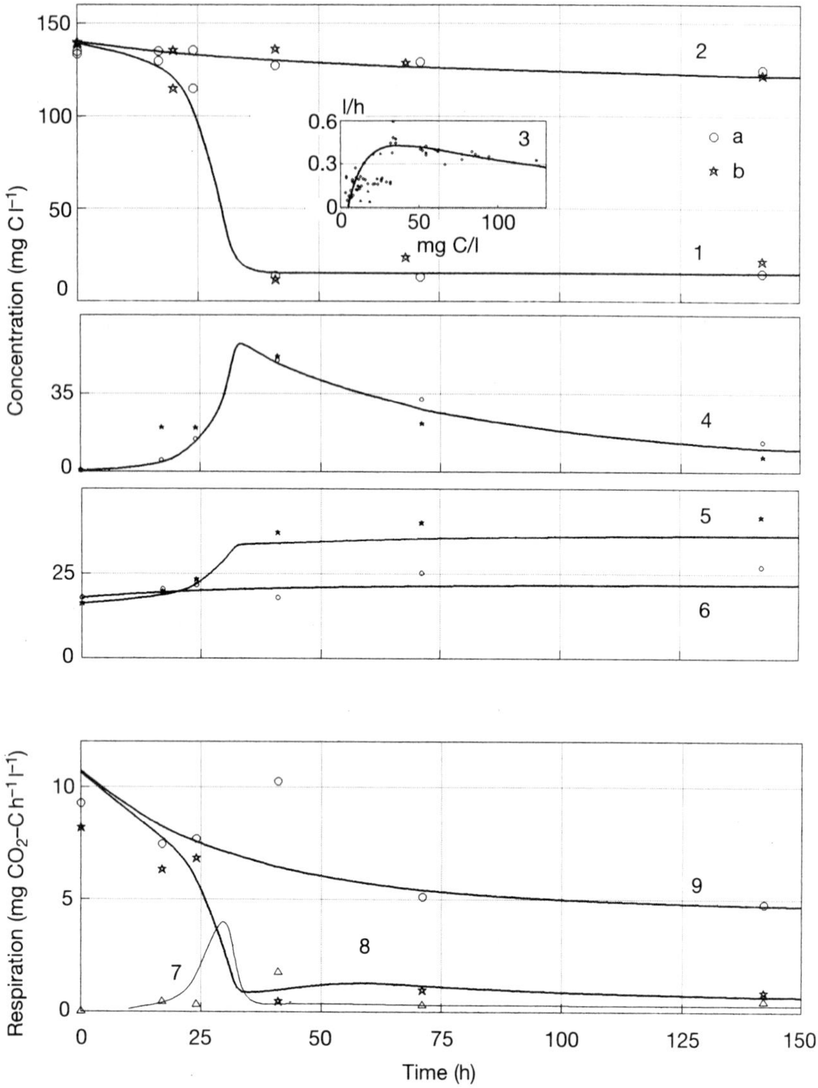

Figure 6.4 Bacterial grazing in a homogeneous system. Batch culture of *Tetrahymena pyriformis* was grown in a shaking flask on mineral solution (medium No 1, section 3.1) containing viable cells of *P. fluorescens* 1472 as the only C source (taken from exponential phase of batch culture and pre-washed by buffer). A parallel run without protozoa was used as control for endogenous decay of bacteria. Recorded dynamic variables: curve 1, bacterial biomass by turbidimetry and kinetic method; curve 2, bacterial biomass in control culture; (insert 3), dependence of *Tetrahymena* sp. growth rate on bacterial biomass concentration, pooled data of Curds and Cuckburn (1968) and Panikov (1991a); curve 4, biomass of protozoa accounted for by carbon balance as difference 'total particulate C' – 'bacterial C'; curve 5, soluble extracellular organic compounds determined by C-analyser; curve 6, the same in control culture; curve 7, respiration of protozoa; curve 8, glucose-induced respiration of bacteria in

Table 6.2 Kinetic model of bacteria grazing by protozoa

Variable	Equation	
Bacterial substrate	$\dot{s} = -(q_s - q'_s)x + \theta a_x r_x x + a_y r_y y$	[A]
Bacteria	$\dot{x} = \mu x - q_s y,\ \mu = Y_s(q_s - m) + a_x r_x$	[B]
r-Variable of bacteria	$\dot{r}_x = Y_s q_s \left(\frac{s}{K_s + s} - r_x \right) - \beta_x r_x$	[C]
Protozoa	$\dot{y} = \eta y, \quad \eta = Y_x q_x - a_y r_y$	[D]
r-Variable of protozoan	$\dot{r}_y = Y_x q_x \left(\frac{x - x^*}{K_x + x - x^*} - r_y \right) - \beta_y r_y$	[E]
Extracellular products	$\dot{H} = (1 - \theta) a_x r_x x + \psi q_x y$	[F]
Specific grazing rate	$q_x = Q_x r_y r_x \frac{x - x^*}{K_x + x - x^* + (x - x^*)^2 / K_{xx}}$	[G]
Actual respiration rate	$v_{resp} = (1 - Y_x - \psi) q_x y + [(1 - Y_s) q_s + q'_s + Y_s m] x$	[H]
Substrate-induced respiration	$V_{resp} = (1 - Y_x - \psi) q_x y + [(1 - Y_s) Q_s + Q'_s + Y_s m] x$	[I]

of spatial factors (escape of bacteria from grazing in soil micropores and cavities). The decline of amoebae after almost the complete consumption of bacteria was quicker than the corresponding decline of *Tetrahymena*, probably due to action of other predators. Specificity of grazing was rather low, i.e. one and the same extinction rate was observed for several bacteria (*Pseudomonas*, *Arthrobacter*, *Azotobacter* and *Azaspirilla*) added to soil.

The procedure for the determination of protozoan biomass involves soil incubation with washed microbial cells in experiments similar to that shown in Figure 6.5. The biomass is calculated according to Eqns [6.2–6.3], where the grazing rate, $v(t)$, is measured by following the decline in

Figure 6.4 *Continued.* the binary system; curve 9, control culture. The curves were calculated according to the model (Table 6.2) with the following parameters: $Y_s = 0.4$, $Y_x = 0.59$ g C g^{-1} C, $a_x = 0.015$ and $a_y = 0.04$ h^{-1}, $Q_s = 0.38$, $m = 0.001$ and $Q_x = 1.2$ mg C h^{-1} mg^{-1} C, $K_s = 1.0$, $K_x = 18.4$ and $K_{xx} = 60.6$ mg C l^{-1}, $x^* = 15.0$ mg C l^{-1}, $\beta_x = 0.03$, $\beta_y = 0.005$, $\theta = 0.7$, $\phi = 0.13$. Initial values of r-variable $r_x = 0.2$, $r_y = 0.5$.

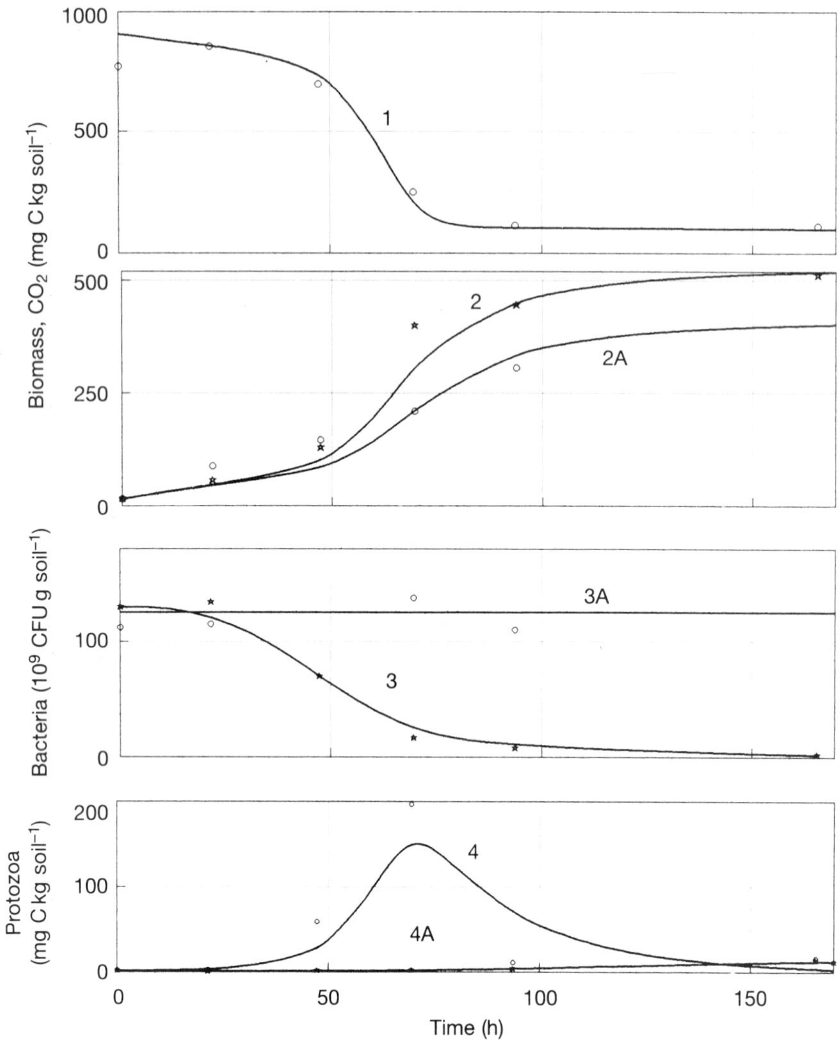

Figure 6.5 Dynamics of bacterial grazing in soil (Panikov, 1991a). The washed cells of *P. fluorescens* were released into grey forest soil and incubated at 25°C and moisture 50% of WHC. In a parallel run the growth of protozoa was suppressed with 200 ppm of actidione (marked by the symbol A). Recorded dynamic variables: curve 1, bacterial biomass by kinetic method; curve 2, CO_2 evolution; curve 3, bacterial population by plating on MPA; curve 4, biomass of protozoa by direct count and cell volume measurement. Curves were calculated according to the model (Table 6.2) with the following parameters: $Y_s = 0.4$, $Y_x = 0.5\,\mathrm{g\,C\,g^{-1}\,C}$, $a_x = 0.015$ and $a_y = 0.01\,\mathrm{h^{-1}}$, $Q_s = 0.38$, $m = 0.001$ and $Q_x = 0.8\,\mathrm{mg\,C\,h^{-1}\,mg^{-1}\,C}$, $K_s = 1.0$, $K_x = 160$ and $K_{xx} = 1000\,\mathrm{mg\,C\,l^{-1}}$, $x^* = 100\,\mathrm{mg\,C\,l^{-1}}$, $\beta_x = 0.03$, $\beta_y = 0.0$, $\theta = 0.7$, $\phi = 0.35$. Initial values of r-variable $r_x = 0.2$, $r_y = 0.1$.

bacterial number or the respiration rate of amended soil less soil treated with actidione, an inhibitor of protozoan growth. The evaluation of the stoichiometric parameter Y_y may be achieved only in time-consuming experiments, so it is better to calibrate kinetic methods against microscopic counting methods.

Other microorganisms

The kinetic methodology is almost uniformly applicable to various physiological groups of soil microorganisms. Details may be found in original publications. Here we will present only a summary.

1. Chemolithotrophic microorganisms (Paleyeva *et al.*, 1990; Popov and Panikov, 1990). For the assessment of the biomass of nitrifying, hydrogen and carboxidotrophic bacteria soil samples were incubated with NH_4^+, under air enriched with H_2 and CO, respectively. The rates of consumption or oxidation of these substrates, $v(t)$, increase for some time according to Eqn. [6.1]. Y evaluation is very important, since growth stoichiometry is very specific and the stoichiometric ratio $\Delta x/\Delta s$ varies in a wide range. The most precise estimation of Δx is provided by the measurement of $^{14}CO_2$ uptake during incubation with respective inorganic substrate, unamended soil serving as a reference to account for heterotrophic CO_2 fixation. The auxiliary use of specific inhibitors allows the differentiation of strictly autotrophic and other components, e.g. autotrophic and heterotrophic nitrifiers.
2. Anaerobic soil microorganisms (Panikov *et al.*, 1993). The recorded variables $v(t)$ are generally the rates of end-product formation. Thus the biomass of denitrifying, methanogenic, acetogenic and sulphate-reducing bacteria was determined from the production rates of $N_2 + N_2O$, CH_4, acetate and S^{2-} respectively. The direct evaluation of Y is rather difficult, as an anaerobic metabolic system is highly branched, so that the easiest course of action is to accept the data obtained for pure cultures.
3. Degraders of xenobiotic compounds (Glagolev *et al.*, 1995). The characteristic feature of the microbial degradation of pesticides (typified by chlorinated anilines) is its rather low specific rate, significant substrate inhibition and the strong adsorption of analysed substances by soil. Hence it was not reasonable to measure the initial rates of substrate consumption at normal soil moisture but instead the incubation was performed in soil suspension. The biomass of pesticides degraders was evaluated from the complete substrate depletion curves.

Limitations

To forestall possible disappointment, we will outline the most obvious forbidden fields for the application of kinetic methods. The general rule is that kinetic techniques should fail in those situations when substrate conversion is not coupled with the unlimited growth of the responsible microbial populations. Examples are pesticide conversion via co-metabolic and partly abiotic chemical reactions, and the decomposition of soil humus, lignocellulose and any polymeric substances which need to be degraded by extracellular enzymes before uptake. The rate of substrate transformation by extracellular enzymes is no longer proportional to the biomass of the microbial populations which produce this enzyme. In addition, often decomposition products are consumed by other micro-organisms, associated with hydrolytics. An additional difficulty is the accumulation of enzymes immobilized on soil particles where functionally they are completely independent of microbial cells.

6.2 MICROBIAL GROWTH IN SOIL AS DEPENDENT ON THE PATTERN OF ORGANIC SUBSTRATE INPUT

This section considers not only the initial microbial response to a soil amendment but also subsequent events associated with successional changes in the composition and functioning of the community. In Chapters 2 and 5 we speculated that the dynamic pattern of substrate supply (continuous, discontinuous) as well as its physical state (volatile, soluble, solid) determines the mode of microbial growth and the specificity of interactions between various populations. Here we will validate these speculations by direct experimental tests.

6.2.1 SINGLE-TERM INPUT OF READILY AVAILABLE COMPOUNDS

Most of the available data use glucose as a model substrate. We shall give two typical examples of incubation experiments when the dynamics of uptake and oxidation of added substrate were followed parallel to measurements of three major components of the soil community, bacteria, fungi and protozoa. In the first experiment (Figure 6.6) the examined soil was recently sampled from a field under barley, while in the second (Figure 6.7) it was maintained in the laboratory at normal field moisture and room temperature for 2 years (an imitation of bare fallow).

Fresh samples displayed a clear-cut sequence of interdependent events of glucose uptake, bacterial growth and multiplication of protozoa. Fungi grew more slowly than bacteria and independently of the protozoan and glucose dynamics. The observed dynamics were simulated by the model (Table 6.3) derived from the SCM and the following assumptions: (a) soil heterogeneity affects only bacteria which are not able to consume some fraction of glucose, s^*, located in remote microsites, while fungi do not encounter any spatial hindrances; (b) protozoa attack bacteria but not fungi and the bacteria have shelter of limited capacity x^* to escape grazing;

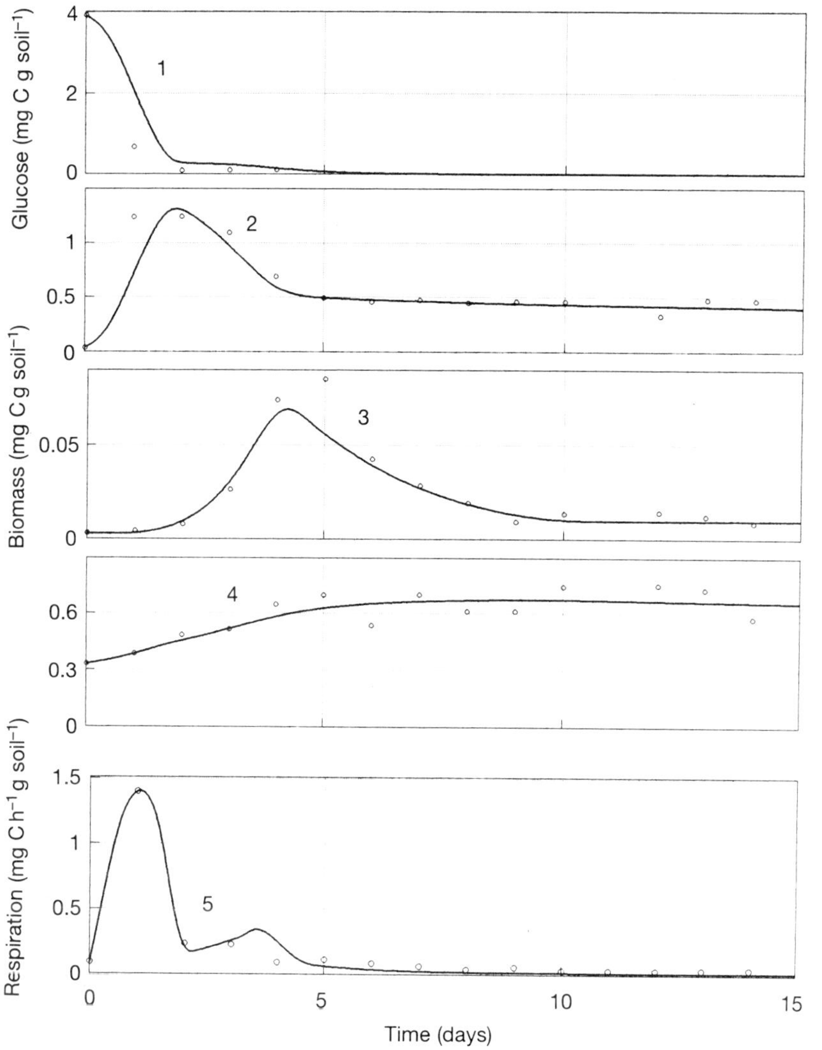

Figure 6.6 Dynamics of microbial succession in glucose-amended freshly sampled soil (Panikov, 1991a). Soil (grey forest soil under barley, Moscow region) was incubated with glucose–mineral mixture as indicated in Figure 6.2. Recorded dynamic variables: curve 1, residual glucose + exometabolites in water extract by dichromate oxidation; curve 2, bacterial population by plating on MPA, biomass being calculated assuming ratio 10^{-12} g C per CFU; curve 3, biomass of protozoa by direct count and cell volume measurement; curve 4, length of fungi mycelium by direct microscopy, biomass-to-length conversion factor 1.52 μg C m^{-1} hypha; curve 5, CO_2 evolution rate. The curves were fitted to model (Table 6.3) with the following parameters: $K_1 = K_2 = K_3 = K_{r1} = K_{r2} = s^* = 0.25$, $K_H = 0.071$, $K_x = 0.18$, and $x^* = 0.5$ mg C g^{-1}, $Y_1 = Y_2 = 0.4$, $Y_x = 0.16$ and $\phi = 0.2$ g C g^{-1} C, $Q_1 = 0.7$, $Q_2 = 0.07$, $Q_3 = 0.003$, $Q_x = 0.47$, $a_1 = 0.014$, $a_2 = 0.0025$ h^{-1}, $\lambda_1 = \lambda_2 = 0.1$, $\beta_1 = 4$ if $s \geq s^*$ else $\beta_1 = 15$, $\beta_2 = 1$.

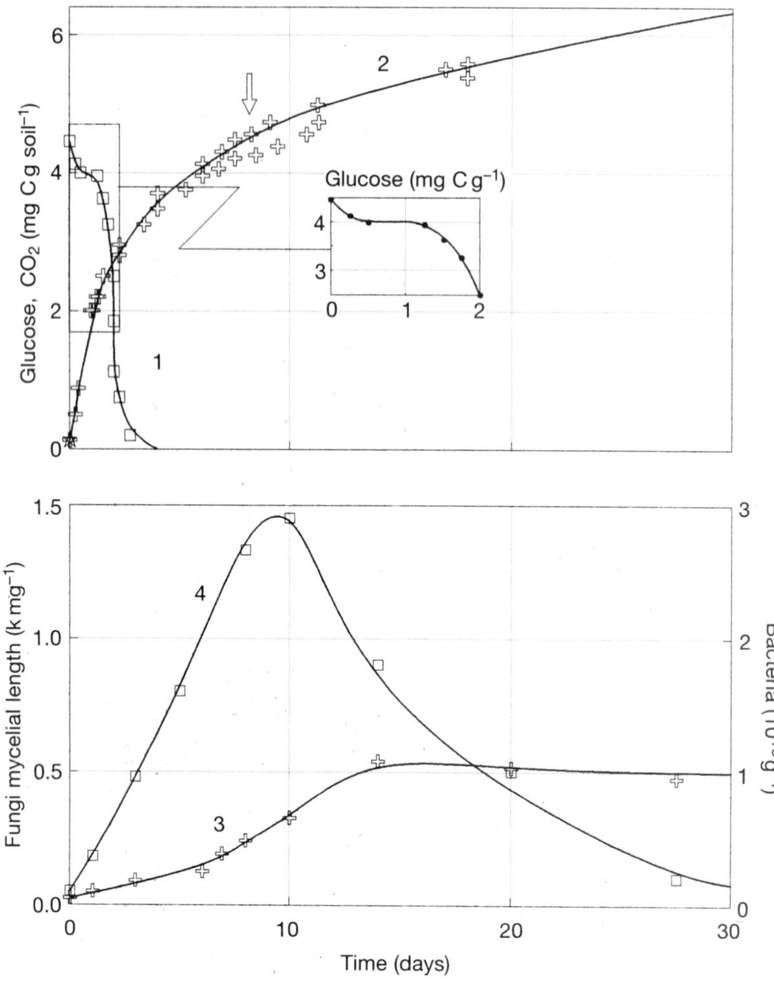

Figure 6.7 Dynamics of microbial succession in glucose-amended samples of bare soil (Panikov, 1991a). Soil samples of typical chernozem (Kursk region) were kept under constant moisture about 50% of water-holding capacity (WHC) during 2 years. Incubation experiments were performed as described in Figure 6.6. Recorded dynamic variables: curve 1, residual glucose with glucose oxidase technique; curve 2, CO_2; curve 3, fungal mycelial length; curve 4, bacteria direct count with acridine orange. The arrow indicates the time when CO_2 evolved has exceeded an amount of glucose-carbon added to the soil. The insert shows the dynamics of residual glucose during the first hours of the experiment.

Table 6.3 Simulation of microbial succession in glucose-amended soil

Variable	Equation	
Glucose	$\dot{s} = -q_1x_1 - q_2x_2 + a_1r_1x_1 + a_2r_2x_2 + a_yr_yy$	[A]
Bacteria	$\dot{x}_1 = (Y_1q_1 - a_1r_1)x_1 + q_xy$	[B]
r-Variable	$\dot{r}_1 = \beta_1Y_1q_1\left(\frac{s - s^*}{K_{r_1} + s - s^*} - r_1\right)$	[C]
Fungi	$\dot{x}_2 = [Y_2(q_2 + q_3) - a_2r_2]x_2$	[D]
r-Variable	$\dot{r}_2 = \beta_2Y_2(q_2 + q_3)\left(\frac{s}{K_{r_2} + s} - r_2\right)$	[E]
Exometabolites	$\dot{H} = \lambda_1q_1x_1 + \lambda_2q_2x_2 + \phi q_xy - q_3x_2$	[F]
Microbovore protozoa	$\dot{y} = (Y_xq_x - a_yr_y)y$	[G]
r-Variable	$\dot{r}_y = Y_xq_x\left(\frac{x - x^*}{K_x + x - x^*} - r_y\right)$	[H]
Specific rates of substrate consumption	$q_1 = r_1Q_1\frac{s - s^*}{(K_1 + s - s^*)(1 + H/K_H)}$, if $s \geqslant s^*$, else $q_1 = 0$ $q_2 = r_2Q_2\frac{s}{K_2 + s}$, $q_3 = r_2Q_3\frac{H}{K_3 + H}$ $q_x = r_yQ_x\frac{x_1 - x_1^*}{K_x + x - x^*}$, if $x \geqslant x_1^*$, else $q_x = 0$	[I]
CO_2 evolution rate	$v_{resp} = (1 - Y_1 - \lambda_1)q_1x_1 + (1 - Y_2 - \lambda_2)q_2x_2 + (1 - Y_2)q_3x_2 + (1 - Y_x - \phi)q_xy$	[J]

and (c) bacteria consume glucose with a concomitant formation of extracellular inhibitory products such as organic acids, which the fungal population utilizes as substrates. These assumptions were sufficient to obtain a reasonable agreement with the experimental points. The simulations revealed two distinct stages in microbial succession. The pioneer stage was dominated by bacteria, which consume most of the added glucose. Intensive r-selection was indicated by very high values of β

(Eqn. [C], Table 6.3), while fungi were not influenced by r-selection ($\beta = 1$, Eqn. [F], Table 6.3). After 7–10 days the microbial community enters the successional stage of K-selection characterized by slow growth and almost zero grazing rates. As the carbon from the added glucose was immobilized in microbial biomass, then the only nutrient source should be derived from the slow decay of bacteria and soil organic matter.

In another soil, which was under 'bare fallow', the first stage of explosive growth of r-selected bacteria and protozoa was absent (Figure 6.7). The non-trivial features of the observed dynamics was that glucose uptake was uncoupled from the apparent microbial growth, i.e. substrate was consumed in 2 days while the maximal biomass accumulation was attained between the 9th and 14th days. Moreover, CO_2 evolution not only lagged glucose uptake but also was considerably higher than may be accounted for solely by the soil amendment (from the instant indicated by the arrow, the cumulative amount of evolved CO_2-C exceeded the total C content of added glucose). Finally, contrary to fresh soil, bacteria revealed slower growth than fungi, attaining a maximal population density toward the end of the experiment.

All the observed unusual details could be easily explained if we were to assume that most of soil bacteria behave like *Arthrobacter* or *Prosthecomicrobium* in a batch culture. The uncoupling of growth from substrate uptake is no more than false diauxie. To perceive the similarity we may not rely on biomass (microscopic counting is too crude) but instead consider the hour-scale of glucose consumption (see insert to Figure 6.7). This was characteristically S-shaped, indicating the filling of storage compartments rather than exponential growth. The absence of protozoan grazing activity is the obvious result of the highly dispersed spatial distribution of K-selected bacteria as potential prey (to travel from one bacterial microcolony to another amoeba should spend more energy than they could gain from the feeding). Thus, 'bare soil' harbours a climax microbial community dominated by K-selected bacteria, while r-selected species have almost completely perished during the 2-year deprivation of fresh plant material.

These results incite a new interpretation of the nature of the 'priming effect' exerted by glucose on C mineralization in soil. Here it was manifested as CO_2 production in excess of added carbon. The traditional explanation of priming as the co-oxidation of humic substances, optimization of the C/N ratio, accumulation of extracellular enzymes, etc. (Kunc, 1975; Parnas, 1976) appears inadequate. Indeed, these mechanisms should be operative in any soils independent of past history, while here we detected considerable priming only in soil containing a climax microbial community. Hence this effect should be immediately related to the physiological features of the K-selected species and in particular to their ability to scavenge volatile organic substances from the air. In addition, there could be significant production of CO_2 resulting from glucose-induced activation of dormant cells which are especially abundant in bare soil.

6.2.2 CONTINUOUS SUPPLY OF READILY AVAILABLE SUBSTRATE

We used aerated columns packed with soil with a continuous delivery of substrate in the form of either glucose-containing nutrient solution or vapours of volatile organic substances

Soil columns continuously fed by solutions were claimed to be an advancing technique for soil microbiological studies (Macura, 1961; Hattori, 1973). During steady-state operation there should be intensive unlimited microbial growth accompanied by r-selection in the column top and substrate-limited growth with K-selection in the bottom. So, the temporal sequence is expected to be resolved in the spatial dimension. However, we found that the long-term performance of such a column is impossible due to the degradation of soil crumbs. A more natural way of continuous substrate delivery was achieved using volatile compounds and gases.

As a matter of fact, there are plenty of diverse volatile organic substances (VOSs) in the soil air. They originate from microbial fermentation (many alcohols, fatty acids, ketones) and plant secondary metabolism. One of the commonest fermentation products is ethanol, and since it is also a major volatile component in root exudates it was selected as a typical VOS. Soil was exposed to the VOS either in a continuously ventilated column (Figure 6.1, b) or in static chambers with soil incubated over VOS solutions (Figure 6.1, c). Microbial growth was followed by recording the respiration rate and the ethanol uptake rate, v_s, by gas chromatography, from the decrease of its concentration in the output air compared with the input air stream (column) or from the loss of its concentration in solution (static chambers). To keep the VOS flux constant, the solutions were replaced as soon as the concentration fell below 90–95% of the initial value.

The results are shown in Figure 6.8. An essential feature of VOS consumption is very strong substrate inhibition (curve 1). The yield factor found from the relationship between q_{resp} and v_s was 0.28 g C g^{-1} C, which is equivalent to a rather low energetic yield, $\eta = 0.19$. It is close to values exhibited by pure cultures slowly growing on glucose (sections 3.2.2, 3.2.3). Another characteristic feature is growth control by the mass-transfer of VOS from gas to liquid phases. We assumed that microbial cells dwell in soil water films and that mass-transfer obeys 'stationary liquid film theory' (Pirt, 1975). The model derived from these assumptions (Table 6.4) was used to estimate the mass-transfer coefficient, Ψ, by recording the dynamics of CO_2 and ethanol in the output air after substrate shutoff

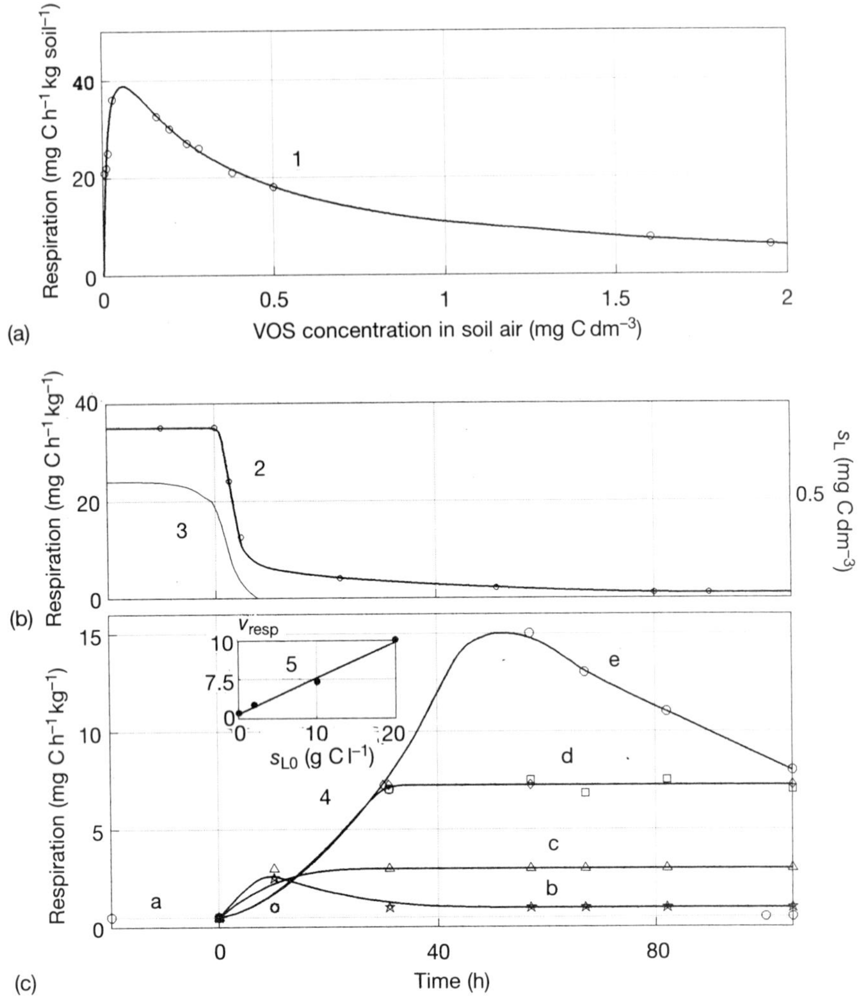

Figure 6.8 Microbial growth in soil continuously supplied with ethanol via gas phase (Panikov, 1991a). (a) Curve 1, dependence of soil respiration on ethanol concentration in soil air. Incubation was performed in a ventilated column; the curve was fitted to Eqn. [F] (Table 6.4) with parameters $K_s = 0.016$ and $K_{ss} = 0.22$ mg C dm^{-3}, $Q_s = 60$ mg C h^{-1} kg^{-1} soil. (b) Curve 2, non-steady-state measurement of ethanol gas to liquid transfer rate. Respiration dynamics was recorded before and after stopping at $t = 0$ of volatile organic substrate (VOS) supply keeping constant air flow rate ($F = 58$ ml h^{-1}). The best-fit value of Ψ_{VOS} was 10 h^{-1} (Table 6.4); curve 3 is simulated ethanol concentration in soil solution. (c) Curve 4, microbial growth followed as respiration dynamics of soil incubated in static chambers (Figure 6.1, 3) over ethanol solutions of various concentration (s_{L0}, g Cl^{-1}): a, 0; b, 0.2; c, 2.0; d, 10; e, 20. Curve 5 shows the relationship between steady-state respiration rate and s_{L0}.

Table 6.4 Simulation of microbial growth on volatile organic compounds entering soil column

Variable	Equation	
VOS in soil air	$\dot{s}_g = \frac{F}{V_g}(s_{0g} - s_g) - \Psi_{VOS}\left(s_g - \frac{s_L}{H_{VOS}}\right)$	[A]
VOS in soil solution	$\dot{s}_L = \Psi_{VOS}\left(s_g - \frac{s_L}{H_{VOS}}\right)\frac{V_g}{V_L} - q_s\frac{x}{W}$	[B]
CO_2 in soil solution	$\dot{p}_L = (1 - Y)q_s\frac{x}{W} + \frac{\phi}{W} - \Psi_p\left(p_L - H_p p_g\right)$	[C]
CO_2 in soil air	$\dot{p}_g = \frac{F}{V_g}(p_{0g} - p_g) + \Psi_p(p_L - H_p p_g)$	[D]
Microbial biomass	$\dot{x} = Y(q_s - a)x$	[E]
Specific rate of substrate (VOS) uptake	$q_s = \frac{Qs_L}{K_s + s_L + s_L^2/K_{ss}}$	[F]

F, air flow rate ($cm^3\ h^{-1}$); V_g and V_L, volume of soil gas and liquid phase respectively ($cm^3\ g^{-1}$); W, water content per unit dry soil ($cm^3\ g^{-1}$); H, Henry's constant (ml/ml); Ψ, mass transfer coefficient (h^{-1}), subscripts VOS and p denote respectively VOS and CO_2; ϕ, background soil respiration (mg C $h^{-1}\ g^{-1}$ soil). VOS, volatile organic substances.

(curves 2 and 3). Surprisingly, the identified Ψ-value was rather high ($\sim 10\ h^{-1}$) and was comparable with that in shaking flasks.

The long-term growth dynamics of ethanol-utilizing microbial populations were followed for various sizes of VOS fluxes (curves 4 and 5). Small fluxes provided a rapid establishment of steady-state respiration rates, which implies slow linear microbial growth. Higher VOS fluxes induced 1–2 days' exponential increase of v_{resp}, indicating unlimited growth of responsive ethanol-resistant populations. Their amount in terms of active biomass (Eqn. [6.3]) was as high as $x_a = 2.43\ \mu g\ C\ g^{-1}$ soil or 4% of the glucose-assimilating microorganisms. Microbial growth was completely suppressed in the air nearly saturated with ethanol ($s_{L0} = 200\ g\ l^{-1}$).

Thus, a continuous low-rate supply of VOS to soil encourages the growth typical for a 'dispersion zone' of soil, i.e. slow and linear, with a steady-state respiration rate, and without biomass elimination (we did not detect protozoan development) since microbial populations remain sparse.

6.2.3 THE POLYMERIC COMPOUNDS

In this section our primary focus was the decomposition of plant litter and its major component, cellulose. In contrast to monomeric substrates, cellulose and other plant polymers must be transformed beyond the cells by extracellular enzymes, giving rise to a high diversity of interactions between the members of the decomposition community. The second specific feature stems from the fact that moribund plant tissue is severely unbalanced with respect to elemental composition, i.e. in decomposition hot-spots the microbial community encounters excessive amount of C-compounds and a deficiency of other elements, N, P, Mg and Fe. This obstacle may be overcome only by the 'cooperative efforts' of the entire community. Our task was to decipher the mechanism of these cooperative interactions in kinetic terms.

Kinetics of cellulose decomposition

During soil incubation with cellulose (cellophane, filter paper, cotton cloth) we followed the dynamics of residual substrate by dry weight, CO_2 evolution rate and microbial biomass determinations (Panikov *et al.*, 1984). Different soils displayed variations not only in the rates of decomposition but also in the dynamic patterns. Paradoxically, the most essential factor which accounts for the greatest part of the variability was not the basic soil properties (texture, vegetation, soil type) but conditions of the soil samples preservation. Fresh samples taken directly from the field or carefully preserved under low temperature at field moisture were characterized by very slow zero-order decomposition kinetics (Figure 6.9, curves 'a').

Soil dried and remoistened before incubation exhibited an explosive type of decomposition dynamics. That is, during the first 10–20 days of incubation, cellulose was decomposed almost completely, and the subsequent release of breakdown products induced intensive growth and respiration of microorganisms. A decline in activity was associated with the depletion of added substrate (Figure 6.9, curves 'b'), although in some soils this occurred earlier and with a rather high residual substrate.

To simulate the observed dynamic pattern we tested several mechanistic considerations (overgrazing, formation of autoinhibitors, autoselection of cellulase producers, differentiation of biomass into active and dormant parts), from which only one was successful. This model (Table 6.5) has two features: (a) it allows for starvation which results from a lack of some non-carbon nutrients (we have tentatively identified this as the available N-

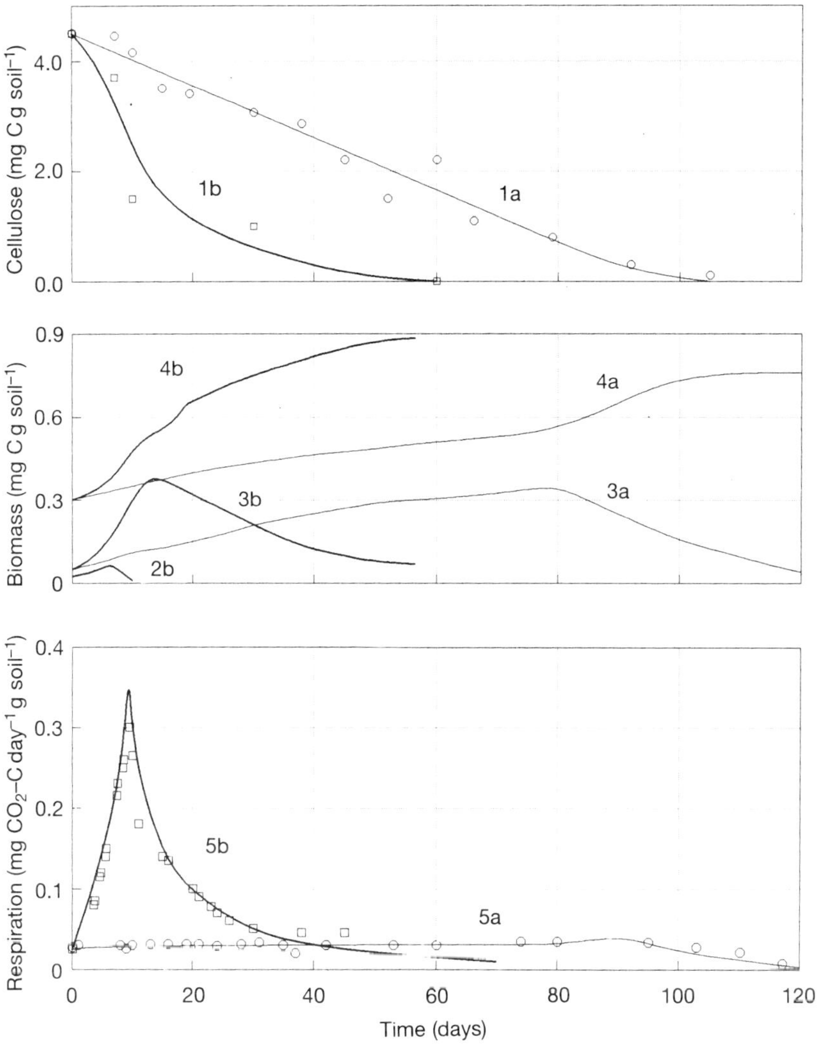

Figure 6.9 Cellulose decomposition in soil as related to population dynamics of microbial populations (Panikov, 1992). Cellulose as cellophane was added to fresh samples of alluvial meadow soil (a) or to soil kept under air-dried state and remoistened 5 days before onset of incubation experiment (b). Dynamic variables: curve 1, residual cellulose; curve 2, simulated biomass of hydrolytic; curve 3, copiotrophic; curve 4, dissipotrophic populations; curve 5, CO_2 evolution rate. Simulation was performed with model (Table 6.5) and the following parameters: specific rates (d^{-1}) $k_{cat} = 40$, $Q_1 = 1.2$, $Q_2 = 2.0$, $Q_3 = 0.5$, $\Psi = 0.5$, $a_1 = 0.1$, $a_2 = 0.25$, $a_3 = a_5 = 0.04$, $a_4 = a_6 = 1.0$, $M = 0.5$; concentration terms (mg kg^{-1} soil), $K_1 = K_2 = 100$, $K_3 = 1.0$, $K_N = 0.1$, $K_s = 300$, $N^* = 0.05$, $E^* = 1.25$; stoichiometric conatants (g g^{-1}), $Y_1 = Y_2 = 0.4$, $Y_3 = 0.2$, $Y_N = 0.05 + 0.1F(N)$, $\sigma_1 = 0.1$, $\sigma_e = 0.16$, $\sigma_2 = 0.05 + 0.1r$, $\theta = 0.2$. $V_1/V_2 = 0.1$.

source); and (b) it discloses the internal structure of the cellulolytic microbial community. It was assumed that the community consists of three microbial populations: (a) hydrolytic microbes producing extracellular cellulases; (b) r-selected copiotrophic organisms capable of explosive growth and N_2 fixation; and (c) K-selected populations capable of scavenging monomeric products diffusing from decomposition hot-spots (so-called dissipotrophic organisms). In addition, there are long-lasting enzymes immobilized on soil particles which provide a background level of depolymerizing activity.

The chain of dramatic events during decomposition is as follows. The addition of cellulose to soil induces the formation of short-living microbial cellulases and a build-up within the hot-spot of degradation products, mono- and oligosaccharides. However, intensive microbial growth on these products is ephemeral and stops as soon as N becomes depleted. The first casualties of N-starvation are hydrolytics which move to a latent state close to 'substrate-accelerated death' (vulnerability to this kind of stress was revealed and examined in section 3.4.1). The second group of the microbial community transiently avoids starvation by switching to N_2 fixation, but this metabolic process is also ephemeral and persists only as long as there is enough available C-substrates to fuel the expensive maintenance of nitrogenase activity. The depletion of accumulated monomeric C-products results in double energy- and N-starvation preventing further activity. The third participant, the group of dissipotrophic microorganisms, is not affected by N-starvation because of their location beyond the decomposition hot-spot. Thus, growth of these organisms proceeds slowly and evenly at the expense of diffusing monomeric products. Thus, the final stages of cellulose decomposition is carried out by the combined action of accumulated short-living and stable extracellular enzymes as well as dissipotrophic microflora. It may proceed rapidly up to complete degradation if residual activity is preserved at a high level. If not (considerable inactivation of cellulases), then cellulose fragments may persist for a longer time. It is a fascinating fact that normally soils are in this state of permanent arrest of cellulase-producing activity.

Verbal narrative is easily translated into mathematical language (Table 6.5). This model is similar to the former one (Table 6.3), being augmented with new variables: enzymes, monomeric products and available N. The threshold concentration N^* triggers the transition from one degradation stage to another. In addition there are two spatial compartments, the degradation loci and the surrounding dispersion zone.

The simulation agrees with experimental data (Figure 6.9), the difference in the response of fresh and air-dried soils being accounted for by the initial conditions, i.e. in fresh soils there are no active hydrolytic organisms. Drying–rewetting activates hydrolytic microbes for some unknown reasons and a similar but much smaller effect is exerted by the addition to

Table 6.5 Kinetic model of cellulose decomposition in soil

Variable	Equation	
Polymeric cellulose	$\dot{s} = -q_s E, \quad q_s = k_{cat} \dfrac{s}{K_s + s}$	[A]
Breakdown products in decay hot-spot	$\dot{L}_1 = q_s E - q_1 x_1 - q_2 x_2 - \Psi \left(L_1 - L_2 \dfrac{V_1}{V_2} \right)$	[B]
	$q_1 = Q_1 \dfrac{L_1}{K_1 + L_1}, \quad q_2 = r Q_2 \dfrac{L_1}{K_2 + L_2}$	
Breakdown products in dispersion zone	$\dot{L}_2 = \Psi \left(L_1 - L_2 \dfrac{V_1}{V_2} \right) - q_3 x_3, \quad q_3 = Q_3 \dfrac{L_2}{K_3 + L_2}$	[C]
Cellulase	$\dot{E} = \theta Y_1 q_1 x_1 - a_e (E - E^*), \quad a_e = \begin{cases} a_1, \text{ if } N \geqslant N^* \\ a_2, \text{ if } N < N^* \end{cases}$	[D]
Available nitrogen	$\dot{N} = \begin{cases} -\sigma_1 \dot{x}_1 - \sigma_2 \dot{x}_2 - \sigma_e \dot{E}, & \text{if } N \geqslant N^* \\ r(a_5 + a_6)\sigma_2 x_2 - q_N x_2, & \text{if } N < N^* \end{cases}$	[E]
	$q_N = \dfrac{r Q_N (N - N^*)}{K_N + N - N^*}$	
Biomass of hydrolytic population	$\dot{x}_1 = \begin{cases} (1 - \theta) Y_1 q_1 x_1 - a_3 x_1, & \text{if } N \geqslant N^* \\ -a_4 x_1, & \text{if } N < N^* \end{cases}$	[F]
Biomass of copiotrophic population	$\dot{x}_2 = \begin{cases} Y_2 q_2 x_2 - a_5 r x_2, & \text{if } N \geqslant N^*, \\ Y_2 (q_2 - M) x_2 \quad a_5 r x_2, & \text{if } N < N^*, q_2 \geqslant M \\ (Y_n q_N - a_6 r) x_2, & \text{if } N < N^*, q_2 < M \end{cases}$	[G]
Biomass of dissipotrophic population	$\dot{x}_3 = Y_3 q_3 x_3$	[H]
r-Variable for copiotrophic	$\dot{r} = \dfrac{d \ln x_2}{dt} \min \begin{cases} F(L_1) = L_1/(K_1 + L_1) \\ F(N) = (N - N^*)/(K_N + N - N^*) \end{cases}$	[I]

k_{cat}, catalytic rate constant of cellulase; Q_1–Q_3, maximum specific uptake rates; K_1–K_3, K_s, K_N, saturation constants; a_1–a_6, specific turnover rates; Y_1–Y_3, Y_N, yields; σ_1, σ_2, σ_e, N content; N^*, E^*, threshold concentrations; Ψ, mass-transfer coefficient for monomeric decay products; V_1, total space volume of decomposition loci; V_2, the same for dispersion zone; θ, cost of cellulase formation; M, maintenance of nitrogenase integrity.

the soil of mineral mixture. Under natural conditions activation and growth of hydrolytics probably takes place sporadically in the manner which was simulated for bacilli at spring-thaw time in tundra soil (Chapter 5).

Plant litter decomposition as related to the latent state of hydrolytic organisms

Simulation of cellulose decomposition prompts a new interpretation of available data on the transformation of plant debris naturally occurring in soils. Many forest arctic and boreal ecosystems are characterized by extremely slow decomposition, resulting in the accumulation of organic soil layers, such as forest litter and peat deposits. The usual eclectic explanation of this phenomenon is the ambiguous 'combined action of various factors', including low temperature, anoxia, toxic effects of plants phenolic substances, lack of active microbial germs, etc. However, careful examination of these factors reveals an inconsistency in the logic, since the adaptive capabilities of microbes to extreme environments are practically unlimited (psychrophily, anaerobic metabolism, resistance to toxins, etc.). The only insuperable obstacle is the low ash content in dying plant tissue. Here microbes fail, since their adaptive strategies typically rely on the intensification of metabolic performance under extreme conditions, which need to be supported by enlarged fluxes of nutrients.

The limited hydrolytic activity of a natural community stems from the long-term coevolution of microbes and higher organisms. Chemically balanced plant tissue would be too vulnerable for microbial invasion. In natural ecosystems all plant organs, including 'fleshy' fruits, sweet berries, juicy rhizomes, etc., preserve immunity towards hydrolytic action by many methods, not omitting the wide ratio between available C-compounds (mainly sugars) and conserved elements, N, P, K and Mg (Guzev, 1988). Under these conditions microbial cells encounter catabolic repression of depolymerase biosynthesis and, moreover, enter a latent state close to the phenomenon of 'substrate-accelerated death' (SAD). Catabolic repression and SAD have some common molecular mechanisms, e.g. they are both controlled by the intracellular content of cAMP, but they are distinctly differentiated in terms of cell physiology and kinetics:

Catabolic repression	*Substrate-accelerated death*
Arrest of biosynthesis of one particular enzyme or cluster of enzymes involved in the transformation of a less preferential substrate	General decline of the entire cell biosynthesis; decrease in the intracellular content of all P-components

Repression is reversible	Repression is almost irreversible
Cells preserve high metabolic activity	Metabolic activity is close to zero
The final result is the optimization of cell performance in a polysubstrate environment (r-type of strategic response)	The final result is the exclusion of cell population from active metabolism for a long time (L-strategic type of survival)

The second problem to be considered is the effect of the heterogeneity of plant litter as compared with individual polymeric substances such as starch or cellulose. It is customary to view decomposition as a number of consecutive steps in microbial attacks, each step being associated with the uptake of less and less preferential substrates. In fact it is too literal, being borrowed from the classic diauxie of *E. coli* on a mixture of sugars, but in any case this idea is welcomed as an explanation of the progressive decline in the plant litter decay rate. This explanation is not consistent for several reasons. Firstly, deceleration is observed in the decomposition not only of complex substrates but also of chemically pure cellulose and other individual compounds. Secondly, chemical analysis usually reveals that structural components of plant litter are decomposed in parallel rather than in a successive manner (Wessen and Berg, 1986). Thirdly, frequently less available components (e.g. lignin) could not be transformed separately from a more accessible moiety (e.g. cellulose) in terms of degradation biochemistry (Golovleva and Leontievsky, 1990).

At the same time we may classify plant litter components with respect to kinetic principles in the following way.

1. Low-molecular weight substances which do not need to be transformed outside the cells before consumption (monosaccharides, amino acids, nuclesides, many phenolic compounds, etc.).
2. Macromolecular chemical species with balanced elemental composition, i.e. having a high contribution of elements other than C (proteins, nucleic acids, lipo- and glycoproteides).
3. Structural insoluble and 'low-ash' macromolecular C-compounds of unbalanced composition (lignocellulose, hemicellulose, pectin, lipopolysaccarides).

Compounds from the first category are readily available and often repress the biosynthesis of extracellular enzymes. Macromolecular compounds of the second group are more available than compounds of the third because they contain the major growth elements and repression of the respective hydrolytic enzymes should not be irreversible.

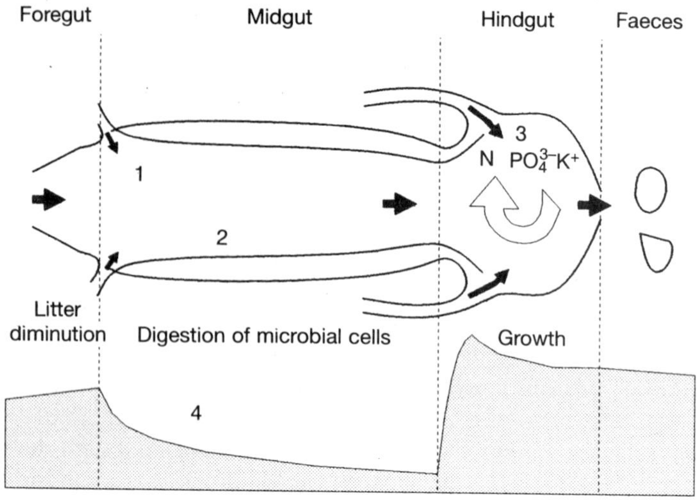

Figure 6.10 Schematic representation of the diplopod gut as a microbial habitat. 1, Site of digestive enzymes secretion; 2, peritrophic membrane; 3, malpighian tubules excreting mineral compounds and uric acid into the hindgut space; 4, hypothetical distribution of microbial biomass along the gut indicating considerable digestion of some microorganisms in the midgut and multiplication of other microbial populations in the hindgut and faeces. (Adapted from Byzov *et al.*, 1993.)

6.3 MICROBIAL GROWTH ASSOCIATED WITH SOIL INVERTEBRATES

Heterotrophic microorganisms and saprotrophic invertebrates partly overlap in their metabolic activity during the decomposition of plant litter. In recent years this fact induced vigorous discussions on the issue of which contribution is greater. Nowadays, it is clear that bacteria and invertebrates in the soil are in no way competitors, rather their interrelationships are of the 'prey–predator' or 'host–parasite' kind. However, the most interesting case is their positive cooperation during the decomposition of plant litter.

The model

The control exerted by animals on microbial growth is illustrated by Figure 6.10. Swallowed by animals, plant litter undergoes mechanical disruption and diminution which greatly accelerates its subsequent digestion in the midgut. A significant proportion of the soil microorganisms that are delivered with the food to this part of the intestine are inactivated and lysed by the animal's digestive enzymes. In contrast to the midgut, the hindgut provides very favourable environmental conditions for microbial growth, i.e. permanently high moisture, neutral pH, elevated concentrations of mineral compounds, uric acid, carbohydrates and other nutrients. It is interesting that the digestive tract of various soil invertebrates harbours not only specific symbionts (they are beyond our scope), but also allows the growth acceleration of free-living microorganisms normally occurring in soil or in plant litter, e.g. *Pseudomonas, Flavobacterium, Vibrio, Enterobacter, Streptomyces* (Szabo, 1974; Marialigeti, 1979; Contreras, 1980).

The growth of these populations in the intestine is continuous because the feeding of animals normally proceeds with very short pauses. Most of the growing cells are subjected to continuous removal with excrements, the probable exception being those microbial cells retained in some pocket (e.g. the peritrophic membrane on Figure 6.10). Finally, due to the peristaltic motion, the content of the gut is more or less homogeneous. The combination of all these features is characteristic for a chemostat type of open cultivation system (1aα, Table 2.1). The closest approximation in the case of diplopods would be chemostats in series with partial biomass feedback (Table 1.3), each stage corresponding to particular section of the gut. In the case of earthworms having a long and narrow digestive tract, an even closer approximation would be a tubular culture. However, for the sake of clarity and in view of the limited experimental data we prefer the conventional one-stage chemostat approximation. Then,

$$\begin{array}{ccccccccc} \text{net increase} & = & \text{input} & - & \text{removal to faeces} & + & \text{growth} & - & \text{digestion} \\ \frac{dx}{dt} & = & \frac{x_0}{\tau} & - & \frac{(1-\Omega)x}{\tau} & + & \mu x & - & \eta x \end{array} \qquad [6.6]$$

where τ is the gut transit time, (note that $1/\tau$ is equivalent to a chemostat dilution rate), x and x_0 are cell quantities in the animal's gut and food respectively, and Ω is the coefficient of food assimilation,

$$\Omega = \frac{F-E}{F} \approx \frac{R}{F}, \qquad [6.7]$$

where F is food consumption, E is excrement production and R is respiration. Under steady state, $dx/dt = 0$, the apparent growth rate, i.e. true growth less digestion is found as

$$\mu - \eta = \frac{(1-\Omega)\bar{x} - x_0}{\bar{x}\tau}. \qquad [6.8]$$

Table 6.6 Microbial growth and elimination in the gut of animals

Recorded group of microorganisms	Animals	x_0	$\bar{x}$	$\mu-\eta$ (day^{-1})
Total biomass by fumigation technique (mg C g^{-1})	*P. hercules*	15.7	7.8	−1.47
	A. officinalis	15.7	10.7	−1.18
	N. caliginosus	0.24	0.26	+30
Bacteria by direct count (10^9 g^{-1})	*P. hercules*	1370	2020	−0.31
	A. officinalis	1370	1930	+0.03
	N. caliginosus	3.3	2.4	−2.00
Bacteria developing on MPA (10^9 g^{-1})	*P. hercules*	89	116	+0.19
	A. officinalis	89	162	+0.29
	N. caliginosus	0.032	0.070	+2.61

MPA, meat–peptone agar.

The experimental test

This was carried out with invertebrates, common in a temperate zone, i.e. millipedes *Pachybrachiulus hercules* (Verhoeff), wood lice *Armadillo officinalis* (Dumeril), and earthworms *Nicodrilus caliginosus* (Sav.). Millipedes and wood lice were fed with fresh leaf litter of *Carpinus orientalis–Quercus pubescens*, while earthworms were reared in unamended meadow soil. The microbial biomass was determined in the food (x_0) and fresh excrements (assumed to be equal to $\bar{x}$) by kinetic methods (glucose-induced respiration), fumigation techniques (Jenkinson and Powlson, 1976), direct counting and plating on meat–peptone agar (MPA). To identify the parameter τ, the animal's food (litter or soil) were tagged with methylene blue and the time the tag appeared in the excrements was recorded. Transit time was 5 h in earthworms and 15–18 h in millipedes and wood lice. Feeding and excretion rates (F and E) were found from the dynamics of food (litter or soil) consumption and excrement formation respectively, by dry weight measurements. Respiration rates of the animals (R) was estimated by gas chromatography as a CO_2 evolution rate. The parameter Ω calculated from [6.7] varied from 0.02 (earthworms) to 0.22–0.27 (wood lice).

The results of the microbial enumerations (Table 6.6) indicate that in millipedes and wood lice the total growth balance of microorganisms is negative (elimination exceeds growth acceleration), the main digested component being fungi. All animals stimulated the growth of plateable bacteria, especially Gram-negative non-spore-forming populations. The most rapid multiplication was found for bacteria identified as *Pseudomonas mendocina* in millipedes and small motile unidentified rods in the intestine of earthworms, the latter having an unusually high specific growth rate,

$0.18\,h^{-1}$, for natural habitats (generation time < 3.8 h). Thus the digestive tract of animals provides conditions for intensive r-selection in favour of motile rapidly multiplying bacteria.

The projection of obtained results to natural habitats

When made in order to account for the true abundance and activity of mesofauna (Gorbenko *et al.*, 1986), this revealed that the overall growth acceleration and elimination of soil microorganisms by soil mesofauna is moderate in terms of the ecosystem's mass balance. That is, at best the production of bacteria in the guts of all invertebrates could be $1\,g\,m^{-2}$ per season as compared with the total seasonal production, $> 10^2\,g\,m^{-2}$. So, control exerted by animals is more essential in qualitative terms, as their guts are a reservoir for r-selected bacteria which rapidly grow and decline.

A much more significant contribution to the mass-flow of ecosystems is made by smaller animals, microarthropods and protozoa. Their respiration may transiently account for as much as 30% of the total respiration of soil biota mineralizing plant litter (Panikov and Simonov, 1986). Oribatides and collembolas are also the most active grazers in the litter (upper soil horizons). Here they have an obvious advantage over unicellular predators (protozoa), in transporting efficiency and the ability to attack large-size prey (fungi mycelium and rhyzomorphs).

6.4 MICROBIAL GROWTH ASSOCIATED WITH PLANT ROOTS

The rhizosphere and rhizoplane are the most densely populated soil quarters, harbouring the main part of heterotrophic microbial activity. This activity is fuelled by so-called rhizodeposition, exudation and root sloughing, which is the invisible component of the primary production. Its quantitative estimations, as well as the assessment of biomass, activity and growth control of root-associated organisms, have been the primary topics of 'rhizosphere biology' for the whole century. However, these basic problems still have not been solved. We will challenge this objective by using a kinetic approach and two methodological principles: (a) to measure the dynamic variables accounting for all the major components of the C-balance of the entire plant–microbe system; and (b) to apply a mathematical analysis to the obtained dynamic data.

Plant growth in mineral solution

This artificial system is considerably simpler than soil culture and was selected as the starting point for our investigations. The main advantage was the possibility of considering separately shoot, root and microbial biomass and their respiration. The isolation of the microbial mass was achieved by the gentle mechanical removal of bacterial films from the root surface with a subsequent determination in cell suspension of the respiratory activity and biomass by three techniques: a kinetic method with glucose as substrate, plating on bean-extract agar, and by a direct count. The combination of measured parameters and dynamic curves during the 2-month experiment is shown in Figure 6.11. The accuracy of the determinations was confirmed by satisfying the C-conservation condition: CO_2 photoassimilation + seed mass = total biomass increase (shoot + root + bacteria) + total respiration of plant and bacteria.

Microorganisms on the roots were mainly small motile rods and immobile cells forming biofilms. The three enumeration techniques that were used gave compatible results. The standing crop of bacterial biomass increased a little, from 0.1 at the start, to 0.6–1.8 mg C per plant at $t = t^*$, and then remained constant. However, bacterial respiration and the r-variable were rather high, indicating that the true growth was much more intensive than the apparent one. An abundant development of infusoria implied that grazing was responsible for this disparity. From independent measurements of the yield factor ($Y \sim 0.4$ g biomass per g glucose consumed) in the cell suspension removed from the roots, and of the respiration rate, we have found that μ, the specific rate of true growth, should be $0.7\ h^{-1}$. Hence, it takes exactly 1 day = $\ln 2/\mu$ for an average bacterial cell on the root surface to divide. This multiplication frequency is probably defined by the circadian rhythms of the plant secretory activity.

To summarize, we may conclude that plants provide a bacterial population with chemostat-type growth conditions (1aα, Table 2.1). Specifically, the observed dynamics may by approximated by the following model:

$$\dot{Z} = (G - R_1 - T)Z' - R_2(Z - Z'),$$

$$\dot{K} = (1 - \theta)TZ' - R_3K,$$

$$\dot{s} = \theta TZ' - q_s x,$$

$$\dot{x} = (Yq_s - E)x,$$

$$Z' = \begin{cases} Z, & \text{if } t \leq t^* \\ Z(t^*), & \text{if } t > t^*, \end{cases} \qquad [6.9]$$

where Z is total shoot phytomass, Z' is the mass of young actively growing shoots, K is root mass, G is true photoassimilation rate, T is the flux of photosynthetic products to the root system, R_1–R_3 are respiration rates of young shoots, older green tissue and roots respectively, θ is the fraction of exuded compounds from total C-flux directed to roots and E is the specific rate of microbial grazing. t^* denotes the time when plant growth becomes limited by some environmental or internal factor. At $t > t^*$

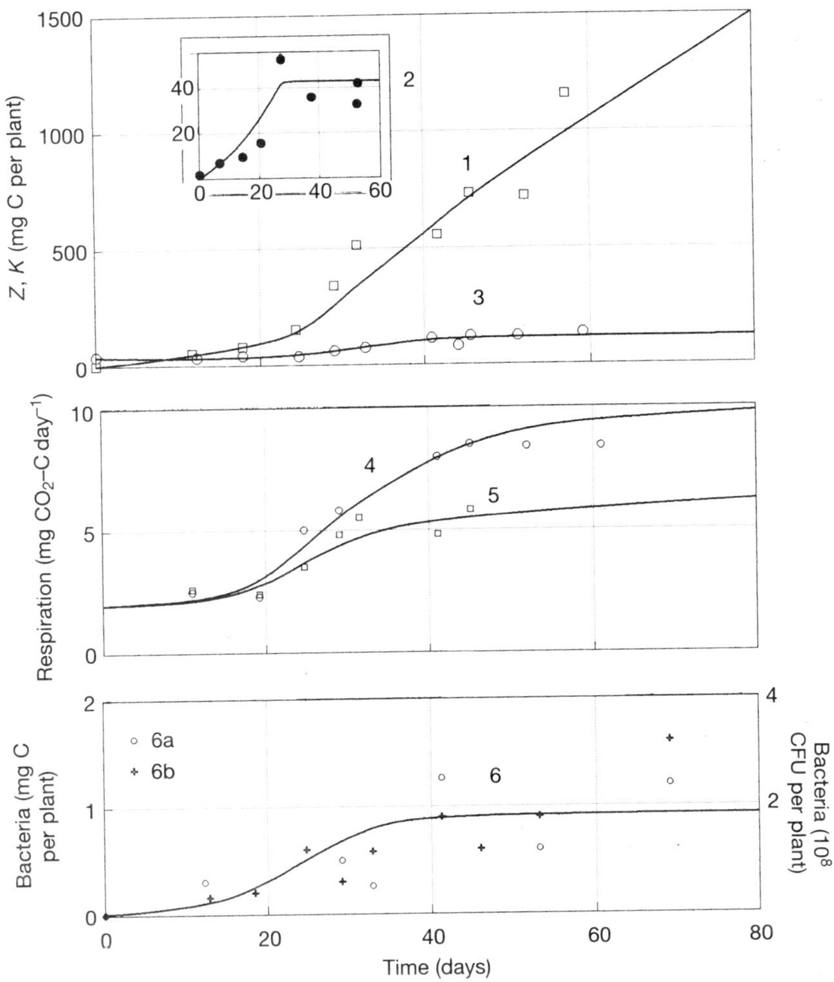

Figure 6.11 Dynamics of plant–microbe system developing on the liquid nutrient medium (Panikov *et al.*, 1987). The plants *Pisum sativum* were grown in the two-compartment chamber allowing estimation of CO_2 exchange rates separately for shoots and root system (Figure 6.1 (h)). The mineral nutrient solution (KNO_3, 1.0, K_2HPO_4, 0.25, KH_2PO_4, 0.25, $MgSO_4 \cdot 7H_2O$, 0.2, $CaCl_2 \cdot 2H_2O$, 0.02 g l^{-1}, total volume 15 l) was replaced every week. Illumination was 7000 lux; air ventilation rate was 2 l min^{-1}. Several identical chambers with 15–18 plants were installed; one was intact and used for continuous monitoring of CO_2 exchange, while 3–5 replicate plants from other chambers were picked each week for biomass determinations and enumeration of microorganisms. Dynamic variables: curve 1, shoots dry weight; curve 2, CO_2 photoassimilation rate; curve 3, roots dry weight; curve 4, respiration of roots + associated microrganisms; curve 5, root respiration after removal of microbial films; curve 6, microbial biomass by kinetic method (a) and plating (b). Curves were calculated according to Eqn. [6.9] with the following parameters: $G = 0.2$, $T = 0.06$, $R_1 = 0.02$, $R_2 = 0.003$, $R_3 = 0.05$ day^{-1}, $t^* = 28$ days, $\theta = 0.5$.

phytomass growth was assumed to be a function of the constant amount of active green mass, $Z'(t^*)$.

Identification of the model's parameters allowed the estimation of the intensity of root exudation. It is interesting that this quantity was strongly dependent on the composition of mineral solution, i.e. a full-strength mixture provided a root exudation equal to 15–20% of G and a deficiency of N, P or K increased this value to 30%.

Soil–plant microcosms

Two types of microcosms were used. The first was soil under barley placed in simple small pots (Figure 6.1(g)). Once a day two replicate pots were sacrificed and used for the determination of microbial biomass, protozoa and soluble organic matter (SOM). The phytomasses of shoots and roots were determined gravimetrically after drying at 70°C. Before drying, the dark respiration of shoots and screen-washed roots (DRS and DRR respectively) was estimated by gas chromatography as the CO_2 evolution rate during incubation at 25°C in 500-ml glass bottles. In parallel, two intact microcosms were covered by Plexiglass enclosures and changes in CO_2 concentration were measured by gas chromatography under alternate dark (v_{dark}) and light conditions (v_{ill}). The soil respiration activity (SRA) and plant photosynthetic activity (PPA) were calculated as:

$$\mathrm{PPA} = v_{\mathrm{dark}} + v_{\mathrm{ill}} - \mathrm{DRS}, \quad \mathrm{SRA} = v_{\mathrm{dark}} - \mathrm{DRS} - \mathrm{DRR}. \qquad [6.10]$$

To minimize chamber artifacts light exposure time was restricted to 3 minutes and PPA values were corrected for elevated CO_2 inside the chamber by the Michaelis–Menten equation.

The second microcosm was planted soil in a two-compartment chamber (Figure 6.1 (h)). Here we used a refined technique based on the continuous automatic infra-red recording of CO_2 concentration in two air streams, one passing through the shoot compartment and another over the constantly shaded root surface. PPA and SRA+DRR were estimated from the p.p.m.-range difference in CO_2 concentration as compared with ambient air.

The results obtained with the first microcosm are shown in Figure 6.12. The most essential features not observed in the experiments with liquid plant culture were oscillations in PPA, SRA, microbial biomass, protozoan numbers and SOM, and a monotonous pattern of phytomass and dark plant respiration. Bare soil did not show any oscillations in CO_2 evolution rates. After cutting the shoots, the soil respiration sharply decreased, damped oscillations being observed for no more than 1–2 weeks.

To simulate the observed dynamics we used a model derived from Eqn. [6.9], taking explicit account of protozoan dynamics and putting $\zeta = \theta T/G$, $\mathrm{f} = \mathrm{gz}$:

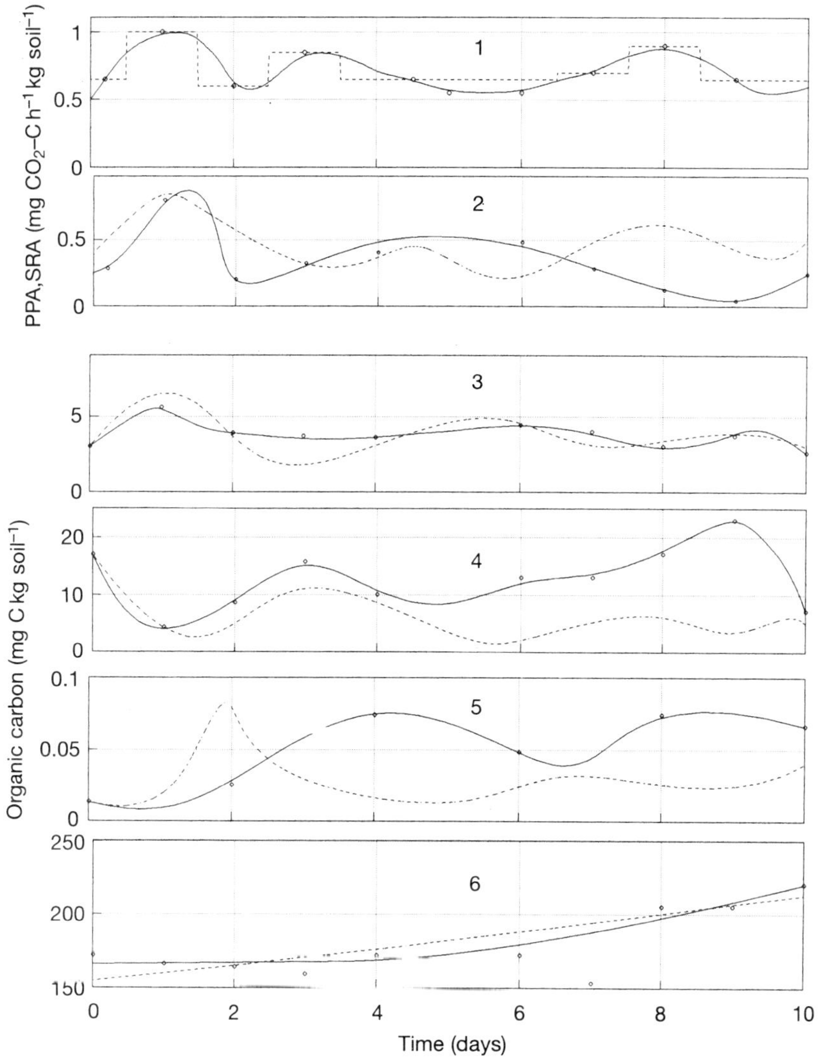

Figure 6.12 Dynamics of plant–soil microcosm (Panikov and Gorbenko, 1992). Barley plants were grown in beakers with 300 g of grey forest soil packed to bulk density 1.2 g cm^{-3} (Figure 6.1 (g)), under illumination of 10 000 lux (light period 14 h), room temperature and moisture 50% of WHC controlled by daily weighing. Dynamic variables: curve 1, plant photosynthetic activity (PPA); curve 2, soil respiration activity (SRA) by static chamber method; curve 3, microbial biomass by kinetic method; curve 4, soluble organic compounds by dichromate oxidation of soil water extract; curve 5, biomass of protozoa by direct count; curve 6, phytomass dry weight. The dotted line is the calculation according to Eqn. [6.10] with the following parameters: $\mu_m = 0.09$ and $a_y = 0.135\ h^{-1}$; $K_s = 13\ \mu g\ C\ g^{-1}$ of soil; $\eta = 0.035$ g $(\mu g\ C)^{-1} h^{-1}$; $\zeta = 0.5$, $Y = 0.2$ and $Y_x = 0.047$.

$$\dot{s} = \zeta F - q_s x\,, \quad q_s = \frac{Qs}{K_s + s}\,,$$

$$\dot{x} = \mu x - q_x y\,, \quad q_x = \frac{Q_x}{K_x} x\,, \quad \mu = Y q_s\,, \qquad [6.11]$$

$$\dot{y} = (\eta - a_y) y\,, \quad \eta = Y_x q_x$$

As shown by qualitative stability analysis, this model may generate autonomous sustained oscillations of x, s and y ('centre') only if $\zeta F > Q a_y K_x / Y_x Q_x$, i.e. in the unrealistic situation when microbial growth is not limited by substrate. Otherwise, oscillations are damped and all variables move towards stable steady-state values (a stable focus). However, sustained oscillations could be induced by external excitation sources. The response of Eqn. [6.11] to F periodic changes was studied in numerical experiments. The main results are summarized as follows: (a) SRA has the same dynamic pattern as PPA and is always delayed; (b) time-delays in response to PPA changes, increase in the sequence SRA ~ SOM < microbial biomass < protozoan number; (c) the amplitude of the induced oscillations depends on the amplitude of the excitation signals; and (d) the steady-state biomass value depends on the grazing activity of protozoa and does not depend on ζF values. All these predictions are in good qualitative agreement with the experimental data. The observed time-delays between PPA and other variables increased in the predicted order. Microbial biomass in the root vicinity and in the unplanted soil was virtually identical (zero 'rhizosphere effect' due to intensive grazing), although microbial respiration activity was 20–30 times higher.

To question the quantitative agreement we performed the numerical integration of Eqn. [6.11] with F-values, determined from day-average experimental points of PPA. As can be seen from Figure 6.12, there is adequate simulation of amplitude and frequencies of oscillations, but no close point-to-point fit. Disparity is obviously the result of the model's simplicity, i.e. it neglects diurnal variations and does not take into account the contribution of non-rhizosphere microorganisms, the input of SOM from sources other than root exudation, etc. Nevertheless, it appeared to be useful as a first-step approximation of the total mass balance and average fluctuating dynamic pattern. This gives us opportunity to make two comments:

1. The microbial growth rate in planted soil turned out to be rather high, with a generation time $g = (\ln 2)/\mu$ equal to 20–40 h. This finding contradicts the deeply-rooted belief in the extremely low growth rates of soil populations (see section 1.5).
2. The residence time of photosynthetic products in the whole of the

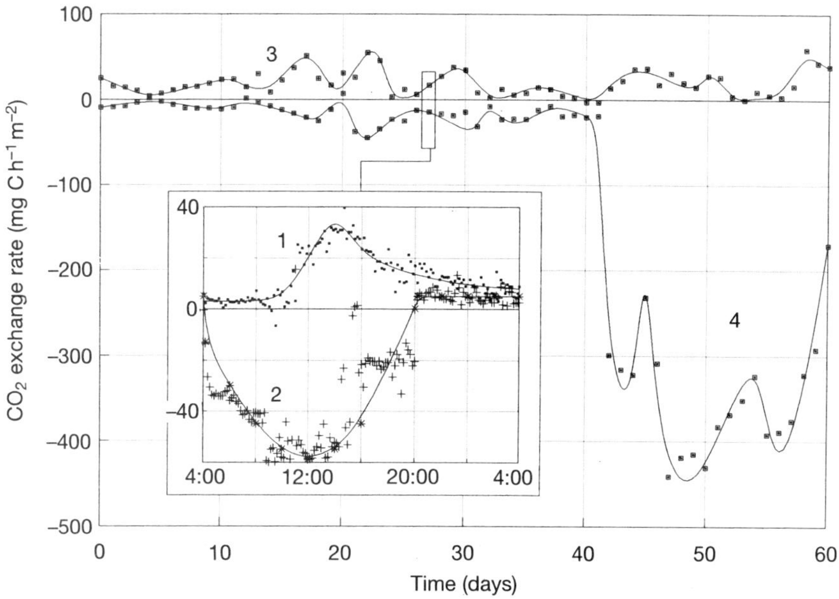

Figure 6.13 Short-term and long-term fluctuations in CO_2 exchange rates between soil–plant microcosm and atmosphere. Five rice plants were grown in two-compartment chamber (Figure 6.1 (h)) under combined artificial and natural illumination in a greenhouse without temperature control. The soil (ordinary chernozem, Voronezh region, 4.5 kg per chamber) was kept moist for the first 2 weeks and then submerged. The root and shoot compartments were continuously flushed with air, $1.5\,l\,min^{-1}$; CO_2 concentration in the output stream was recorded every 10 min by infra-red analyser (Infralyt4). CO_2 exchange rates were calculated as product of air flow rate and difference in CO_2 concentrations between input and output streams. Examples of diurnal cycles in CO_2 exchange rate for root (curve 1) and shoot compartments (curve 2). Long-term dynamics of mean daily rates of CO_2 evolution from root compartment (curve 3) and mean daily net photosynthetic activity of plants (curve 4).

studied system includes the time of (a) CO_2 uptake, (b) the transport of products from leaves to roots, (c) exudation, (d) uptake and oxidation by rhizosphere microorganisms, and finally (e) CO_2 emission from soil to atmosphere. Stage ‘a’ takes seconds, stages ‘b’ and ‘c’ should take several hours, while ‘d + e’ are expected to be longer. Calculated according to Eqn. [6.11], the turnover time of the variable s is $s/(q_s x) \sim$ 35 h. The very close value of 40 h was obtained by Warembourg (1975) who used $^{14}CO_2$-impulse labelling of plants and then followed label release to atmosphere from soil.

The results of the second experiments with the automatic recording of CO_2 exchange rates are presented in Figure 6.13. Diurnal cycles (curves 1 and 2) were characterized by a close relationship between SRA and PPA with 3–6-h time delays. These cycles were clearly observed early in the morning when PPA jumps from zero to submaximal value while SRA remained for 5–6 h at nocturnal low level. Maximal PPA was observed at mid-day while the highest SRA value was at 2 p.m. In the evening the decrease of PPA was quicker than the decline of SRA.

The daily average time-series for PPA and SRA exhibited fluctuations of roughly the same frequency as in the first experiment. Again, it was obvious that the two time-series are closely correlated in terms of the frequency of the quasi-periodic fluctuations. To analyse them quantitatively we used time-series statistics (Gorbenko and Kryshev, 1985). The contributions of the periodic components from two time-series was approximated by equation

$$z(t) = \bar{z} + \sum_{k=1}^{n} R_k \cos 2\pi f_k(t - \phi_k) + L_t , \qquad [6.12]$$

where $\bar{z}$ is the mean, R_k, ϕ_k and f_k are amplitude, phase and frequency, repectively, of a harmonic with index k, while L_t is residual noise. The fitting of data to this equation by a non-linear least-square procedure revealed an extremely complicated harmonic composition which included 31 significant ($P < 0.05$) components for PPA and 78 for SRA. The largest contribution to the total variance of the two time-series was provided by the following harmonics:

Period (days)	Fraction of total variance (%)	
	PPA	SRA
31.4	19.01	2.1
7.9	0.71	6.1
2.9	0.54	1.7
1.0	31.6	6.0

The next step in the statistical analysis was the estimation of time-delays between the PPA and SRA series by the calculation of the cross-correlation functions r_{xy} (τ):

$$r_{xy} = \frac{C_{xy}(\tau)}{S_x S_y}, \quad c_{xy} = \frac{1}{N} \sum_{t=1}^{N-\tau} (x(t) - \bar{x})(y(t + \tau) - y^*) , \qquad [6.13]$$

where x and y are means, $x(t)$ and $y(t + \tau)$ are the values of x at moments t and y at moments $t{+}\tau$ respectively.

The calculated correlation functions varied at best from 0.2 to 0.3, low values resulting from spectrum complexity. This analysis also confirmed the statistical significance of SRA *versus* PPA time-delays, which ranged from 8 hours to 3.4 days ($P < 0.05$).

Thus, we have confirmed the existence of a close relationship between PPA, on the one hand, and parameters related to microbial activity, on the other. However, to substantiate this conclusion we took a further step and analysed the field data.

6.5 FIELD OBSERVATIONS ON MICROBIAL DYNAMICS

Site and methods

The study was carried out at a field station in Pushchino, in the Moscow region (55°32′N, 37°39′E). The soil was arable grey forest silty clay loam, C_{org} 0.92%, N_{org} 0.11%, pH_{KCl} 5.5, tilled in early spring by a rotary cultivator to the depth of 10 cm, and seeded with barley (cultivar Trumph).

Over the course of 44 days, 15 dynamic variables were measured (Figure 6.14). Most of these determinations were made with the same techniques as in laboratory microcosm experiments, so here we shall outline only new elements. CO_2 fluxes were determined using static chambers (Conrad *et al.*, 1983) which consisted of two parts: a permanent square 40 × 40-cm stainless steel base permanently inserted into the soil and a removable Plexiglas cover installed only during the 15–30 min of flux measurements. The measurements were made under alternate dark and light conditions as described above. However, as well as PPA, calculated according to Eqn. [6.10], we considered separately the dark respiration of the whole soil–plant community (DRSPC), and measured soil respiration (SRA) directly between plant rows.

The CO_2 concentration in the soil air as well as other soil gases were determined by means of membrane probes (hollow perforated stainless steel tubes covered by polythene membrane) inserted into the soil at a depth of 10 cm. Bearing in mind the existence of diurnal cycles, all measurements were made at a fixed time, from 9.00 to 9.45 a.m.

Time-series analysis of observation data

Results are shown in Figure 6.14 as a time-series for the 15 variables. Similar to the microcosm experiment, all variables, except phytomass, exhibited frequent fluctuations with rises and falls every 3–6 days. At the same time the general trend observed for PPA, DRSPC and SRA was an increase parallel to plant growth.

The dynamics of each of the 15 time-series were fitted to Eqn. [6.12], statistically unreliable (at the level $P = 0.5$) harmonics were discarded, and the resulted curves plotted in Figure 6.14. Equation [6.12] explained

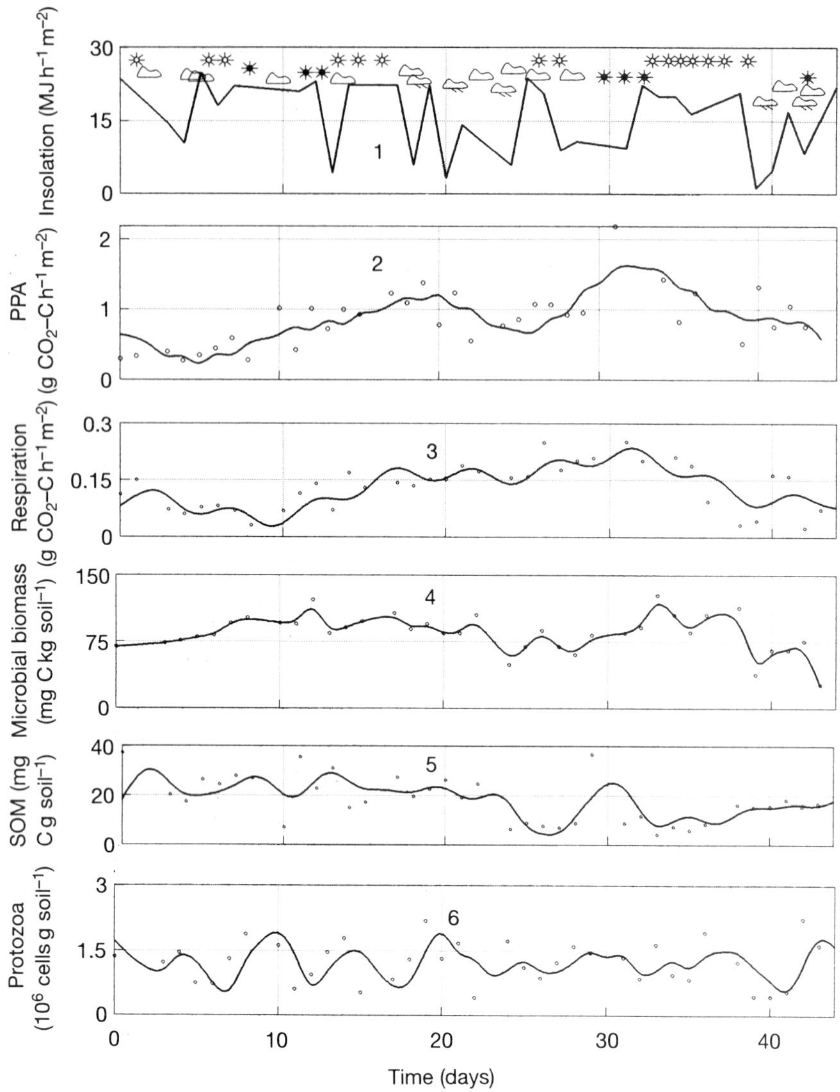

Figure 6.14 Seasonal dynamics of CO_2 exchange rates and related variables of arable grey forest soil under barley (Panikov and Gorbenko, 1992). Curve 1, insolation; curve 2, plant photosynthetic activity (PPA, static chamber technique); curve 3, dark respiration of soil–plant community (DRSPC, static chamber technique); curve 4, soil microbial biomass in soil (kinetic method); curve 5, soluble organic matter (SOM, dichromate oxidation of soil water extract); curve 6, protozoan population (direct count); curve 7, shoots phytomass dry weight;

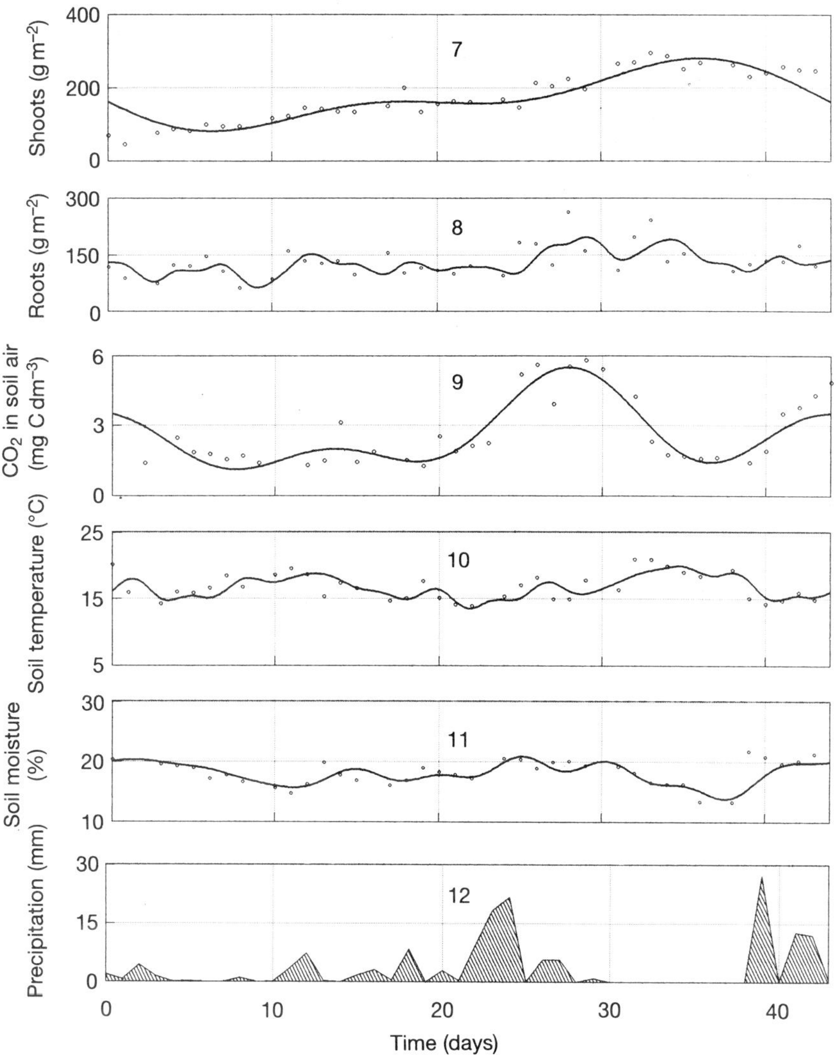

Figure 6.14 *Continued.* curve 8, roots phytomass dry weight; curve 9, CO_2 in soil air at depth of 10 cm (membrane probe technique); curve 10, soil temperature at depth of 2 cm (with thermoresistors); curve 11, soil moisture at depth 0–10 cm (gravimetrically); curve 12, precipitation (mm). Curves are calculated according to Eqn. [6.12] taking into account only statistically valid ($P = 0.05$) periodic components.

Table 6.7 Summary of time-series statistical analysis

Cluster	Variables	Correlation/delay*			
		1	2	3	4
1	1. SRA	×			
	2. CO_2 in soil air	0.83/0	×		
	3. DRSPC	0.65/−3	0.58/−3	×	
	4. PPA	0.55/−2	0.49/−3	0.60/0	×
2	1. Shoots	×			
	2. DRS	−0.83/0	×		
	3. SOM	−0.62/0	0.54/4	×	
	4. Roots	0.52/1	−0.42/−1	−0.55/−2	×
3	1. Microbial biomass	×			
	2. Moisture	−0.83/0	×		
	3. Temperature	0.51/0	−0.54/0	×	

* Negative delay, when row parameter leaves behind the heading parameters of the column. SRA, soil respiratory activity; DRS, dark respiration of shoots; PPA, plant photosynthetic activity; DRSPC, dark respiration of soil–plant community; SOM, soluble organic matter.

70–85% of the total variance, the rest being attributed to noise. The variables could be divided, with respect to harmonic composition, into two groups. One group is characterized by the predominance of a low-frequency component (periods 22 and 44 days), and the other by high-frequency components (periods from 2 to 14.7 days). To the first group belong all variables which are related to plant growth and CO_2-exchange, i.e. PPA, DRSPC, SRA, DRS and DRR, CO_2 in soil air and phytomasses. The second group includes microbial biomass, protozoan population and SOM.

The cross-correlation functions (Eqn. [6.13]) for all possible paired combinations of the 15 time-series were then calculated. The results of the cluster analysis are presented in Table 6.7. There are three clusters, the first formed by parameters PPA, DRSPC, SRA and CO_2, in soil air. PPA and DRSPC were correlated with each other without time delay, and both these variables were lagged by SRA and CO_2 concentration in soil air by 2–3 days. This is in excellent accord with laboratory data, confirming the operation of a functional relationship between the plant and microbial community under field conditions. It is interpreted as a chain of events. A temporary increase of PPA is accompanied by the simultaneous increase of root deposition with a subsequent acceleration of microbial respiratory activity. The time delay in this case corresponds to the residence time of CO_2 in the plant–soil system and is very close to that estimated from Eqn. [6.11] and the $^{14}CO_2$ data of Warembourg (1975).

The second cluster was formed by parameters which characterize the physiological state and biomass of the plant. The negative correlation of dark plant respiration with phytomass is the result of plant senescence. The inverse relation between phytomass and SOM gives evidence for a decrease in root exudation during the later stages of plant development. It is essential that SOM is linked with plant-derived parameters, but not with those of soil.

The third cluster was formed by microbial biomass, temperature and moisture of soil. The negative correlation between temperature and moisture was expected, due to increased evaporation under warm conditions. The inverse relation between microbial biomass and soil moisture content could be tentatively explained by a higher grazing activity of protozoa under moist conditions.

The origin of fluctuations

Now we can outline the main results of the time-series analysis of the field data. This confirmed the existence of periodic components in the ecosystem's dynamics, supporting our premise that it is plant activity which induces rhythms of microbial growth. Contrary to earlier hypotheses explaining microbial fluctuations by specific metabolites and grazing, this concept does allow for the **synchronization** of jumps and falls in microbial activity in the numerous spatially isolated soil microhabitats. Their interconnection is provided by the highly branched root system which penetrates the entire soil mass. Diurnal cycles of PPA and root exudation force rhizosphere bacteria to divide synchronously approximately once a day. Day-to-day variations in rhizodeposition probably have an effect on the amount of soil microsites with elevated content of available organic substrates. The final result is the generation of fluctuations in biomass and activity of microorganisms and subsequently of protozoa and other microbovores.

The last question to be answered is, what is the cause of PPA fluctuations? These factors are presently not well understood. Firstly we noted that oscillations of SRA (and therefore PPA) take place synchronously in distinct locations (Figure 6.15). So oscillations of PPA are probably not autonomous but are induced in different places by a common 'master' oscillator. Secondly, as was indicated by time-series analysis, none of the known environmental parameters (moisture, temperature, illumination) has a significant effect on the **periodicity** of CO_2-fluxes (not to be confused with the absolute values of the fluxes). And thirdly, fluctuations were observed not only outside, but also in laboratory experiments under con-

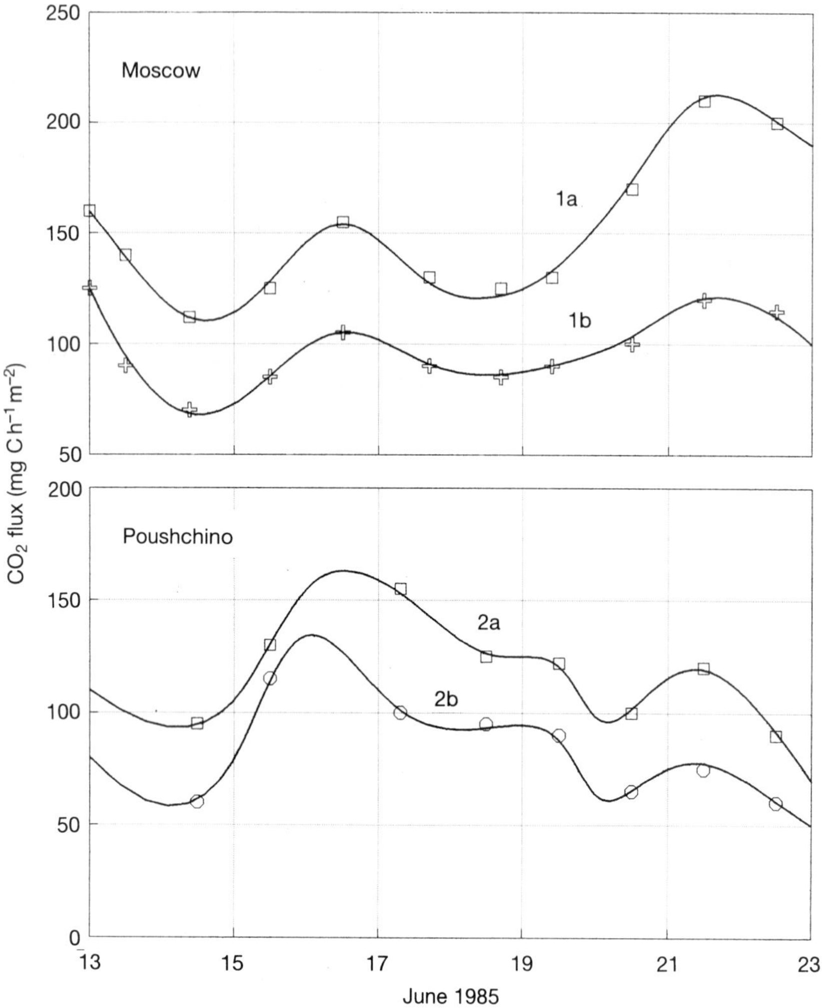

Figure 6.15 Evidence for temporal synchronization of CO_2 evolution from soils of different geographical locations. The day-to-day variation of soil respiration was followed by static chamber technique simultaneously at two sites 120 km apart: curve 1, on sod-podzolic soil under grass, field experimental station of Moscow State University; curve 2, on arable grey forest soil under barley in Poushchino. The small letters a and b refer to individual chambers 3 m apart, fixed in space by non-removable bases.

Table 6.8 Calculations of seasonal production of microbial biomass in arable grey forest soil

As estimated from	Production (kg C (ha^{-1}) per 44 days)	μ (day^{-1})	$\bar{g}$ (days)
Frequencies of fluctuations in microbial count	290	0.077	9.0
Flux of plant organic material to soil (Eqn. [1.77])*	520	0.14	5.0
Respiration rate of soil microorganisms†	530	0.14	5.0

* Ecosystem's parameters: gross primary production, 3740; net primary production, 1100; plant respiration, 390; microbial respiration, 1580; protozoan respiration, 80; input of organic matter to soil, 2240 kg C ha^{-1}; mean standing crop of soil microorganisms, 86 kg C ha^{-1}.
† Calculation according to relationship $\mu x = v_{resp}/Y_{p/x}$, where average v_{resp} was 36 kg C day^{-1} ha^{-1} and $Y_{p/x} = 3.0$ was taken from laboratory microcosm experiments.

trolled conditions. One of the speculations is that the 'master' oscillator is a component of solar insolation with enhanced penetrative ability. Indirect evidence for this comes from the surprisingly perfect reiteration of 27.3-day cycles in long-term records of SRA, the period of 27.3 days corresponding to the Sun's rotation time with respect to the Earth (Panikov *et al.*, 1988). However, for the moment substantial direct evidence is absent.

Carbon balance of a soil ecosystem and seasonal production of microbial biomass

In Chapter 1 we outlined three different techniques used for productivity estimation:

- from the frequency of x fluctuations;
- from the flux of organic matter to soil; and
- from the respiration of soil microorganisms.

We now have enough data to compare these techniques with respect to the studied ecosystem.

The results of the calculations (Table 6.8) indicate that the first method significantly underestimates microbial production as compared with the second and third methods. The difference is attributed to that part of microbial production which is concealed by grazing. Hence, the frequency of oscillations could not be regarded as adequate. However, it does not mean that oscillations themselves contradict the conservation law and the C-balance of the ecosystem.

The mean generation time of microbial populations in grey forest soil was 5 days. Bearing in mind that the rhizosphere provides more intensive growth with $g \approx 1$ day, one would expect very slow microbial growth beyond the root zone. This is supported by low average values of the r-variable for unplanted soils.

6.6 CONCLUSION

The main optimistic outcome of our studies is the evidence that a kinetic approach can be efficiently applied to microbial growth under natural conditions. Heterogeneity of soil and soil community, interactions between microbial populations and higher organisms, a changeable environment and other complications are not absolutely insuperable obstacles to quantitative analysis. Let us just outline the results of the soil kinetic studies.

The initial response of the microbial community to a soil amendment was uniform for the various types of amendments (glucose, oxidizable minerals, CO_2 under illumination, washed microbial cells, etc.). In each case this was observed as the exponential growth of responsive populations, with a lag phase dependent on their physiological state in soil before the substrate addition. This finding led to development of a new category of kinetic methods for microbial biomass estimation, which are exact, rapid and unbiased.

The long-term effects of soil amendments was shown to be associated with successional changes in the composition of the microbial community. The mechanistic interpretation of these events was obtained for three particular cases when successions were launched by addition of glucose, cellulose or a continuous supply of ethanol vapour. The understanding of the observed dynamics was achieved only by an account of competition or (and) cooperative interactions between at least two to four functional groups of soil microorganisms. Under different circumstances they were either bacteria–fungi, or r-K-L-selected species, or diazotrophic–dissipotrophic–hydrolytic populations, or microbial prey–protozoan predators. The most complicated interactions were identified in the case of cellulose decomposition, when we had to consider multiple and alternate C- and N-limitation of microbial activity associated with non-trivial growth-suppression effects.

The interactions of microbial populations with higher organisms (animals and plants) were also accessible to interpretation in quantitative terms. The animal's gut was treated as a small-scale continuous fermenter propagating a particular sort of soil bacteria. The plant–microbe association was portrayed as an open system like a chemostat with a fluctuating substrate feed in the form of root exudation and protozoan grazing instead of wash-out.

It evolved that microbial growth *in situ* and *ex situ* have many common features. We found in soil basically the same dynamic phenomena as those described earlier in pure cultures, i.e. growth limitation and auto-inhibition, multiple limitation, false diauxie, substrate-accelerated death, exponential growth preceded by a lag phase dependent on past history. On the other hand, in pure laboratory culture it was possible to imitate several features which beforehand were considered as unique natural phenomena (priming effects, appearance of unculturable forms, severe growth retardation).

Finally, field observations revealed dynamic patterns very similar to those observed in a soil–plant microcosm. A close relationship was found between microbial growth and primary production via rhizodeposition, manifested as quasi-harmonious oscillations in the measured variables with characteristic time-delays. It was possible to accommodate the microbial domain within the total carbon budget of the entire ecosystem without any violation of the conservation law and to estimate the average values of true growth rates of the microbial populations *in situ*.

The future development of kinetic studies of soil microorganisms inevitably requires us to abandon microbiology as a particular autonomous branch of natural history. Progress should be dependent upon the interdisciplinary synthesis of basic knowledge in the constellation of biospheric sciences. This is probably the only way to quantitatively describe, understand and control microbial functioning in nature.

Appendix: Derivation of mathematical models and identification of kinetic parameters

Even conventionally trained microbiologists with rather low mathematical competence may efficiently use the simple mathematical models described in this book. Here we will present several practical recommendations on mathematical techniques to those readers who are still amateurs in this esoteric field.

A.1 SOME INSTRUCTIVE RULES ON THE MODEL'S DERIVATION

HOW TO START THE FORMULATION OF THE MODEL, THE ROLE OF THE FLOW DIAGRAM

It is very helpful to start from the construction of the flow diagram of simulated processes (see example in Table 1.5). The rules are very simple: you have to assign all desired variables X_i using short and clear notations, sources ($\ldots \rightarrow X_i$) and sinks ($X_i \rightarrow \ldots$) should be shown for every such variable. Then you have to construct a set of ordinary differential equations (ODE) bearing in mind that: a) the number of ODE is to be equal to the number of assigned variables, b) the right side of every ODE should contain as many terms as there are arrows before and after respective variable, c) the sources-terms being positive, and sinks terms being negative.

DIMENSIONS

Possible errors in derived equations could be efficiently revealed by analysis of dimensions (Cornish-Bowden, 1981). You should remember that it is forbidden to add or subtract terms of different dimensions, although you may multiply or divide them. For example, the expression '$\ldots (s + x)$' in your ODE is not correct. If s is mg of substrate per l, and x is mg of biomass per l, the correct expression should be '... $(Ys + x)$', where Y has dimension g of biomass per g of substrate.

You cannot compare kinetic constants of different dimension, this is as absurd as asking which is higher, 3 miles or 6°C. Some operations (rising to power) are allowed for only simple dimensionless numbers, e.g. the expression $e^{2.5t}$ where t is time, only make sense if 2.5 has dimension $(\text{time})^{-1}$.

CONSERVATION CONSTRAINTS

Suppose we have accounted in the growth model for all the components of the C budget: biomass (x); organic substrate (s); CO_2 (p) and exometabolites (m); and chose one, and the same, units for all assigned variables (e.g. mg C l^{-1}). Then we may inspect the correctness of derived ODE by the following reliable test: the sum of all derivatives should be equal to zero in a closed system (e.g. in a batch culture) or the difference between fluxes entering and leaving the open system (e.g. chemostat culture):

$$\frac{dx}{dt} + \frac{ds}{dt} + \frac{dp}{dt} + \frac{dm}{dt} = 0 \quad \text{(batch culture)},$$

$$\frac{dx}{dt} + \frac{ds}{dt} + \frac{dp}{dt} + \frac{dm}{dt} = D(s_0 - s - x - p - m) \quad \text{(chemostat)}.$$

Similar conservation conditions may be derived for other elements (N, P, ...), available electrons, etc. It is strongly recommended to calculate the sum of the derivative during numeric integration of the set of ODE, then deviation from expected constant value is a clear indication of some unnoticed errors in the equation's right-hand side.

STRUCTURED MODELS, AN ACCOUNT OF 'INTERNAL VARIABLES'

These are a particular case of the use of the conservation constraints described above. All structured models contain two types of variables: 'external', measured as g l^{-1} of cultural liquid (exosubstrate, total biomass, concentration of exometabolites), and 'internal' which have dimension g or mole of component per g of total cell mass (intracellular concentration of RNA, reserved substances, pool of intermediates, etc.). As an example we will consider the derivation of ODE for intermediate L (see flow chart, Table 3.4). Let us denote by lower case 'l' intermediate concentration in terms of g per unit of culture volume (g l^{-1}) and by capital 'L' its fraction of total biomass (g g^{-1}). Then $L = l/x$, and $l = Lx$. According to the flow chart we derive ODE for l by account of all sources and sinks:

$$\frac{dl}{dt} = q_s x - q_{resp} x - q_L x + ax' - Dl .$$

The relationship between dl/dt and dL/dt is then found from differentiation rules:

$$\frac{dl}{dt} = \frac{d(Lx)}{dt} = L\frac{dx}{dt} + x\frac{dL}{dt},$$

$$\frac{dL}{dt} = \frac{1}{x}\left[\frac{dl}{dt} - L\frac{dx}{dt}\right] = q_s - q_{resp} - q_L + a(1 - L) - \mu L .$$

Notice that ODE for every internal variable X_i should contain the term '$-\mu X_i$', which is interpreted as the dilution of this component due to cell growth.

A.2 IDENTIFICATION OF THE KINETIC PARAMETERS*

This is one of the most difficult technical problems in biokinetic studies and may annoy even a normally dispassionate investigator. Below we will outline our experience in this field.

Sometimes growth parameters are found graphically by manually drawing a straight line through linear forms of data. A well-known example is the case when parameters of a Monod equation are found from the plot of $1/\mu$ *vs.* $1/s$ as y- and x-intercepts corresponding to $1/\mu_m$ and $1/K_s$ respectively. However, manual line drawing is very arbitrary, and obtained parameters could not be assessed statistically for confidence limits. Thus this procedure should be avoided (Cornish-Bowden, 1976) in favour of the analytical techniques described next.

SIMPLE LINEAR REGRESSION

Suppose we have a kinetic equation which specifies the relationship between experimental data x and y as well as a set of parameters (constant coefficients) $a_0, a_1, \ldots, a_n$:

$$y = f(x, a_0, a_1, \ldots, a_n) . \qquad [A.1]$$

If measurements are independent of each other, and errors are normally distributed with mean zero, then the best estimate of parameters is found by minimizing the sum of square residuals, S:

$$S = \sum_{i=1}^{N} [y_i - f(x_i, a_0, a_i, \ldots, a_n)]^2 . \qquad [A.2]$$

* This section has been written jointly with Michael V. Glagolev, who is also the author of most of the developed computer routines.

If some observations are more precise than others, then we minimize the weighted sum of squares, i.e. multiply every i-term by the value of $w_i = 1/(ey_i)^2$, where e is relative experimental error (0.05–0.10 in most of our determinations).

Obviously the easiest way of minimization is accomplished when Eqn. [A.1] is linear (e.g. zero kinetic order of reactions, see Eqns [2.10] and [4.5] in the main text). In some cases we have a possibility of making a relatively simply linear transformation of non-linear functions (Table A.1). If the transformation is $y = g(Y)$, then weights should be chosen for kth result as

$$W_k = \left[\frac{1}{g'(Y_k)}\right]^2,$$

where g' is the first derivative of the g-function.

However this type of linearizable function is also rather narrow and encompasses only some simple cases. So we need to make one additional step.

GENERAL LINEAR MODEL (GLM)

Often the non-linear function of independent variables and parameters is a linear combination of several functions of the same variable (they are called basis functions; Ebert and Ederer, 1985). This is a case when GLM may be used. Surprisingly, this approach was found to be very efficient and almost uniformly applicable in microbiological kinetics. Let us demonstrate it by the following examples.

In Chapter 3 we frequently used Eqn. [3.29] containing three parameters. At the first glance it is not susceptible to linear transformation and needs to be treated by non-linear regression. However, we have a much better alternative. Multiplying both sides of Eqn. [3.29] by $(K_s + s + s^*)$ produces the linear form:

$$y = a_0 + a_1 f_1 + a_2 f_2 , \qquad \text{[A.3]}$$

where $y = q_s s$, $f_1 = s$, $f_2 = q_s$, $a_0 = -Q_s s^*$, $a_1 - Q_s$, and $a_2 = s - K_s$. Now minimization of square residuals may be attained by use of multiple linear regression. Obtained coefficients $a_0 - a_2$ are then easily recalculated to required parameters Q_s, K_s, and s^*.

The second example is a double reduced SCM equation which describes batch growth with lag-phase (Eqn. [3.66]). The most interesting case is when we consider $r_0 < 1$ and $\beta = 1$, then

$$x = x_0 r_0 \exp(\mu t) + (1 - r_0) x_0 . \qquad \text{[A.4]}$$

Integration of this equation gives

Table A.1 Linear transformation of some equations used in microbiological kinetics

Original equation	Linear form	Weight	Parameters	Equation number
$\mu = \dfrac{\mu_m s}{K_m + s}$	$\dfrac{1}{\mu} = \dfrac{K_s}{\mu_m}\dfrac{1}{s} + \dfrac{1}{\mu m}$	μ^4	μ_m, K_s	[1.18]
	$\dfrac{s}{\mu} = \dfrac{K_s}{\mu_m} + \dfrac{s}{\mu_m}$	$\dfrac{\mu^4}{s}$	μ, K_s	[1.25]
$N = N_0 \exp(\mu t)$	$\ln N = \ln N_0 + \mu t$	N^2	N_0, μ	[1.7]
$N = N_0 \exp(-kC^n t)$	$\ln(\ln(N/N_0)/t) = \ln(-k) + n \ln(C)$	$N^2 \ln^2 \dfrac{N}{N_0}$	k, n	[1.10]
$x = x_0 (t+1)^{-a}$	$\ln(x) = \ln(x_0) - a \ln(t+1)$	x^2	x_0, a	[3.27]
$r = \dfrac{s^n}{K_r + s^n}$	$\ln\left(\dfrac{1}{r-1}\right) = \ln K_r - n \ln s$	$(r - r^2)^2$	K_r, n	[3.57]

$$x(t) - x_0 = \mu \int_0^t x(t)\,dt - (1 - r_0)x_0\mu t\,. \qquad [A.5]$$

Eqn. [A.5] is already linear and equivalent to [A.3] if we put $y = x(t) - x$, $a_0 = 0$, $a_1 = \mu$, $a_2 = (1 - r_0)x_0\mu$, while basis function $f_1 = \int x(t)\,dt$ and $f_2 = t$.

Certainly it is not usual to see an integral as a basis function. However it is quite justifiable, since integrals may found numerically for any t similar to analytically expressed elementary functions. Other linear transformations are shown in Table A.2. In practical terms, multiple regression and calculations of basis functions including integrals were performed with standard routines available in Lotus 1-2-3 and similar software.

NON-LINEAR REGRESSION

This approach is uniformly applied to any data but there are serious obstacles associated with lack of convergency in the iteration procedure, discussed elsewhere (Afifi and Azen, 1979; Zhiglhiazsky and Zhilinskas, 1991). The computer routine used in present work was based on algorithms POISK2 and POISK4 (Krut'ko, Maximov and Skwortsov, 1988) and always provided the required convergence. However non-linear regression frequently fails in obtaining the global minimization of square residuals, and produces instead some local minimum of error. This is illustrated by Figure A.1, where the residual sum of squares (obtained for Eqn. [3.29]) is plotted *vs.* parameter K_s. The non-linear model produces two local minima, one of which is global, while the linear one is characterized by one and only one minimum (curve 2). However, the best estimates of K_s for linear and original non-linear forms are not absolutely identical because linearization itself introduces a bias. Therefore GLM could not be the only procedure used. The approach we have accepted as most appropriate is as follows: (a) GLM is used to find the global minimum of parameters for the linear form of a kinetic equation, and (b) this set of parameters is applied as a starting point for non-linear regression. A combination of GLM and non-linear regression provides global minimization of residuals.

CONFIDENCE LIMITS OF KINETIC PARAMETERS

An excellent description of this technique based on traditional statistics and linear kinetic equations was given by Cornish-Bowden (1976) and Robinson (1985). Here we may demonstrate the results of this analysis as applied to Eqn. [3.29] and experimental data from Figure 3.28: $Q_s = 0.218 \pm 0.217$ h^{-1}, $s^* = 66 \pm 8$ mg l^{-1} and $K_s = 12 \pm 101$ mg l^{-1}

Table A.2 Transformation of some kinetic equations in the form suitable for GLM (General Linear Model)

Original equation	Linear form	y	Parameters	Basis functions
$q = Q\dfrac{s^{n}}{K + s^{n}}$	$y_1 = k_1 f_1(s) + k_2 f_2(s)$	$\int_{s_0}^{s} q^2\,ds$	$k_1 = Q$	$f_1(s) = \int_{s_0}^{s} q\,ds$
			$k_2 = -Q/n$	$f_2(s) = s_0 q_0 - sq + \int_{s_0}^{s} q\,ds$
	$y_2 = k_3 f_3(s)$	$(Q - q)s^{n}$	$k_3 = K$	$f_3(s) = q(s)$
$K_R = K_R^{max}\dfrac{s}{K_s + s + s^2/K_{ss}}$	$y = k_1 f_1(s) + k_2 f_2(s) + k_3 f_3(s)$	$K_R s$	$k_1 = K_R^{max}$	$f_1(s) = s$
			$k_2 = -1/K_{ss}$	$f_2(s) = K_R s^2$
			$k_3 = -K_s$	$f_3(s) = K_R$
$\mu = \mu_m\dfrac{s - s^*}{K_s + s}$	$y = k_1 f_1(s) + k_2 f_2(s) + k_3$	μs	$k_1 = \mu_m$	$f_1(s) = s$
			$k_2 = -K_s$	$f_2(s) = \mu$
			$k_3 = -\mu_m s^*$	$f_3(s) = 1$
$\mu = \mu_1\dfrac{s}{K_1 + s} + \mu_2\dfrac{s}{K_2 + s}$	$y = k_1 f_1(s) + k_2 f_2(s) + k_3 f_3(s) + k_4 f_4(s)$	μs^2	$k_1 = \mu_1 K_2 + \mu_2 K_1$	$f_1(s) = s$
			$k_2 = -K_1 K_2$	$f_2(s) = \mu$
			$k_3 = \mu_1 + \mu_2$	$f_3(s) = s^2$
			$k_4 = -(K_1 + K_2)$	$f_4(s) = \mu s$

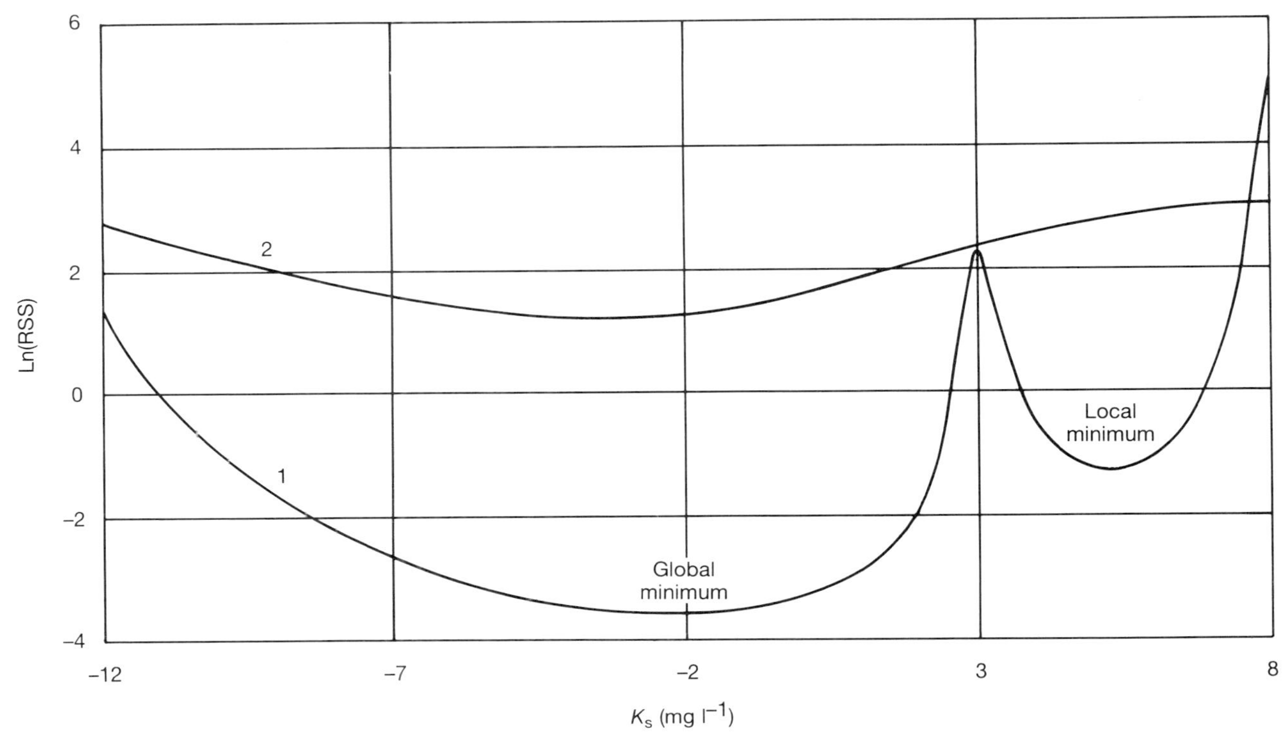

Figure A.1 The residual sum of squares (RSS) for original (1) and linear forms of (2) of Eqn. [3.29] calculated with different values of parameter K_s. Experimental points are taken from Figure 3.28. Please note that there are two local minima for non-linear equation and one single minimum of RSS for the linear form.

($P = 0.68$). These results turned out to be disappointing; relative errors in kinetic parameters were found to be too large (up to 863%! in the case of K_s). The main reason is the rather close correlation between particular basis functions. However, some particular combinations of required parameters ($Q_s s^*$ and $s^* - K_s$) were identified with low residuals. This implies that presented experimental data were not sufficient for exact identification, and one of the three required parameters should be determined in independent (additional) experiments. For instance, we may measure the parameter $Q_s = 0.22\ h^{-1}$ directly in a chemostat culture as specific uptake rate of substrate at subcritical D. Then confidence limits for the rest of the parameters became sensibly low; $s^* = 68 \pm 4\ mg\ l^{-1}$, and $K_s = 7.6 \pm 5.4\ mg\ l^{-1}$.

However, this type of sensitivity analysis is still not satisfactory because of statistical bias; we had to make an assumption that all measurement errors are normally distributed with a mean of zero. This bias may be eliminated in an approach which belongs to one or more modifications of the so-called 'bootstrap' methods (from the idiomatic expression 'pulling yourself up by your own bootstrap', i.e. relying on yourself). Actually it relies on the power of modern computers. Having only one set of experimental data from n pairs of measurements, y_i and x_i, we may produce by a pseudo-random number generator many other sub-sets of data (say, 1000 n) confined within the range $[y_i - q, y_i + q]$, where q is the experimental error. There are other techniques of this kind, e.g. jack-knife methods, which differ from the described approach in detail (Diaconis and Efron, 1983), but we will restrict ourselves to classic bootstrap.

We have found it to be reliable and easy to use with an ordinary personal computer for relatively simple kinetic expressions, when non-linear regression may be executed for several iterations. An example of bootstrap-generated distributions of identified parameters is shown on Figure A.2 for reduced SCM. The reliability of the identification procedure turned out to be rather high, which may be envisaged as a narrow range of parameter variation. The confidence limits are characterized explicitly as mean ± standard deviation: $\mu_m = 0.366 \pm 0.100$ and $x_0 = 23.46 \pm 0.66$.

A.3 NUMERIC INTEGRATION OF ORDINARY DIFFERENTIAL EQUATIONS (ODE) AND IDENTIFICATION OF THEIR PARAMETERS

ODEs are the main instruments of mathematical modelling. The principal technical problems associated with their use are (a) stability analysis, (b) the solving of direct problems (analytical or numeric integration), and (c) the reverse problem of parameter identification. The first aspect was considered in details in many books (Bazin, 1982; Romanovsky, Stepanova and Chemovski, 1984), so we will concentrate on the points (b) and (c).

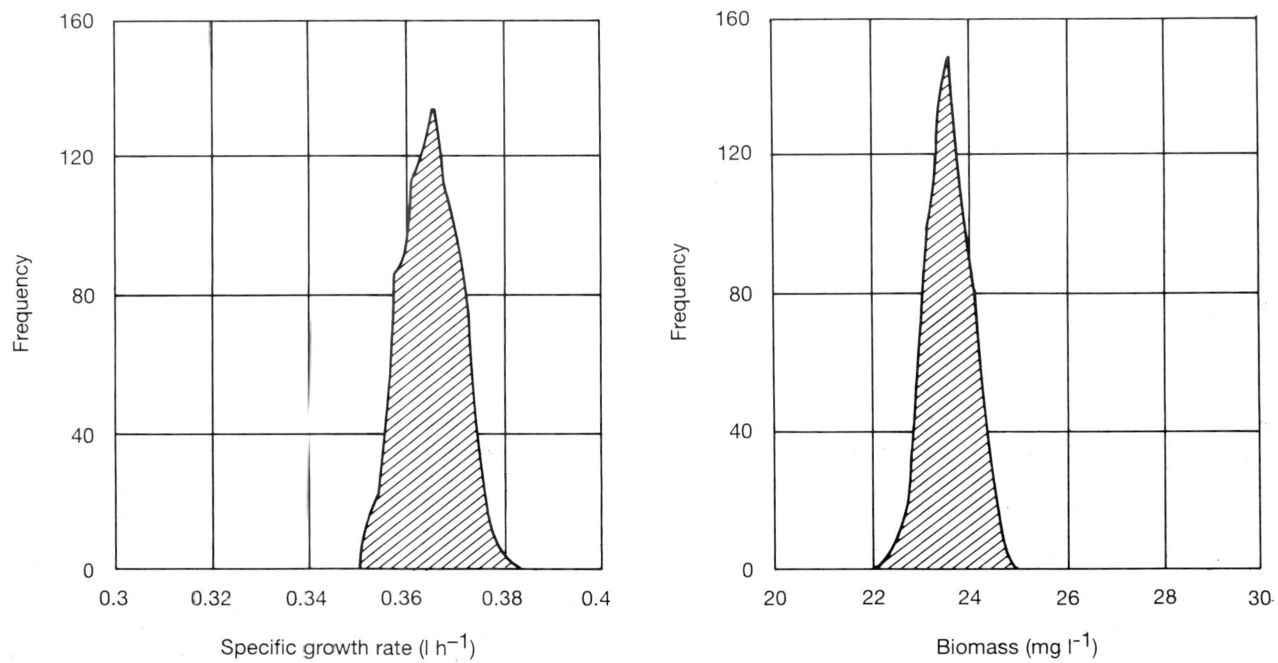

Figure A.2 An example of 'bootstrap' distribution for parameters of Eqn. [3.65] (particular case of $r_0 = 1$) fitted to experimental data (Figure 3.17, curve 1) by non-linear regression. The number of bootstrap-generated sub-sets of data was $n = 1000$, experimental error was assumed to be 5%.

NUMERIC INTEGRATION, STIFF ODE

There are two components of global error which are associated with the procedure of numeric integration (Samarskii and Goolin, 1989): one is originated from the substitution of an infinitesimally small increment, dt, by some small but finite value, Δt, and the second one results from an error in the set-up of initial conditions. Additional difficulty appears when the set of ODE contains both quick and slow variables. For instance all versions of chemostat models include ODE for quick variables (limiting substrate concentration, pool of intermediate) and slow variables (biomass, index of physiological state). To prove it let us rewrite Monod's chemostat model in dimensionless form by substitution of variables: $\tau = Dt$, $\alpha = \mu_m/D$, $\chi = x/Ys_0$, and $\zeta = s/K_s$. Then Eqn. [1.28] is modified as follows:

$$\frac{d\chi}{d\tau} = \chi\left(\frac{\alpha\zeta}{1+\zeta} - 1\right), \qquad [A.7]$$

$$\epsilon\frac{d\zeta}{d\tau} = 1 - \chi\frac{\alpha\zeta}{1+\zeta} - \zeta\epsilon,$$

where new state variables ζ and χ are now of the same order of magnitude and $\delta = K_s/s_0 \ll 1$ is the so-called small parameter. Such kinds of ODE are called stiff equations. Their numeric solution needs to be made by special methods, since standard explicit techniques (Euler, Runge–Kutta) require small integration steps for extremely large time intervals. We used our own program written in Turbo-Pascal and based on a phenomenally effective algorithm devised by Novikov, Novikov and Yumatova (1987) which belongs to the Rosenbrok type of implicit Runge–Kutta methods.

IDENTIFICATION OF PARAMETERS OF ODE BY 'SHORT-CUT' METHOD

The straightforward method to find the best estimate of parameters is based on local minimization of square residuals, which implies the necessity to solve many times the direct problem (numeric integration of ODE). This way is not efficient, since numeric integration requires significant computation time. To overcome this difficulty it was suggested to approximate experimental curves by spline-functions and then to calculate residuals from simpler algebraic expressions (Marchuk, 1991). It is important to avoid numerous numeric integrations of original ODE, although the integration of spline-functions still remains to be done, often this problem may be solved analytically.

Such an approach belongs to so-called 'short-cut' methods of parameter identification. Unfortunately it has not always resulted in achieving the global minimum of residuals; therefore we tried, where possible, to perform preliminary linearization of ODE and use GLM. This technique is demonstrated by the following example, in which we will identify the parameters of the Monod model from the experimental data on batch growth. The ODE [1.28] at $D = 0$ is

$$\frac{dx}{dt} = \mu_m \frac{sx}{K_s + s},$$

$$\frac{ds}{dt} = -\frac{\mu_m}{Y}\frac{sx}{K_s + s}. \qquad [A.8]$$

Let us divide the first equation by the second and take an integral within the boundary from x_0 to $x(t)$ and from s_0 to $s(t)$, then

$$x(t) - x_0 = Y[s_0 - s(t)]. \qquad [A.9]$$

Equation [A.9] is equivalent to the linear form $y = k_1 f_1(t)$, where $y = x(t) - x_0 k_1 = Y$, and basis function $f_1 = s_0 - s(t)$. Equation [A.9] is used for identification of parameter Y.

To find other parameters we will modify the second ODE of the set [A.8] in the following way: multiply both sides of ODE by (K_s+s), insert s under the differential and integrate the obtained equation:

$$K_s(s - s_0) + 0.5(s^2 - s_0^2) = -\frac{\mu_m}{Y}\int_{t_0}^{t} sx \, dt \qquad [A.10]$$

It is equivalent to the linear form $y = k_1 f_1(t) + k_2 f_2(t)$, where $y = 0.5 (s^2 - s_0^2)$, $k_1 = K_s$, $k_2 = -\mu_m/Y$, and basis functions are

$$f_1(t) = s_0 - s(t), \quad f_2(t) = \int_{t_0}^{t} sx \, dt.$$

The integral $f_2(t)$ is estimated numerically by a simple trapezoid method (it may be performed with standard software like Lotus 1-2-3) or with the use of spline-functions. GLM gives the first approximation of desired parameters, which provides the global minimization of square residuals for the linear form. Afterwards these parameters are used in non-linear regression as a starting point in the iteration procedure. It results in obtaining global minimization of residuals for the original non-linear set of ODE. Thus the problem of parameter identification may be solved quite quickly and accurately.

Linear transformation of some other ODEs are presented in Table A.3. Similar transformations may be performed for a wide spectrum of models containing between one and three ODEs. More complicated models like SCM are less suited to such procedures because linearization produces too bulky analytical expressions. For these types of models it is advisable to identify parameters step by step, e.g. from well-defined experimental data like those illustrated on Figures 3.5, 3.8 and 3.12.

Table A.3 Transformation of some differential equations in the form suitable for GLM (General Linear Model)

Original equation	Linear form	y	Parameters	Basis functions
$\frac{dN}{dt} = rN\frac{\ln N/K}{\ln K}$	$y = k_1f_1(t) + k_2f_2(t)$	$\ln\frac{N}{N_0}$	$k_1 = r/\ln K$	$f_1(t) = \int_{t_0}^{t} N\, dt$
			$k_2 = -r$	$f_2(t) = t - t_0$
$\frac{dx}{dt} = ax - bx^2$	$y = k_1f_1(t) + k_2f_2(t)$	$x(t) - x_0$	$k_1 = a$	$f_1(t) = \int_{t_0}^{t} x\, dt$
			$k_2 = -b$	$f_2(t) = \int_{t_0}^{t} x^2\, dt$
$\frac{dR}{dt} = K_R - kR$	$y = k_1f_1(t) + k_2f_2(t)$	$R(t) - R_0$	$k_1 = K_R$	$f_1(t) = t - t_0$
			$k_2 = -k$	$f_2(t) = \int_{t_0}^{t} R\, dt$

References

Abrosov, N.S. and Kovrov, B.G. (1977) *Analysis of the Structure of Community of Unicellular Organisms*, Nauka Publ., Novosibirsk, 190 p.

Adams, J., Kinney, T., Thompson, S. *et al.* (1979) Frequency-dependent selection for plasmid-containing cells of *Escherichia coli*. *Genetics*, **91**(4), 627–37.

Afifi, A.A. and Azen, S.P. (1979) *Statistical analysis. A Computer Oriented Approach*, Academic Press, New York, San Francisco, London.

Agrawal, P., Lim, H.C. and Ramkrishna, D. (1983) An extended bottleneck model: accounting for the metabolic turnover effect in microorganisms. *J. Chem. Technol. Biotechnol.*, **33B**, 155–63.

Akagi, Y., Taga, N. and Simidu, U. (1977) Isolation and distribution of oligotrophic marine bacteria. *Can. J. Microbiol.*, **23**, 981–7.

Akimenko, V.K., Trutko, S.M., Medentsev, A.G. and Korobov, V.P. (1983) Distribution of cyanide-resistant respiration among yeasts and bacteria and its relation to oversynthesis of metabolites. *Arch. Microbiol.*, **136**, 234–41.

Aminov, R.I. and Golovlev, E.L. (1987) Growth kinetics of *Rhodococcus minimus* in batch and continuous culture. *Microbiology* (*Mikrobiologiya*), **56**(1), 64–70.

Anderson, J.P.I. and Domsch, K.H. (1973) Quantification of bacterial and fungal contribution to soil respiration. *Archiv für Mikrobiologie*, **93**, 113–27.

Anderson, T.-H. and Domsch, K.H. (1985) Maintenance carbon requirements of actively-metabolizing microbial populations under *in situ* conditions. *Soil Biol. Biochem.*, **17**(2), 197–203.

Andrews, J.F. (1968) A mathematical model for the continuous culture of microorganisms utilizing inhibitory substrates. *Biotechnol. Bioeng.*, **10**(4), 707–23.

Andrews, J.H. and Harris, R.F. (1986) r- and K-selection and microbial ecology. *Adv. Microbial Ecol.*, **9**(1), 99–147.

Angerer, K. von (1919) Über die Arbeitsleistung eigenbeweglicher Bakterien. *Archiv für Hygiene*, **88**(1), 139–55.

Aristovskaya, T.V. (1965) *Microbiology of Podzolic Soils*, Nauka Publ., Moscow, Leningrad.

Aristovskaya, T.V. (1972) The theoretic aspects of assessment of microbial abundance, biomass and productivity in soil. In: *Problems of Assessment of Quantity, Biomass and Productivity of Soil Microorganisms*, Nauka Publ., Leningrad, pp.7–20.

Aristovskaya, T.V. (1975) Quantity, biomass and productivity of soil bacteria. In: *Resources of the Biosphere (Synthesis of the Soviet Studies for the International Biological Programme)* (eds L.E. Rodin and N.N. Smirnov), Nauka Publ., Leningrad, pp.241–59.

Bader, F.G., Meyer, J.S., Frederickson, A.G. and Tsuchiya, H.M. (1975) Comments on microbial growth rate. *Biotechnol. Bioeng.*, **17**(2), 279–83.

Bauchop. T. and Elsden, S.R. (1960) The growth of microorganisms in relation to their energy supply. *J. Gen. Microbiol.*, **23**(2), 457–69.

Bazin, M.J. (ed.) (1982) *Microbial Population Dynamics*, CRC Press, Boca Raton, Florida.

Beijerinck, M.W. and van Delden, A. (1903) Über ein farblos Bakterie deren Kohlenstoffndhrung aus der Luft herrbhrt atmosphdrischen. *Zentralblatt fbr Bakteriologie, Parasitenkunde und Infektionskrankheiten. Abt. II*, **10**, 33–47.

Bell, R.T. (1986) Thymidine incorporation as a measure of bacterial production in lakes. *Acta Univ. Upsal. Compr. Summ. Uppsala Diss. Fac. Sci.*, No. 43, 2–31.

Belova, S.E., Dorofeyev, A.G. and Panikov, N.S. (1995) Kinetics and stoichiometry of colonial growth on agar medium of bacteria *Pseudomonas fluorescens* and *Alcaligenes* sp. *Microbiology (Mikrobiologiya)*, **64**.

Berezin, I.V. and Varfolomeyev, S.D. (1979) *Biokinetics*, Nauka Publ., Moscow.

Bergter, F. (1969) Zwei stabile stationdre Zustdnde bei der Lichtemission von Photobacterium pierantonii. *Studia Biophysica*, **14**(1), 63–6.

Blackman, F.F. (1905) Optima and limiting factors. *Ann. Bot. London*, **19**(74), 281–93.

Bondarenko, T.F., Panikov, N.S. and Dobrovolskaya, T.G. (1985) Changes in the morphology of soil bacteria as a function of their growth rate. *Microbiology (Mikrobiologiya)*, **54**(5), 798–803.

Broadbent, F.E. (1947) Nitrogen release and carbon loss from soil organic matter during decomposition of added plant residues. *Proc. Soil Sci. Soc. Am.*, **12**(2), 246–9.

Brock, T.D. (1971) Microbial growth rates in nature. *Bact. Rev.*, **35**(1), 39–58.

Brock, T.D. and Brock, M.L. (1968) Measurement of steady-state growth rates of a thermophilic alga directly in nature. *J. Bacteriol.*, **95**, 811–5.

Brookes, P.C., Tate, K.R. and Jenkinson, D.S. (1983) The adenylate energy charge of the soil microbial biomass. *Soil Biol. Biochem.*, **15**(1), 9–16.

Brown, J. *et al.* (eds) (1980) *An Arctic Ecosystem (the Coastal Tundra at Barrow, Alaska)*, Stroudsberg, Pennsylvania, 440 pp.

Brunner, W. and Focht, D.D. (1984) Deterministic three-half-order kinetic model for microbial degradation of added carbon substrates in soil. *App. Environ. Microbiol.*, **47**(1), 167–72.

Bryson, V. (1952) The turbidostat selector – a device for automatic isolation of bacterial variants. *Science*, **116**(1), 48–50.

Buchanan, R.E. (1918) Life phases in a bacterial culture. *J. Infect. Dis.*, **23**(1), 109–25.

Burmaster, D.E. (1979) The unsteady continuous culture of phosphate-limited *Monochrysis lutheri* Droop: experimental and theoretical analysis. *J. Exp. Mar. Biol. Ecol.*, **39**(2), 167–86.

Burton, A.C. (1936) The basis of the principle of the master reaction in biology. *J. Cell Comp. Physiol.*, **9**(1), 1–15.

Button, D.K. (1969) Thiamine limited steady state growth of the yeast

Cryptococcus albidus. *J. Gen. Microbiol.*, **58**(1), 15–21.

Button, D.K. (1978) On the theory of control of microbial growth kinetics by limiting nutrient concentrations. *Deep-Sea Res.*, **25**(12), 1163–77.

Button, D.K. (1985) Kinetics of nutrient-limited transport and microbial growth. *Microbiol. Rev.*, **49**(3), 270–97.

Byzov, B.A., Vu Nguyen Thanh and Babjeva, I.P. (1993) Interrelationships between yeasts and soil diplopods. *Soil Biol. Biochem.*, **25**(8), 1119–26.

Caperon, J. (1968) Population growth response of *Isochrysis galbana* to nitrate variation at limiting concentrations. *Ecology*, **49**(4), 866–72.

Caperon, J. and Meyer, J. (1972) Nitrogen-limited growth of marine phytoplankton. II. Uptake kinetics and their role in nutrient limited growth of phytoplankton. *Deep-Sea Res.* **19**(9), 619–32.

Campbell, A. (1957) Synchronization of cell division. *Bacteriol. Rev.*, **21**(2), 263–72.

Campbell, R. (1985) *Plant Microbiology*, Arnold, London, 191 p.

Carpenter, S.R. (1981) Decay of heterogeneous detritus: a general model. *J. Theor. Biol.*, **89**(4), 539–47.

Catler, D.W., Crump, L.M. and Sandon, H.A. (1923) A quantitative investigation of the bacterial and protozoan population of the soil with an account of the protozoan fauna. *Phil. Trans. Roy. Soc. Ser. B*, **211**, 317–50.

Chapman, S.J. and Gray, T.R.G. (1986) Importance of cryptic growth, yield factors and maintenance energy in models of microbial growth in soil. *Soil Biol. Biochem.*, **18**(1), 1–4.

Chernavsky, D.S. and Ierusalimsky, N.D. (1966) The principle of minimum in kinetics of enzymatic reactions. In: *Controlled Biosynthesis*, Nauka Publ., Moscow, pp.19–24.

Chesbro, W.R., Evans, T. and Eifert, R. (1979) Very slow growth of *Escherichia coli*. *J. Bacteriol.*, **139**, 625–38.

Clarholm, M. and Rosswall, T. (1980) Biomass and turnover of bacteria in a forest soil and a peat. *Soil Biol. Biochem.*, **12**(1), 49–57.

Collins, J.F. and Richmond, M.H. (1962) Rate of growth of *Bacillus cereus* between divisions. *J. Gen. Microbiol.* **28**(1), 15–33.

Collins, S.H., Jarvis, A.W., Lindsay, R.J. and Hamilton W.A. (1976) Proton movements coupled to lactate and alanine transport in *Escherichia coli*: isolation of mutants with altered stoichiometry in alanine transport. *J. Bacteriol.*, **126**(7), 1232–44.

Conn, H.J. (1948) The most abundant groups of bacteria in soil. *Bacteriol. Rev.*, **12**(3), 257–73.

Conrad, R. and Seiler, W. (1981) Decomposition of atmospheric hydrogen by soil microorganisms and soil enzymes. *Soil Biol. Biochem.*, **13**(1), 43–50.

Conrad, R., Seiler, W. and Bunse, G. (1983) Factors influencing the loss of fertilizer into the atmosphere as N_2O. *J. Geophys. Res.*, **88**(C11), 6709–18.

Contois, D.E. (1959) Kinetics of bacterial growth, relationship between population density and specific growth rate of continuous culture. *J. Gen. Microbiol.*, **21**(1–2), 40–9.

Contreras, E. (1980) Studies on the intestinal actinomycete flora of *Eisenia lucens* (Annelida, Oligochaeta). *Pedobiologia*, **20**(6), 411–6.

Cooney, C.L. (1979) Continuous culture of microorganisms: an overview and perspective. *Ann. N.Y. Acad. Sci.*, **326**, 295–314.

Cooper, S. (1988) What is the bacterial growth law during the division cycle? *J. Bacteriol.*, **170**(11), 5001–5.

Cooper, S. (1990) The *Escherichia coli* cell cycle. *Res. Microbiol.*, **141**(1), 17–29.

Cornish-Bowden, A. (1976). *Principles of Enzyme Kinetics*, Butterworths, London, Boston, 206 p.

Cornish-Bowden, A. (1981) *Basic Mathematics for Biochemists*, Chapman & Hall, London, New York.

Curds, C.R. and Cuckburn, A. (1968) Studies on the growth and feeding of *Tetrahymena pyriformis* in axenic and monoxenic culture. *J. Gen. Microbiol.*, **54**(3), 343–58.

Dagley, S. and Hinshelwood, C.N. (1938) Physicochemical aspects of bacterial growth. I. Dependence of growth of *Bact. lactis aerogenes* on concentration of medium. *J. Chem. Soc.*, 1938 (12), 1930–6.

D'Ari, R., Maguin, E., Bouloc, P. *et al.* (1990) Aspects of cell cycle regulation. *Research in Microbiology*, **141**(1), 9–16.

Dawson, P.S.S. (1965) Continuous phased growth, with a modified chemostat. *Can. J. Microbiol.*, **11**, 893–903.

Dedysh, S.N., Paleyeva, M.V. and Panikov, N.S. (1991) A kinetic method for the determination of algal biomass in soil. *Microbiology* (*Mikrobiologiya*), **60**(2), 387–94.

Dennis, J.E. and Schnabel, R.B. (1983) *Numerical Methods for Unconstrained Optimization and Nonlinear Equations*. Prentice-Hall, Inc., Englewood Cliffs, New Jersey.

Dicks, J.W. and Tempest, D.W. (1966) The influence of temperature and growth rate on the quantitative relationship between potassium, magnesium, phosphorus and ribonucleic acid of *Aerobacter aerogenes* growing in a chemostat. *J. Gen. Microbiol.*, **45**(3), 547–57.

Domach, M.M., Leung, S.K., Cahn, R.E. *et al.* (1984) Computer model for glucose-limited growth of a single cell of *Escherichia coli* B/r-A. *Biotech. Bioeng.*, **26**(2), 203–16.

Donachie, W.D. (1968) Relationship between cell size and time of initiation of DNA replication. *Nature (London)*, **219**, 1077–9.

Dorofeyev, A.G. and Panikov, N.S. (1988) Growth efficiency in batch cultures for microorganisms with different ecological strategies. *Microbiology* (*Mikrobiologiya*), **57**(4), 491–8.

Dorofeyev, A.G. and Panikov, N.S. (1991a) The quantitative description of microbial growth in a batch culture depending on the physiological state of inoculate. *Microbiology* (*Mikrobiologiya*), **60**(4), 652–60.

Dorofeyev, A.G. and Panikov, N.S. (1991b) Dynamics of the dying off of starved microorganisms as a function of the initial growth rate. *Microbiology (Mikrobiologiya*), **60**(5), 560–6.

Dorofeyev, A.G. and Panikov, N.S. (1994) Mass-balance technique for determining of intracellular pool of respiratory substrates. *Microbiology* (*Mikrobiologiya*), **63**(5).

Dorofeyev, A.G. and Panikov, N.S. (1995) Endogeneous respiration and turnover in growing and nongrowing culture of *Pseudomonas fluorescens*. *Microbiology* (*Mikrobiologiya*), **64**(1).

Dorofeyev, A.G., Glagolev, M.V., Bondarenko, T.F. and Panikov, N.S. (1992) Observation and explanation of the unusual kinetics of *Arthrobacter globifor-*

mis. Microbiology (Mikrobiologiya), **61**(1), 24–31.

Droop. M.R. (1974) The nutrient status of algal cells in continuous culture. *J. Mar. Biol. Assoc. U.K.*, **54**, 825–55.

Dykhuizen, D. and Davies, M. (1980) An experimental model: bacterial specialists and generalists competing in chemostats. *Ecology*, **61**(5), 1213–27.

Dykhuizen, D.E. and Hartl, D.L. (1983) Selection in chemostats. *Microbiol. Rev.*, **47**(2), 150–68.

Eakman, J.M., Fredericson, A.G. and Tsuchija, H.M. (1966) Statistics and dynamics of microbial cell population. *Chem. Eng. Progr.*, Symp. Ser. No 69, **62**, p. 37–49.

Ebert, K. and Ederer, H. (1985) *Computeranwendungen in der Chemie*, VCH Verlagsgesellschaft mbH, Weinheim.

Egunov, M.A. (1914) *The laws of microbial colonies growth and reproduction*, M. Kvara Printing House, Petrograd.

Esener, A.A., Veerman, T., Roels, J.A. and Kossen, N.W.F. (1982) Modeling of bacterial growth; formulation and evaluation of a structured model *Biotechnol. Bioeng.*, **24**(8), 1749–64.

Evdokimova, G.A. (1976) Dynamics of biological productivity of bacterial communities in the rhizosphere of cereals. *Soviet Soil Sci.* (*Pochvovedenie*), No. 12, 97–102.

Evdokimova, N.V. and Panikov, N.S. (1994) Growth and sporulation of *Bacillus subtilis* in batch and dialysis cultures: experiments and mathematical simulation. *Microbiology* (*Mikrobiologiya*), **63**(5).

Evdokimova, N.V., Dorofeyev, A.G. and Panikov, N.S. (1994) Dynamics of survival and transition to dormant state of nitrogen starving bacteria *Pseudomonas fluorescens*. *Microbiology* (*Mikrobiologiya*), **63**(2), 194–203.

Fantechi, R., Maracchi, G. and Almeida-Teixeira, M.E. (eds) (1991) *Climatic Change and Impacts: A General Introduction*, Office for Official Publications of the European Communities, Luxembourg.

Ferroni, G.D., Leduc, L.G. and Winterhalder, E.K. (1985) The measurement of mineralization activity in soils. *Soil Biol. Biochem.*, **17**(5), 727–8.

Fichman, B.A. (1967) *Microbiological Refractometry*, Meditzyna Publ., Moscow, 280 p.

Flanagan, P.W. and Veum, A.K. (1974). Relationship between respiration, weight loss, temperature, and moisture in organic residues on tundra. In: *Soil Organisms and Decomposition in Tundra*, (eds A.J. Holding, O.W. Heal, S.F. MacLean, Jr. and P.W. Flanagan), Tundra Biome Steering Committee, Stockholm, pp.249–77.

French, D.D. (1974) Classification of IBP tundra biome sites based on climate and soil properties. *In: Soil Organisms and Decomposition in Tundra*, (eds A.J. Holding, O.W. Heal, S.F. McLean, Jr and P.W. Flanagan), Tundra Biome Steering Committee, Stockholm, pp.3–25.

Freter, R., Freter, R.R. and Brickner H. (1983) Experimental and mathematical models of *Escherichia coli* plasmid transfer *in vitro* and *in vivo*. *Infect. Immun.*, **39**(1), 60–84.

Galadzhiev, M.A. (1932) On the problem of protozoa immortality. *Proc. USSR Acad. Sci.* (*Izvestiya Akademii Nauk*), No. 9, 1269–1300.

Gaudy, A.F., Jr, Obayashi, A. and Gaudy, E.T. (1971) Control of growth rate by initial substrate concentration at values below maximum rate. *Appl.*

Microbiol., **22**(5), 1041–7.
Gause, G.F. (1934) *The Struggle for Existence*, Williams and Wilkins, Baltimore (Reprint, 1964), 163 p.
Gause, G.F. and Vitt, A.A. (1934) Periodic oscillations in populations: the mathematical theory of interaction between predators and preys and its application to two protozoan populations. *Proc. USSR Acad. Sci.* (*Izvestiya Akademii Nauk*), No.10, 1551–9.
Georgiou, G. and Shuler, M.L. (1986) A computer model for the growth and differentiation of a fungal colony on solid substrate. *Biotech. Bioeng.*, **28**(3), 405–16.
Gerson, U. and Chet, I. (1981) Are allochtonous and autochthonous soil microorganisms r- and K-selected? *Revue d'écologie et de biologie du sol*, **18**(3), 285–9.
Gilmanov, T.G. and Bazilevitch, N.I. (1983) Conceptual balance model of organic matter turnover in ecosystem as a theoretical basis of monitoring. In: *Theoretical Basis and Implementation of Ecological Monitoring*, Nauka Publ., Moscow, pp.7–57.
Glagolev, M.V., Vasiljeva, G.K. and Panikov, N.S. (1995) Kinetic method for enumeration of soil microorganisms degrading chloranilines. *Microbiology* (*Mikrobiologiya*), **64** (in press).
Golle, H.A. (1953) Theoretical considerations of continuous culture system. *Agric. Fd. Chem.*, **1**, 1789–98.
Golovlev, E.L. (1978) Coryneform bacteria and related microorganisms. *Adv. Microbiol.* (*Uspechi mikrobiologii*), **13**, 73–83.
Golovlev, E.L. (1983) Biology of saprophytic mycobacteria. DSc Thesis., Institute of Microbial Physiology and Biochemistry, Poushchino, 300 p.
Golovleva, L.A. and Leontievsky, A.A. (1990) Biodegradation of lignin. *Adv. Microbiol.* (*Uspechi mikrobiologii*), **24**, 128–55.
Gompertz, B. (1825) On the nature of the function expressive of the law of human mortality. *Phyl. Trans.*, **115**, 513–85.
Goodwin, B.C. (1969) Synchronization of *E. coli* B in chemostat by periodic phosphate feeding. *Eur. J. Biochem.*, **10**(4), 511–4.
Gorbenko, A.J., Panikov, N.S. and Zvyagintsev, D.G. (1986) Effect of invertebrates on growth of soil bacteria. *Microbiology* (*Mikrobiologiya*), **55**(3), 403–8.
Gorbenko, J.A. and Kryshev, I.I. (1985) *Statistical Analysis of Microbial Dynamics in Marine Ecosystems*, Naukova Dumka Publ., Kiev, 143 p.
Gray, T.R.G. (1976) Survival of vegetative microbes in soil. In *The Survival of Vegetative Microbes*, 26th Symp. Soc. Gen. Microbiol., Cambridge University Press, 327–64.
Gray, T.R.G. and Williams, S.T. (1971) *Microbial Productivity in Soil*, 21st Symp. Soc. Gen. Microbiol., Cambridge University Press, pp.255–86.
Grdzer-Lampart, S.D., Egli, T. and Hamer, G. (1986) Growth of Hyphomicrobium ZV620 in the chemostat: regulation of NH_4^--assimilating enzymes and cellular composition. *J. Gen Microbiol.*, **132**(12) 3337–47.
Grover, N.B., Woldringh, C.L. and Koppes, L.J. (1987) Elongation and surface extension of individual cells of *Escherichia coli* B/r: comparison of theoretical and experimental size distributions. *J. Theor. Biol.*, **129**(2), 337–48.
Gurevich, U.L. (1984) *Stability and Regulation of Reproduction in Microbial*

Populations, Nauka Publ., Novosibirsk, 160 p.
Guthke, R. and Knorre, W.A. (1980) Bistability in a model of microbial product formation. *Z. Allgem. Mikrobiol.*, **20**(7), 441–7.
Guzev, V.S. (1988) The ecological assessment of anthropogenic effects upon soil microbial system. DSc Thesis, Moscow University, Moscow.
Guzev, V.S. and Ivanov, P.I. (1986) Functional structure of the zymogenic part of soil microbial community. *Proc. USSR Acad. Sci.* (*Izvestiya Akademii Nauk USSR*), No.5, 739–46.
Haddon, E.C. (1928) Apparatus for obtaining a continuous bacterial growth. *Trans. Roy. Soc. Trop. Med. Hyg.*, **21**(2), 299–300.
Hagström, A., Larsson, U., Hörstedt, P. and Normark S. (1979) Frequency of dividing cells, a new approach to the determination of bacterial growth rates in aquatic environments. *Appl. Environ. Microbiol.*, **37**(5), 805–12.
Haldane, J.B.S. (1930) *Enzymes*, Longmans Green, London. 430 p.
Hanson, B.J., Cummins, K.W., Barnes, J.R. and Carter, M.W. (1984) Leaf litter processing in aquatic systems: a two variable model. *Hydrobiologia*, **111**(1), 21–9.
Harrison, M.J., Wright, R.T. and Morita, R.Y. (1971) Method for measuring mineralization in lake sediments. *Appl. Microbiol.*, **21**(4), 698–702.
Hattori, T. (1973) *Microbial Life in the Soil*, Marcel Dekker, N.Y.
Hattori, T. (1981) Enrichment of oligotrophic bacteria at microsites of soil. *J. Gen. Appl. Microbiol.*, **27**(1), 43–55.
Hattori, T. (1985) Kinetics of colony formation of bacteria: an approach to the basis of the plate count method. The Reports of the Institute for Agricultural Res. Tohoku Univ., **34**, pp.1–36.
Helmstetter, C.E., Cooper, S., Pierucci, O. and Revelas E. (1968) On the bacterial life cycle sequence. *Cold Spring Harbor Symposium on Quantitative Biology*, **33**, 809–22.
Henrici, A.T. (1928) *Morphologic Variation and the Rate of Growth of Bacteria*, C.C. Thomas, Springfield, Illinois, 194 p.
Herbert, D. (1958) Continuous culture of microorganisms; some theoretical aspects. In: *Continuous Cultivation of Microorganisms. A symposium*, Publishing House of Czech. Acad. of Sci., Prague, pp.45–52.
Herbert, D. (1959) Some principles of continuous culture. In: *Recent Progress in Microbiology*, Blackwell, Oxford, pp.81–96.
Herbert, D. (1961a) A theoretical analysis of continuous culture systems. In: *Continuous Culture of Microorganisms*, Monograph No. 12, Society of Chemistry and Industry, London, pp.21–53.
Herbert, D. (1961b) The chemical composition of microorganisms as a function of their environment. 11th Symp. Soc. Gen. Microbiol., Cambridge University Press, pp.391–416.
Herbert, D., Elsworth, R. and Telling, R.C. (1956) The continuous culture of bacteria: a theoretical and experimental study. *J. Gen. Microbiol.*, **14**(3), 601–22.
Hueting, S., de Lange, T. and Tempest, D.W. (1979) Energy requirement for maintenance of the transmembrane potassium gradient in *Klebsiella aerogenes* NCTC 418: a continuous culture study. *Arch. Microbiol.*, **123**(2), 183–8.
Hungate, R.E. (1962) Ecology of bacteria. In: *The bacteria* (eds I.C. Gunsalus and R.Y. Stanier), Academic Press, N.Y., vol. 4, pp.95–119.

Ierusalimsky, N.D. (1967) Bottle-necks in metabolism as growth rate controlling factors. In: *Continuous Cultivation of Microorganisms* (ed. E.O. Powell *et al.*), H.M. Stationery Office, Salisbury, U.K., pp.23–33.

Ierusalimsky, N.D. (1963) *The Fundamentals of Microbial Physiology*, Publ. House of USSR Akademy of Sciences, Moscow, 146 p.

Ierusalimsky, N.D. and Neronova, N.M. (1965) Quantitative relationship between metabolic products concentration and growth rate of microorganisms. *Ann. USSR Acad. Sci.* (*Doklady Akademii Nauk*), **161**(6), 1437–40.

Ivanov, M.V. (1955) The method for determination of production of bacterial biomass in water reservoir. *Microbiology* (*Mikrobiologiya*), **24**(1), 79–89.

James, T.W. (1978) Selection and evolution of yeast cells in a chemostat. In: *Cell Reproduction*, Academic Press, New York, pp.127–37.

Jenkinson, D.S. and Ladd, J.N. (1981) Microbial biomass in soil: measurement and turnover. In: *Soil Biochemistry* (eds E.A. Paul and J.N. Ladd), Marcel Dekker, New York, vol. 5, pp.415–71.

Jenkinson, D.S. and Powlson, D.S. (1976). The effect of biocidal treatments on metabolism in soil. V. A method for measuring soil biomass. *Soil Biol. Biochem.*, **17**(1), 57–63.

Jenny, H., Gessel, S.P. and Bingham, F.T. (1949) Comparative study of decomposition rates of organic matter in temperate and tropical regions. *Soil Sci.*, **63**(3), 419–32.

Jeong, J.W., Snay, J. and Ataai, M.M. (1990) A mathematical model for examining growth and sporulation processes of *Bacillus subtilis*. *Biotechnol. Bioeng.*, **35**, 160–84.

Jones, G.L., Jansen, F. and McKay, A.J. (1973) Substrate inhibition of the growth of Bacterium NCIB 8250 by phenol. *J. Gen. Microbiol.*, **74**(1), 139–48.

Jordan, R.C. and Jacobs, S.E. (1944) The growth of bacteria with a constant food supply. *J. Bacteriol.*, **48**(5), 579–98.

Kaprelyants, A.S., Gottschal, J.C. and Kell, D.B. (1993.) Dormancy in nonsporulating bacteria. *FEMS Microbial Rev.*, **104**, 271–86.

Karl, D.M. (1979) Measurement of microbial activity and growth in the ocean by rates of stable ribonucleic acid synthesis. *Appl. Environ. Microbiol.*, **38**(5), 850–60.

Karl, D.M. (1980) Cellular nucleotide measurements and applications in microbial ecology. *Microbiol. Rev.*, **44**(4), 739–96.

Khmel, I.A. (1970) The regulation of growth rate and associated modifications in microbial cells. *Adv. Microbiol.* (*Uspechi microbiologii*), **6**, 58–85.

Khudyakov, J.P. (1972) Fluctuations in microbiological processes in soil and its origin. In: *Problems of Assessment of Quantity, Biomass and Productivity of Soil Microorganisms*, Nauka Publ., Leningrad, pp.20–37.

Kluyver, A.J. and Van Niel, C.B. (1956) *The Microbe's Contribution to Biology*, Harvard University Press, Cambridge, Massachusetts.

Koch, A.L. (1975) Kinetics of mycelial growth. *J. Gen. Microbiol.*, **89**(2), 209–16.

Kompala, D.S., Ramkrishna, D., Jansen N.B. and Tsao, G.T. (1986) Investigation of bacterial growth on mixed substrates: experimental evaluation of cybernetic models. *Biotechnol. Bioeng.*, **28**(7), 1044–55.

Kostitzin, V.A. (1934) *Simbios, Parasitism et Evolution*, Hermann, Paris.

Kostychev, S.P., Sheloumova, A. and Shulgina, O. (1926) Studies on soil biodynamics. II. Microbiological survey of southern soils. Report of Dept. of

Agricultural Microbiology, 15–39.
Kozheshnik, Ya. (1971) On the dynamic modelling of microbiological industrial processes. *Biophysics* (*Biofisika*), **16**(2). 270–84.
Krassilnikov, N.A. (1958) *Soil Microorganisms and Higher Plants*. Publ. House of USSR Akademy of Sciences, Moscow, 463 p.
Krut'ko, P.D., Maximov. A.I. and Skwortsov, L.M. *Algorithms and computer routines for design of automatic systems*, Radyo i Svyaz Publ., Moscow, 306 p. (In Russian).
Kunc, F. (1975) Biological mechanisms of the 'priming effect' in soil. In: *Studies about Humus*, Prague, pp.241–6.
Kuznetsov, S.I. (1981) The prospects of the microbiological studies of continental reservoirs. *Adv. Microbiol.* (*Uspechi mikrobiologii*), **16**, 159–74.
Kuznetsov, S.I., Dubinina, G.A. and Lapteva, N.A. (1979) Biology of oligotrophic bacteria. *Annu. Rev. Microbiol.*, **33**, 377–87.
Larson, R.J. (1980) Role of biodegradation kinetics in predicting environmental fate. In: *Biotransformation and the Fate of Chemicals in the Aquatic Environment*, (eds A.W. Maki *et al.*), Amer. Soc. for Microbiol. Washington, DC, pp.67–86.
Laws, E.A. (1984) Improved estimates of phytoplankton carbon based on 14C incorporation into chlorophyll a *J. Theor. Biol.*, **110**(3), 425–34.
Laws, E.A., Jones, D. and Karl, D.M. (1986) Method for assessing heterogeneity in turnover rates within microbial communities. *Appl. Environ. Microbiol.*, **52**(4), 866–74.
Li, W.K.W. (1984) Microbial uptake of radiolabeled substrates: estimates of growth rates from time course measurements. *Appl. Environ. Microbiol.*, **47**(1), 184–92.
Litvinenko, L.A., Kintana, M.E. and Petrikevich, S.B. (1980) The effect of nitrogen input concentrations on growth of *Candida utilis* in a continuous culture with pulsed carbon feed. *Microbiology* (*Mikrobiologiya*), **49**(6), 945–51.
Lotka, A.J. (1925) *Elements of Physical Biology*, Williams and Wilkins, Baltimore, 320 p.
Luscombe, B.M. and Gray, T.R.G. (1974) Characteristics of *Arthrobacter* growth in continuous culture. *J. Gen. Microbiol.*, **82**(2), 213–22.
Lynch, J.M. and Panting L.M. (1980) Cultivation and the soil biomass. *Soil Biol. Biochem.*, **12**(1), 29–33.
Maaloe, O. and Kjeldgaard, N.O. (1966) *Control of Macromolecular Synthesis*, W.A. Benjamin, New York.
MacArthur, R.H. and Wilson, E.O. (1967) *The Theory of Island Biogeography*, Princeton Univ. Press, Princeton, New Jersey, 203 p.
MacCracken, M.C. and Luther, F.M. (eds) (1985) Projecting the climate effects of increasing carbon dioxide. US Department of Energy, DOE/ER-0237, Washington, DC.
MacKendrick, A.G. (1926) Applications of mathematics to medical problems. *Proc. Edinburgh Math. Soc.*, **44**(1), 98–130.
Macura, I. (1961) Continuous flow method in soil microbiology. *Folia Microbiol.*, **6**(5), 328–34.
Makarova, N.A., Panikov, N.S., Marfenina, O.E. and Zvyagintsev, D.G. (1957) Fungal growth in porous media composed of small glass beads. *Microbiology*

(*Mikrobiologiya*), **56**(1), 84–90.

Malek, I. (1958) The physiological state of microorganisms during continuous culture. In: *Continuous Cultivation of Microorganisms. A Symposium*, Publ. House of Czech. Acad. Sci., Prague, pp.11–28.

Malthus, T.R. (1798) *An Essay on the Principle of Population*, Johnson, London.

Manakov, M.N. (1981) Stoichiometry and energetics of biosynthesis. *Appl. Biochem. and Microbiol.* (*Prikladnaya biokhimiya i mikrobiologiya*) **17**(3), 375–9.

Mandelstam, J. (1963) Protein turnover and its function in the economy of the cell. *Ann. N.Y. Acad. Sci.*, **102**(3), 621–6.

Marayama, S. (1984) Decomposition kinetics of straw saccharides and synthesis of microbial saccharides under field conditions. *J. Soil Sci.*, **35**, 231–42.

Marchuk, G.I. (1991) *Mathematical Models in Immunology. Numerical Methods and Experiments*, Nauka Publ., Moscow, 300 p.

Marialigeti, K. (1979) On the community-structure of the gut-microbiota of *Eisenia lucens* (Annelida, Oligochaeta). *Pedobiologia*, **19**(3), 213–20.

Markvichev, N.S. and Manakov, M.N. (1985) A method for cultivation of micro-organisms at constant concentrations of medium components. *Appl. Biochem. Microbiol.* (*Prikladnaya biokhimiya i mikrobiologiya*), **21**(4), 510–15.

Marr. A.G., Nilson, E.H. and Clark, D.J. (1963) The maintenance requirement of *Escherichia coli*. *Ann. N.Y. Acad. Sci.*, **102**(3), 536–48.

Mayberry, W.R., Prochazka, G.J. and Payne, W.J. (1967) Growth yields of bacteria on selected organic compounds. *Appl. Microbiol.*, **15**(6), 1332–8.

McClung, L.S. (1949) Recent developments in microbiological techniques. *Ann. Rev. Microbiol.*, **3**, 395–422.

McLaren, A.D. (1980) Nitrification reactions in a column of soil with a flowing nutrient solution. *Folia Microbiol.*, **25**(2), 86–9.

Meyerhof, O. (1917) Untersuchungen über den Atmungsvorgang nitrifizierender Bakterien. IV. Die Atmung des Nitritbildners und ihre Beeinflussung durch Chemische Substanzen. *Pflüger's Archiv*, **166**, 240–62.

Meynell, G.G. (1959.) Use of superinfecting phage for estimating the division rate of lysogenic bacteria in infected animals. *J. Gen. Microbiol.*, **21**, 421–37.

Michaelis, N. and Menten, M.I. (1913) Die Kinetik der Invertinwirkung. *Biochem. Z.*, **49**, 333–69.

Minkevich, I.G. and Eroshin, V.K. (1976) Principles of mass- and energy balance of microbial growth. *Adv. Modern Biol.* (*Uspekhi sovremennoy biologii*), **82**(1), 103–16.

Minkevich, I.G., Krynitskaya, A.J. and Eroshin V.K. (1987) The bistat method of cultivation combining the features of chemostat and pH-auxostat. *Microbiology* (*Mikrobiologiya*), **56**(2), 254–8.

Mirchink, T.G. and Babjeva, I.P. (1981) Mycelium forming fungi and yeasts in natural ecosystems. *J. Gen. Biol.* (*Zhurnal obshchei biologii*), **42**(3), 390–8.

Monakhova, E.V. (1984) The biology of coryneform bacteria isolated from fresh-water reservoir. PhD Thesis, Institute of Physiology and Biochemistry of Microorganisms, Puschino.

Monod, J. (1942) *Recherches sur la Croissance des Cultures Bacteriennes*, Hermann, Paris.

Monod, J. (1949) The growth of bacterial cultures. *Ann. Rev. Microbiol.*, **3**, 371–94.

Monod, J. (1950) La technique de culture continué: theorie et applications. *Ann. Instit. Pasteur*, **79**, 390–410.

Monod, J., Wyman, J. and Changeux, J. (1965) On the nature of allosteric transition: a plausible model. *J. Mol. Biol.*, **12**(1), 88–118.

Morita, R.Y. (1982) Starvation–survival of heterotrophs in the marine environment. *Adv. Microbial Ecol.*, **6**, 117–98.

Morita, R.Y. (1990) The starvation–survival state of microorganisms in nature and its relationship to bioavailable energy. *Experientia*, **46**, 813–7.

Moser, H. (1958) The dynamics of bacterial populations maintained in the chemostat. Carnegie Inst. Wash., Pub. 614, Washington DC, 136 p.

Müller, M. (1895) Ueber den Einfluss von Fiebertemperaturen auf die Wachsthumsgeschwindigkeit und die Virulenz des Typhus-bacillus. *Z. Hyg. Infektionskrankh.*, **20**, 245–80.

Müller, P.J., von Frommannshausen, B., Knorre, W.A. and Bergter, F. (1979) More than one stationary state in the cell concentration of thiamine-limited *Escherichia coli* 2000 chemostat cultures. *Z. Allgemeine Mikrobiol.*, **19**(9), 667–70.

Nakas, J.P. and Klein, D.A. (1981) Use of an amino acid mixture to estimate the mineralization capacity of grassland soils. *Soil Biol. Biochem.*, **13**(3), 427–8.

Nelson, L. and Parkinson, D. (1978) Effect of starvation on survival of three bacterial isolates from an arctic soil. *Can. J. Microbiol.*, **24**(12), 1460–7.

Nelson, L.M. and Visser, A.S. (1978) Effects of spring thaw on microorganisms in the arctic meadow site. *Arc. Alp. Res.*, **10**, 679–88.

Newman, E.I. and Watson, A. (1977) Microbial abundance in the rhizosphere: a computer model. *Plant and Soil*, **48**(1), 17–56.

Nikitin, D.I. and Nikitina, E.S. (1978) *The Processes of Environment Selfremediation and Bacterial Parasites of Genera Bdellovibrio*, Nauka Publ., Moscow,
205 p.

Nikitin, D.I., Vasiljeva, L.V. and Lochmatcheva, R.A. (1966) *New and Rare Forms of Soil Microorganisms*, Nauka Publ., Moscow, 70 p.

Nion, I. and Furusaka, C. (1968) Growth of bacteria in the heat-killed cell suspensions of the same bacteria. *J. Gen. Appl. Microbiol.*, **14**(4), 373–85.

Novick, A. (1959) Experimentation with chemostat. In: *Recent Progress in Microbiology*, Blackwell, Oxford, pp.403–15

Novick, A. and Szilard, L. (1950a) Description of the chemostat. *Science*, **112**(2921), 715–6.

Novick, A. and Szilard, L. (1950b) Experiments with the chemostat on spontaneous mutations of bacteria. *Proc. Natl Acad. Sci.*, **36**(12), 708–19.

Novikov, V.A., Novikov, E.A. and Yumatova, L.A. (1987) Freezing of a Jacobian matrix in the Rosenbrok type method of the second order of accuracy. *J. Computing Math. Phys.* (*Zhurnal vychislitel'noy matematiki i matematicheskoy fisiki*), **27**(3), 385–90.

Nyholm, N. (1976) A mathematical model for microbial growth under limitation by conservative substrates. *Biotechnol. Bioeng.*, **18**(8), 1043–56.

Nyholm, N. (1977) Kinetics of phosphate limited algal growth. *Biotechnol. Bioeng.*, **19**(3), 467–92.

Overbeck, J. and Daley, R.J. (1973) Some precautionary comments on the Romanenko technique for estimating heterotrophic bacterial production.

Ecol. Res. Bull., **17**, 342–4.

Paleyeva, M.V., Svetlov, S.V. and Panikov, N.S. (1990) Measurement of the biomass of soil microorganisms utilizing molecular hydrogen. *Microbiology* (*Mikrobiologiya*), **59**(3), 344–9.

Palumbo, S.A., Johnson, M.G., Rieck, V.T. and Witter, L.D. (1971) Growth measurements on surface colonies of bacteria. *J. Gen. Microbiol.*, **66**(2), 137–43.

Panikov, N.S. (1976) The nucleic acids and their transformations by microorganisms. PhD Thesis, Moscow University.

Panikov, N.S. (1979) Steady state growth kinetics of *Chlorella vulgaris* under double substrate (urea and phosphate) limitation. *J. Chem. Tech. Biotechnol.*, **29**(4), 442–50.

Panikov, N.S. (1984a) Effect of substrate concentration on growth of *Debaryomyces formicarius* in continuous culture. *Microbiology* (*Mikrobiologiya*), **53**(3), 318–24.

Panikov, N.S. (1984b) Rate of endogenous respiration as an indicator of maintenance energy in a chemostat culture of *Debaryomyces formicarius. Microbiology-Engl.Tr.* (*Mikrobiologiya*), **53**(4), 476–83.

Panikov, N.S. (1991a). *Kinetics of Microbial Growth: General Principles and Ecological Applications*, Nauka Publ., Moscow.

Panikov, N.S. (1991b) A synthetic chemostat model as a means of describing the complex dynamic behaviour of microorganisms. *Microbiology* (*Mikrobiologiya*), **60**(3), 299–307.

Panikov, N.S. (1992) Kinetic mechanisms of organic matter decomposition in soil. In: *Problems of Soil Science*. Soviet Pedologists to the 14th International Congress of Soil Science (eds V.A. Kovda and M.A. Glasovskaya), Nauka Publ., Moscow, pp.103–9.

Panikov, N.S. and Bondarenko, T.F. (1981) Measurement of the rate of microbial respiration in the continuous culture. *Appl. Chem. Microbiol.* (*Prikladnaya Biokhimiya i Mikrobiologiya*), **17**(5), 728–33.

Panikov, N.S. and Evdokimova, N.V. (1989) The effect of mineral salts on the growth and amylase production in a batch culture of *Penicillium funiculosum. Appl. Biochem. Microbiol.* (*Prikladnaya biokhimia i mikrobiologia*), **25**(1), 84–90.

Panikov, N.S. and Gorbenko, A.J. (1992) The dynamics of gas exchange between soil and atmosphere in relation to plant–microbe interactions: fluxes measuring and modelling. *Ecol. Bull. (Copenhagen)*, **42**, 101–9.

Panikov, N. and Pirt, J.S. (1978) The effects of cooperativity and growth yield variation on the kinetics of nitrogen or phosphate limited growth of *Chlorella* in a chemostat culture. *J. Gen. Microbiol.* **108**(2), 295–303.

Panikov, N.S. and Simonov, J.V. (1986) Effect of microfauna on the rate of plant litter decomposition. *Ecology* (*Ekologia*), No. 4, 10–17.

Panikov, N.S. and Zyvagintsev, D.G. (1983) Kinetic principle for estimating diversity of microbial habitat types in soil. *Ann. Russian Acad. Sci.* (*Doklady Akademii Nauk USSR*), **268**(5), 49–51.

Panikov, N.S. Afremova, V.D. and Aseyeva, I.V. (1981) The growth kinetics of *Mucor plumbeus* and *Mortierella ramanniana* on solid media containing glucose. *Microbiology* (*Mikrobiologiya*), **50**(1), 55–61.

Panikov, N.S., Afremova, V.D. and Aseyeva, I.V. (1984) Kinetics of cellulose decomposition in soil. *Soviet Soil Sci. (Pochvovedenie)*, No. 1, 56–63.

Panikov, N.S., Afremova, V.D. and Aseyeva, I.V. (1987) Growth of plants and root associated microorganisms on mineral nutrient solutions of different composition. *Agric. Chem.* (*Agrokhimiya*), No. 3, 56–63.

Panikov, N.S., Afremova, V.D., Aseyeva, I.V. and Vshivtsev, V.S. (1983) Respiration of fungi *Trichoderma koningii* and *Mucor plumbeus* during growth on solid agar medium with glucose. *Microbiology* (*Mikrobiologiya*), **52**(2), 230–7.

Panikov, N.S., Aseyeva, I.V. and Chistyakova, I.K. (1980) Kinetics of continuous growth of the yeasts *Debaryomyces formicarius* in chemostat and in continuous-flow columns with a solid phase. *Microbiology* (*Mikrobiologiya*), **49**(5), 794–803.

Panikov, N.S., Dedysh, S.N. and Zenova, G.M. (1989a) Use of approaches of nonsteady-state kinetics for quantitative determination of biological productivity of algae in soil. *Microbiology* (*Mikrobiologiya*), **58**(1), 93–101.

Panikov, N.S., Dobrovolskaya, T.G. and Lysak, L.V. (1989b) Ecology of corynebacteria. *Adv. Microbiol.* (*Uspechy mikrobiologii*), **24**, 51–91.

Panikov, N.S., Dorofeyev, A.G., Bondarenko, T.F. and Zyvagintsev, D.G. (1987) Effect of substrate concentration on the growth and respiration of microorganisms with different life strategies. *Microbiology* (*Mikrobiologiya*), **56**(6), 963–72.

Panikov, N.S., Dorofeyev, A.G. and Zvyagintsev, D.G. (1982) Maintenance energy of *Debaryomyces formicarius* in chemostat and continuous dialysis cultures. *Microbiology* (*Mikrobiologiya*), **51**(5), 581–6.

Panikov, N.S., Evdokimova, N.V. and Andreyev, S.V. (1989c) The growth and amylase production in a chemostat culture of *Penicillium funiculosum*. *Appl. Biochem. and Microbiol.* (*Prikladnaya biokhimia i mikrobiologia*), **25**(1), 84–90.

Panikov, N.S., Gorbenko, A.J. and Zvyagintsev, D.G. (1988a) Oscillative pattern of microbial growth in soil and its possible explanation. *Bull. of Moscow University, Ser. 17* (*Vestnik Mokovskogo Universiteta*), No. 1, 34–9.

Panikov, N.S., Paleyeva, M.V., Dedysh, S.N. and Dorofeyev A.G., (1992) Kinetic methods for determining the biomass and activity of various groups of soil microorganisms. *Eurasian Soil Sci.*, **24**(8), 96–111.

Panikov, N.S., Shekhovtsova, N.V., Dorofeyev, A.G. and Zvyagintsev, D.G. (1988b) Quantitative investigation of the dynamics of death in starving microorganisms. *Microbiology* (*Mikrobiologiya*), **57**(6), 786–94.

Parinkina, O.M. (1989) *Microflora of Tundra Soils*, Nauka Publ., Leningrad.

Paris, D.F., Steen, W.C., Baughman, G.L. and Barnett, J.T. Jr (1981) Second-order model to predict microbial degradation of organic compounds in natural waters. *Appl. Environ. Microbiol.*, **41**(3), 603–9.

Parkinson, D., Domsch, K.H. and Anderson, J.P.E. (1978) Die Entwicklung mikrobieller Biomassen im organischen Horizont eines Fichten standortes. *Oecologia Plantarum*, **13**(3), 355–66.

Parnas, H. (1976) A theoretical explanation of the priming effect based on microbial growth with two limiting substrates. *Soil Biol. Biochem.*, **8**(2), 139–44.

Parr, J.F., Parkinson, D. and Normann, A.G. (1967) Growth and activity of soil microorganisms in glass microbeads. II. Oxygen uptake and direct observations. *Soil Sci.*, **103**(5), 303–10.

Parsons, T.R. and Strickland, J.D.H. (1962) On the production of particulate organic carbon by heterotrophic processes in sea water. *Deep-Sea Res.*, **8**(2), 211–22.

Pawlowsky, U., Howell, J.A. and Chi, C.T. (1973) Mixed culture biooxidation of phenol. III. Existence of multiple steady states in continuous culture with wall growth. *Biotechnol. Bioeng.*, **15**(5), 905–16.

Pearl, R. (1925) *The Biology of Population Growth*, A.A. Knopf, New York, 260 p.

Pechurkin, N.S. (1978) *Populational Microbiology*, Nauka Publ., Novosibirsk, 277 p.

Pechurkin, N.S. and Terskov, I.A. (1973) *Autoselective Processes in Continuous Culture of Microorganisms*, Nauka Publ., Novosibirsk, 64 p.

Peitersen, N. and Ross, E.W. (1979) Mathematical model for enzymatic hydrolysis and fermentation of cellulose by *Trichoderma*. *Biotechnol. Bioeng.*, **21**(6), 997–1017.

Penfold, W.J. (1914) On the nature of bacterial lag. *J. Hygiene*, **14**, 215–33.

Penfold, W.J. and Norris, D. (1912) Influence of concentration of food supply on generation time of bacteria. *J. Hygiene*, **12**, 527–41.

Perfil'ev, B.V. and Gabe, D.R. (1969) *Capillary Methods of Investigating Microorganisms*, Oliver and Boyd, Edinburgh.

Pfeffer, W. (1904) *Pflanzenphysiologie*, Engelmann, Leipzig, 885 p.

Pirt, S.J. (1965) The maintenance energy of bacteria in growing cultures. *Proc. Roy. Soc., B*, **163**, 224–31.

Pirt, S.J. (1972) Introductory lecture. Prospects and problems in continuous flow culture of micro-organisms. *J. Appl. Chem. Biotechnol.*, **22**(1), 55–64.

Pirt, S.J. (1975) *Principles of Microbe and Cell Cultivation*, Blackwell, Oxford.

Pirt, S.J. (1982) Maintenance energy: a general model for energy-limited and energy-sufficient growth. *Arch. Microbiol.* **133**(4), 300–2.

Pirt, W.M. and Pirt, S.J. (1977) Photosynthetic production of biomass and starch by *Chlorella* in chemostat culture. *J. Appl. Chem. Biotechnol.*, **27**(4), 643–50.

Pirt, S.J., Panikov, N.S. and Lee, Y.-K. (1979) The miniloop: a small-scale air-lift microbial culture vessel and photobiological reactor. *J. Chem. Technol. Biotechnol.*, **29**(3), 437–41.

Poindexter, J.S. (1981a) The Caulobacters: ubiquitous unusual bacteria. *Microbiol. Rev.*, **45**(1), 123–79.

Poindexter, J.S. (1981b) Oligotrophy. Fast and famine existence. *Adv. Microbial Ecol.*, **5**, 63–89.

Popov, A.I. and Panikov, N.S. (1990) A kinetic method for the measurement of the numbers of nitrifying bacteria in soil. *Microbiology* (*Mikrobiologiya*), **59**(4), 482–6.

Postgate, J.R. (1967) Viability measurements and the survival of microbes under minimum stress. *Adv. Microbial Physiol.*, **1**, 1–23.

Postgate, J.R. and Gray, T.R.G. (eds) (1976) *The Survival of Vegetative Microbes*, 26th Symp. Soc. Gen. Microbiol., Cambridge University Press.

Powell, E.O. (1958) Discussion. In: *Recent Progress in Microbiology*, Blackwell, Oxford, pp.22–3.

Powell, E.O. (1967) The growth rate of microorganisms as a function of substrate concentration. In: *Continuous Cultivation of Microorganisms*, (eds E.O. Powell *et al.*), H.M. Stationery Office, Salisbury, pp.34–55.

Powell, E.O. (1969) Transient changes in the growth rate of microorganisms. In: *Continuous Cultivation of Microorganisms*, Academia, Prague, pp.275–84.

Powell, E.O., Evans, G.G.T., Strange, R.E. and Tempest, D.W. (eds) (1967) *Continuous Cultivation of Microorganisms*, Third International Symp., H.M. Stationery Office, Salisbury, UK.

Powell, G.E. (1985) Stable coexistence of syntrophic associations in continuous culture. *J. Chem. Technol. Biotechnol.*, **B35**(1), 46–50.

Pozmogova, I.N. (1983) *The Cultivation of Microorganisms under Perturbed Conditions*, Nauka Publ., Moscow, 103 p.

Rabotnova, I.L. (1974) Physiological variability of microorganisms and its regulation. *Adv. in Microbiol.* (*Uspekhi mikrobiologii*), **10**, 120–31.

Rafales-Lamarka, E.E. and Nikolaev, V.G. (1971) *Some Procedures of Optimization and Mathematical Analysis of Biological Experiments*, Naukova Dumka Publ., Kiev, 119 p.

Rahn, O. (1932) *Physiology of Bacteria*, P. Blackiston's Son and Co., Philadelphia, 438 p.

Ramkrishna, D., Frederickson, A.G. and Tsuchiya, H.M. (1967) Dynamics of microbial population: models considering inhibitors and variable cell composition. *Biotechnol. Bioeng.*, **9**(2), 129–70.

Raulin, M.J. (1869) Etudes chimiques sur la vegetation. *Ann. des Sci. Naturelles, Serie 5*, **11**, 93–299.

Rayner, A.D.M. and Boddy, L. (1988) Fungal communities in the decay of wood. *Adv. Microbial Ecol.*, **10**, 115–66.

Reeve, C.A., Amy, P.S. and Matin, A. (1984) Role of protein synthesis in the survival of carbon-starved *Escherichia coli* K-12. *J. Bacteriol.*, **160**(3), 1041–6.

Reyrolle, J. and Letellier, F. (1979) Autoradiographic study of the localization and evolution of growth zones in bacterial colonies. *J. Gen. Microbiol.*, **111**, 399–406.

Řičica, J. (1973) Continuous cultivation of microorganisms. A review. *Folia Microbiol.*, **18**(3), 418–48.

Řičica, J., Nečinova, S., Stejskalova, E. and Fencl, Z. (1967) Properties of microorganisms grown in excess of the substrate at different dilution rates in continuous multistream culture systems. In: *Continuous Cultivation of Microorganisms* (eds E.D. Powell *et al.*), H.M. Stationery Office, Salisbury, pp.196–208.

Rijn, J. van and Shilo, M. (1986) Nitrogen limitation in natural populations of cyanobacteria (*Spirulina* and *Oscillatoria* spp.) and its effect on macromolecular synthesis. *Appl. Environ. Microbiol.*, **52**(2), 340–4.

Robertson, B.R. and Button, D.K. (1979) Phosphate-limited continuous culture of *Rhodotorula rubra*: kinetics of transport, leakage, and growth. *J. Bacteriol.*, **138**(3), 884–95.

Robertson, T.B. (1924) Allelocatalytic effect in cultures of *Colpidium* in hay-infusion and in synthetic media. *Biochem. J.*, **18**(6), 1240–7.

Robinson, J.A. (1985) Determining microbial kinetics parameters using nonlinear regression analysis. Advantages and limitations in microbial ecology. *Adv. Microbial Ecol.*, **8**, 61–114.

Robinson, J.A. and Tiedje, J.M. (1982) Kinetics of hydrogen consumption by rumen fluid, anaerobic digestor sludge, and sediment. *Appl. Environ. Microbiol.*, **42**(12), 1374–84.

Rodin, V.B. and Panikov, N.S. (1995) Two-dimensional model of colonial growth of unicellular microorganisms and its experimental verification. *Microbiology (Mikrobiologiya)*, **64**, 000–000.

Roels, J.A. (1980) Application of macroscopic principles to microbial metabolism. *Biotechnol. Bioeng.*, **22**(12), 2457–514.

Romanenko, V.I. (1964) Heterotrophic CO_2 assimilation by bacterial flora of water. *Microbiology (Mikrobiologiya)*, **33**(4), 679–83.

Romanovsky, J.M., Stepanova, N.V. and Chernavski, D.S. (1984) *Math. Biophys.*, Nauka Publ., Moscow, 304 p.

Rubner, M. (1903/4) Energieverbrauch im Leben der Mikroorganismen. *Archiv für Hygiene*, **48**(3), 260–311.

Samarsky, A. and Goolin, A. (1989) *Numerical Methods*, Nauka Publ., Moscow, 432 p.

Saunders, P.T. and Bazin, M.J. (1973) Non-steady state studies of nitrification in soil: theoretical considerations. *Soil Biol. Biochem.*, **5**(5), 545–57.

Scow, K.M., Simkins, S. and Alexander, M. (1986) Kinetics of mineralization of organic compounds at low concentrations in soil. *Appl. Environ. Microbiol.*, **51**(5), 1028–35.

Semenov, A.M. and Vasiljeva, L.V. (1985) Morpho-physiological characteristics of budding prosthecobacterium *Labrys monahos* with radial symmetry of cell grown in batch and continuous culture. *Proc. USSR Acad. Sci. (Izvestiya Akademii Nauk)*, No. 2, 288–93.

Semenov, A.M. and Vasiljeva, L.V. (1986) The morphological and physiological characteristics of the oligotrophic prosthecobacterium *Prosthecobacterium hirschii* grown in batch and continuous culture. *Microbiology (Mikrobiologiya)*, **55**(2), 248–52.

Seto, M. and Tazaki, T. (1973) Growth and respiratory activity of mold fungus (*Trichoderma lignorum*). *Bot. Mag. (Tokyo)*, **88**(1012), 255–6.

Shapiro, J.A. (1986) Control of *Pseudomonas putida* growth on agar surface. In: *The Bacteria*, (ed. J.R. Sokatch), Academic Press, New York, vol. 10, pp.27–69.

Shapiro, J.A. (1987) Organization of developing *Escherichia coli* colonies viewed by scanning electron microscopy. *J. Bacteriol.*, **169**(1), 142–56.

Shapiro, J.A. and Hsu, C. (1989) *Escherichia coli* K-12 cell–cell interactions seen by time-lapse video. *J. Bacteriol.*, **171**(11), 5963–74.

Shapiro, L. (1976) Differentiation in the *Caulobacter* cell cycle. *Annu. Rev. Microbiol.*, **30**, 377–407.

Shehata, T.E. and Marr, A.G. (1971) Effect of nutrient concentration on the growth of *Escherichia coli*. *J. Bacteriol.*, **107**(1), 210–6.

Sherman, G.M. and Albus, W.R. (1923) Physiological youth of bacteria. *J. Bacteriol.*, **8**, 127–9.

Sherman, J.M. and Albus, W.R. (1924) The function of lag in bacterial cultures. *J. Bacteriol.*, **9**(3), 303–5.

Shvytov, I.A. (1974) Some principles of mathematical simulation of microbial population dynamics. *J. Gen. Biol. (Zhurnal obshchei biologii)*, **35**(6). 904–10.

Simkins, S. and Alexander, M. (1984) Models for mineralization kinetics with the variables of substrate concentration and population density. *Appl. Environ. Microbiol.*, **47**(6), 1299–306.

Skurida, G.I., Mironov, V.A. and Drozdov-Tikhomirov, L.N. (1983) Characteristic tables of monomers biosynthesis in microbial cell. *Microbiology (Mikrobiologiya)*, **52**(5), 702–8.

Skurida, G.I., Mironov, V.A. and Drozdov-Tikhomirov, L.N. (1984) Calculation of minimal substrate requirements for energetic metabolism of microorganisms. *Microbiology (Mikrobiologiya)*, **53**(2), 187–94.

Slator, A. (1916) The rate of growth of bacteria. *Trans. Chem. Soc.*, **109**, 2.

Smith, K.A. (1980) A model of the extent of anaerobic zones in aggregated soils, and its potential application to estimates of denitrification. *J. Soil Sci.*, **31**(2), 263–77.

Söderström, B.E. (1977) Vital staining of fungi in pure cultures and in soil with fluorescein diacetate. *Soil Biol. Biochem.*, **9**(1), 59–63.

Soina, V.S., Dorofeyev, A.G. and Panikov, N.S. (1984) The particular features of the cell structures of soil bacteria under growth limitation. In: *Cytology of Microorganisms*, Nauka Publ., Poushchino, pp.62–3.

Sommer, U. (1984) The paradox of the plankton: fluctuations of phosphorus availability maintain diversity of phytoplankton in flow-through cultures. *Limnol. Oceanogr.*, **29**(3), 633–6.

Sorokina, T.A., Mishustin, E.N. and Pisarchuk, E.N. (1984) Distribution of non-sporeforming bacteria of genera *Pseudomonas* and *Arthrobacter* in soils of different climatic zones. In: *Soil Organisms as a Component of Biogeocenosis*, (ed. E.N. Mishustin), Nauka Publ., pp.199–217.

Staley, J.T. (1971) Growth rates of algae determined *in situ* using an immersed microscope. *J. Phycol.*, **7**(1), 13–7.

Stanier, R.Y. (1977) Obituary. Jacques Monod, 1910–1976. *J. Gen. Microbiol.*, **101**(1), 1–12.

Stanier, R.Y., Adelburg, E.A. and Ingraham, J.L. (1979) *General Microbiology*, 4th edn, Macmillan Press Ltd, London and Basingstoke, 273 p.

Stepanova, N.V., Chernavsky, D.S. and Ierusalimsky, N.D. (1967) Kinetics of microbial cultivation and biological inertia. In: *The Oscillatory Processes in Biological and Chemical Systems*, Nauka Publ., Moscow, pp.317–25.

Stepanova, N.V., Gevixman, H.N., Basnak'yan, I.A. and Dubinina, G.P. (1980) Theoretical and experimental studies of size distribution of bacteria grown in chemostat. In: *Limitation and Inhibition of Microbial Growth*, Nauka Publ., Poushchino, pp.137–52.

Stephenson, M. (1949) *Bacterial Metabolism*, M.I.T. Press, Cambridge, Massachusetts and London, 398 p.

Stotzky, G. and Norman, A.G. (1961) Factors limiting microbial activities in soil. I. The level of substrate nitrogen and phosphorus. *Arch. Microbiol.*, **40**(4), 344–69.

Stouthamer, A.H. (1973) A theoretical study on the amount of ATP required for synthesis of microbial cell material. *Ant. van Leeuvenhoek J. Microbiol. Serol.*, **39**(3), 545–65.

Stouthamer, A.H. (1977) Energetic aspects of the growth of microorganisms. In: *Microbial Energetics*, 27th Symp. Soc. Gen. Microbiol., Cambridge University Press, pp.285–315.

Svensson, B.H. and Rosswall, T. (1980) Energy flow through the subarctic mire at Stordalen. In: *Ecology of a Subarctic Mire*, (ed. M. Sonesson), *Ecol. Bull. (Stockholm)*, **30**, 283–301.

Svirezhev, J.M. (1976) *Vito Volterra and Modern Mathematical Ecology.*

Introduction to Russian edition of Volterra, 1931. Nauka Publ., Moscow, pp.245–86.

Szabo, I.M. (1974) *Microbial Communities in a Forest–Rendzina Ecosystem. The Pattern of Microbial Communities*, Akad. Kiado, Budapest.

Takamatsu, T., Shioya, S. and Okuda, K. (1981) A comparison of unstructured growth models of microorganisms. *J. Ferment. Technol.*, **59**(2), 131–6.

Tarapchak, S.J. and Herche, L.R. (1986) Phosphate uptake by microorganisms in lake water: deviations from simple Michaelis–Menten kinetics. *Can. J. Fish. Aquat. Sci.*, **43**(2), 319–28.

Tauson, V.O. (1933) Calorimetric technique to study microbial bioenergetics. *Microbiology* (*Mikrobiologiya*), **2**(1), 19–50.

Tauson, V.O. (1934) On the basic metabolism of microfungi. *Microbiology* (*Mikrobiologiya*), **3**(4), 523–55.

Tauson, V.O. (1937) The method of retarded culture and rate of basic metabolism of microfungi. *Microbiology* (*Mikrobiologiya*), **6**(4), 517–44.

Teissier, G. (1956) Kinetic behaviour of heterogeneous populations in completely mixed reactors. *Ann. Physiol. Physiochem. Biol.*, **12**, 527–86.

Tempest, D.W., Dicks, J.W. and Hunter, J.R. (1966) The interrelationship between potassium, magnesium and phosphorus in potassium-limited chemostat culture. *J. Gen. Microbiol.*, **45**(1), 135–46.

Tempest, D.W., Herbert, D. and Phipps, P.J. (1967) Studies on the growth of *Aerobacter aerogenes* at low dilution rates in a chemostat. In: *Continuous Cultivation of Microorganisms*, (eds E.O. Powell *et al.*), H.M. Stationery Office, Salisbury, pp.240–53.

Tempest, D.W., Hunter, J.R. and Sykes, J. (1965) Magnesium-limited growth of *Aerobacter aerogenes* in a chemostat. *J. Gen. Microbiol.*, **39**(3), 355–66.

Tempest, D.W. and Neijssel, O.M. (1977) Are maximum yield values (Y_{maxsub}, Y_{maxO2}, Y_{maxATP}) physiologically interpretable parameters? *Proc. Soc. Gen. Microbiol.*, **4**(2), 69–70.

Tempest, D.W. and Neijssel, O.M. (1984) The status of Y_{ATP} and maintenance energy as biologically interpretable phenomena. *Annu. Rev. Microbiol.*, **38**, 459–86.

Tempest, D.W., Meers, J.L. and Brown, C.M. (1970). Influence of environment on the content and composition of microbial free amino acid pools. *J. Gen. Microbiol.*, **64**, 171–85.

Terroine, E.F. and Wurmser, R. (1922) L'energie de croissance. I. Le développement de l'*Aspergillus niger*. *Bull. Soc. Chem. Biol.*, **4**, 519–67.

Tiezen, L.L. (1974) Photosynthesis in relation to primary production. In: *Tundra Biome*, (eds F.E. Wiegolaski and T. Rosswall), Tundra Biome Steering Committee, Stockholm, pp.52–62.

Tijhuis, L., van Loosdrecht, M.C.M. and Heijnen, J.J. (1993) A thermodynamically based correlation for maintenance Gibbs energy requirements in aerobic and anaerobic chemotrophic growth. *Biotechnol. Bioeng.*, **42**(4), 509–19.

Trinci, A.P.J. (1971) Influence of the width of the peripheral growth zone on the radial growth rate of fungal colonies on solid media. *J. Gen. Microbiol.*, **67**, 325–43.

Turner, B.G., Ramkrishna, D. and Jansen, N.B. (1989) Cybernetic modeling of bacterial cultures at low growth rates: single-substrate systems. *Biotechnol. Bioeng.*, **34**, 252–61.

Utenkov, M.D. (1941) *Microgenerating*, Sovetskaya Nauka Publ., Moscow, Leningrad.

Van Niel, C.B. (1949) The kinetics of growth of microorganisms. In: *The Chemistry and Physiology of Growth*, Princeton University Press, Princeton, NJ, pp.91–105.

Varfolomeyev, S.D. (1987) The kinetic principles of analysis of microbial populations. In: *Contemporary Problems of Biokinetics*, Moscow University Publ., Moscow, pp.6–77.

Veldkamp, H. (1977) Ecological studies with the chemostat. *Adv. in Microbial Ecol.*, **1**(1), 59–94.

Verhoff, F.H., Sundaresan, K.R. and Tenney, M.W. (1972) A mechanism of microbial cell growth. *Biotechnol. Bioeng.*, **15**(3), 411–36.

Verhulst, P.F. (1838) Notice sur la loi que la population suit dans son accroissement. *Corresp. Math. Phys.*, **10**(1), 113–21.

Volterra, V. (1976) *Lezons sur la Theorie Mathematique de la Lutte pour la Vie*, Gauthier-Villars, Paris, 214 p.

Von Foester, H. (1959) Some remarks on changing populations. In: *The Kinetics of Cellular Proliferations*, New York, pp.382–407.

Ward, H.M. (1895) On the biology of *Bacillus ramosus* (Fraenkel), a schizomycete of the River Thames. *Proc. Roy. Soc. (London)*, **58**, 276–463.

Warembourg, F.R. (1975) Application de techniques radioisotopiques a l'étude de l'activité biologique dans la rhizosphere des plantes. *Revue d'Ecol. Biol. Sol*, **12**(1), 261–72.

Wessen, B. and Berg, B. (1986) Long-term decomposition of barley straw: chemical changes and ingrowth of fungal mycelium. *Soil Biol. Biochem.*, **18**(1), 53–9.

Williams, P.J.leB. (1973) The validity of the application of simple kinetic analysis to heterogeneous microbial populations. *Limnol. Oceanogr.*, **18**(1), 159–65.

Winogradsky, S.N. (1925) Etudes sur la microbiologie du sol. Premier memoire: sur la méthode. *Ann. Inst. Pasteur*, **39**, 299–300.

Winogradsky, S.N. (1949) *Microbiologie du Sol*, Masson et Cie Editeurs, Paris, 861 p.

Woodruff, L.L. and Baitsell, G.A. (1911a) The reproduction of *Paramaecium aurelia* in a 'constant' culture medium of beef extract. *J. Exp. Zool.*, **11**(1), 135–42.

Woodruff, L.L. and Baitsell, G.A. (1911b) Rhythms in the reproductive activity of infusoria. *J. Exp. Zool.*, **11**(4), 339–59.

Wright, R.T. and Hobbie, J.E. (1966) Use of glucose and acetate by bacteria and algae in aquatic ecosystems. *Ecology*, **47**(3), 447–64.

Zavarzin, G.A. (1970) On the role of 'dispersive' microflora in carbon cycle. *J. Gen. Biol.* (*Zhurnal obshei biologii*), **31**(4), 386–93.

Zavarzin, G.A. (1972) *Lithotrophic Microorganisms*, Nauka Publ., Moscow, 323 p.

Zavarzin, G.A. (1976) Extensive Microbiology. *Proc. of USSR Academy of Sciences* (*Izvestiya Academii Nauk*), Ser. biol., No. 1, 121–34.

Zavarzin, G.A. (1978) *Hydrogen Bacteria and Carboxydobacteria*, Nauka Publ., Moscow, 207 p.

Zhiglhiavsky, A.A. and Zhilinskas, A.G. (1991) *Methods of Global Optimization*, Nauka Publ., Moscow, 248 p.

Zvyagintsev, D.G. (1987) *Soil and Microorganisms*, Moscow University Publ., Moscow, 256 p.

Zvyagintsev, D.G., Kozhevin, P.A. and Malachov, V.V. (1976) Ecological problems of soil microbiology. *J. Gen. Biol.* (*Zhurnal obshei biologii*), **37**(5), 691–706.

Index